STATISTICAL MODELS AND CAUSAL INFERENCE

A Dialogue with the Social Sciences

David A. Freedman presents here a definitive synthesis of his approach to causal inference in the social sciences. He explores the foundations and limitations of statistical modeling, illustrating basic arguments with examples from political science, public policy, law, and epidemiology. Freedman maintains that many new technical approaches to statistical modeling constitute not progress, but regress. Instead, he advocates a "shoe-leather" methodology, which exploits natural variation to mitigate confounding and relies on intimate knowledge of the subject matter to develop meticulous research designs and eliminate rival explanations. When Freedman first enunciated this position, he was met with skepticism, in part because it was hard to believe that a mathematical statistician of his stature would favor "low-tech" approaches. But the tide is turning. Many social scientists now agree that statistical technique cannot substitute for good research design and subject matter knowledge. This book offers an integrated presentation of Freedman's views.

David A. Freedman (1938–2008) was Professor of Statistics at the University of California, Berkeley. He was a distinguished mathematical statistician whose theoretical research included the analysis of martingale inequalities, Markov processes, de Finetti's theorem, consistency of Bayes estimators, sampling, the bootstrap, and procedures for testing and evaluating models and methods for causal inference. Freedman published widely on the application—and misapplication—of statistics in works within a variety of social sciences, including epidemiology, demography, public policy, and law. He emphasized exposing and checking the assumptions that underlie standard methods, as well as understanding how those methods behave when the assumptions are false—for example, how regression models behave when fitted to data from randomized experiments. He had a remarkable talent for integrating carefully honed statistical arguments with compelling empirical applications and illustrations. Freedman was a member of the American Academy of Arts and Sciences, and in 2003 he received the National Academy of Science's John J. Carty Award for his "profound contributions to the theory and practice of statistics."

David Collier is Robson Professor of Political Science at the University of California, Berkeley. He is co-author of *Rethinking Social Inquiry: Diverse Tools, Shared Standards* (2004) and co-editor of *The Oxford Handbook of Political Methodology* (2008) and *Concepts and Method in Social Science* (2009). He is a member of the American Academy of Arts and Sciences and was founding president of the Organized Section for Qualitative and Multi-Method Research of the American Political Science Association.

Jasjeet S. Sekhon is Associate Professor of Political Science at the University of California, Berkeley. His research interests include elections, applied and computational statistics, causal inference in observational and experimental studies, voting behavior, public opinion, and the philosophy and history of science. Professor Sekhon received his Ph.D. in 1999 from Cornell University and was a professor at Harvard University in the Department of Government from 1999 to 2005.

Philip B. Stark is Professor of Statistics at the University of California, Berkeley. His research centers on inference (inverse) problems, primarily in physical science. He is especially interested in confidence procedures tailored for specific goals and in quantifying the uncertainty in inferences that rely on simulations of complex physical systems. Professor Stark has done research on the Big Bang, causal inference, the U.S. Census, earthquake prediction, election auditing, the geomagnetic field, geriatric hearing loss, information retrieval, Internet content filters, nonparametrics (confidence procedures for function and probability density estimates with constraints), the seismic structure of the Sun and Earth, spectroscopy, and spectrum estimation.

Cover illustration: The data are from the Intersalt study of the relationship between salt intake and blood pressure, discussed in Chapter 9. The horizontal axis is urine salt level. The vertical axis is systolic blood pressure. Each dot represents the median value among subjects at one of 52 research centers in 32 countries. The four red dots correspond to two centers in Brazil that studied Indian tribes (Yanomamo and Xingu), a center in Papua New Guinea, and a center in Kenya. The two lines are least-squares regression lines. The purple line is fitted to all the data—the red dots and the blue dots. The blue line is fitted only to the blue dots. If all the data are included, median blood pressure is positively associated with median excreted salt. If only the blue dots are included, median blood pressure has a weak negative association with median salt. These data have been considered evidence that increasing salt intake increases blood pressure. The difference between the two regression lines suggests that any link between salt intake and blood pressure is weak. Chapter 9 discusses this and other shortcomings of the Intersalt study.

Statistical Models and Causal Inference

A Dialogue with the Social Sciences

David A. Freedman

Edited by

David Collier
University of California, Berkeley

Jasjeet S. Sekhon
University of California, Berkeley

Philip B. Stark
University of California, Berkeley

CAMBRIDGE
UNIVERSITY PRESS

32 Avenue of the Americas, New York NY 10013-2473, USA

Cambridge University Press is part of the University of Cambridge.

It furthers the University's mission by disseminating knowledge in the pursuit of
education, learning and research at the highest international levels of excellence.

www.cambridge.org
Information on this title: www.cambridge.org/9780521195003

First published 2010

A catalogue record for this publication is available from the British Library

Library of Congress Cataloguing in Publication data
Freedman, David, 1938–
Statistical models and causal inference : a dialogue with the social sciences / David
A. Freedman ; edited by David Collier, Jasjeet Sekhon, Philip B. Stark.
   p.   cm.
Includes bibliographical references and index.
ISBN 978-0-521-19500-3
1. Social sciences – Statistical methods.   2. Linear models (Statistics)   3. Causation.
I. Collier, David, 1942–   II. Sekhon, Jasjeet Singh, 1971–   III. Stark, Philip B.   IV. Title.
HA29.F6785   2009
519.5–dc22   2009043216

ISBN 978-0-521-19500-3 Hardback
ISBN 978-0-521-12390-7 Paperback

Contents

Part I
Statistical Modeling: Foundations and Limitations

Bayesians and frequentists disagree on the meaning of probability and other foundational issues, but both schools face the problem of model validation. Statistical models have been used successfully in the physical and life sciences. However, they have not advanced the study of social phenomena. How do models connect with reality? When are they likely to deepen understanding? When are they likely to be sterile or misleading?

Statistical inference with convenience samples is risky. Real progress depends on a deep understanding of how the data were generated. No amount of statistical maneuvering will get very far without recognizing that statistical issues and substantive issues overlap.

Regression models are used to make causal arguments in a wide variety of applications, and it is time to evaluate the results. Snow's work on cholera is a success story for causal inference based on nonexperimental data, which was collected through great expenditure of effort and shoe leather. Failures are also discussed. Statistical technique is seldom an adequate substitute for substantive knowledge of the topic, good research design, relevant data, and empirical tests in diverse settings.

Part II
Studies in Political Science, Public Policy, and Epidemiology

The U.S. Census is a sophisticated, complex undertaking, carried out on a vast scale. It is remarkably accurate. Statistical adjustments are likely to introduce more error than they remove. This issue was litigated all the way to the Supreme Court, which in 1999 unanimously supported the Secretary of Commerce's decision not to adjust the 2000 Census.

Gary King's book, *A Solution to the Ecological Inference Problem*, claims to offer "realistic estimates of the uncertainty of ecological estimates." Applying King's method and three of his main diagnostics to data sets where the truth is known shows that his diagnostics cannot distinguish between cases where estimates are accurate and those where estimates are far off the mark. King's claim to have arrived at a solution to this problem is premature.

King's method works with some data sets but not others. As a theoretical matter, inferring the behavior of subgroups from aggregate data is generally impossible: The relevant parameters are not identifiable. King's diagnostics do not discriminate between probable successes and probable failures.

Statistical ideas can clarify issues in qualitative analysis such as case selection. In political science, an important argument about case selection evokes Hempel's Paradox of the Ravens. This paradox can be resolved by distinguishing between population and sample inferences.

Making sense of earthquake forecasts is surprisingly difficult. In part, this is because the forecasts are based on a complicated mixture of geological maps, rules of thumb, expert opinion, physical models, stochastic models, and numerical simulations, as well as geodetic, seismic, and paleoseismic data. Even the concept of probability is hard to define in this context. Other models of risk for emergency preparedness, as well as models of economic risk, face similar difficulties.

Experimental evidence suggests that the effect of a large reduction in salt intake on blood pressure is modest and that health consequences remain to be determined. Funding agencies and medical journals have taken a stronger position favoring the salt hypothesis than is warranted, demonstrating how misleading scientific findings can influence public policy.

Epidemiologic methods were developed to prove general causation: identifying exposures that increase the risk of particular diseases. Courts of law often are more interested in specific causation: On balance of probabilities, was the plaintiff's disease caused by exposure to the agent in question? There is a considerable gap between relative risks and proof of specific causation because individual differences affect the interpretation of relative risk for a given person. This makes specific causation especially hard to establish.

Proportional-hazards models are frequently used to analyze data from randomized controlled trials. This is a mistake. Randomization does not justify the models, which are rarely informative. Simpler methods work better. This discussion matters because survival analysis has introduced a new hazard: It can lead to serious mistakes in medical treatment. Survival analysis is, unfortunately, thriving in other disciplines as well.

Part III
New Developments: Progress or Regress?

Regression adjustments are often made to experimental data to address confounders that may not be balanced by randomization. Since randomization does not justify the models, bias is likely. Neither are the usual variance calculations to be trusted. Neyman's non-parametric model serves to evaluate regression adjustments. A bias term is isolated, and conditions are given for unbiased estimation in finite samples.

The logit model is often used to analyze experimental data. Theory and simulation show that randomization does not justify the model, so the usual estimators can be inconsistent. Neyman's non-parametric setup is used as a benchmark: Each subject has two potential responses, one if treated and the other if untreated; only one of the two responses can be observed. A consistent estimator is proposed.

A number of algorithms purport to discover causal structure from empirical data with no need for specific subject-matter knowledge. Advocates have no real success stories to report. These algorithms solve problems quite removed from the challenge of causal inference from imperfect data. Nor do they resolve long-standing philosophical questions about the meaning of causation.

Causal relationships cannot be inferred from data by fitting graphical models without prior substantive knowledge of how the data were generated. Successful applications are rare because few causal pathways can be excluded a priori.

The use of propensity scores to reduce bias in regression analysis is increasingly common in the social sciences. Yet weighting is likely to increase random error in the estimates and to bias the estimated standard errors downward, even when selection mechanisms are well understood. If investigators have a good causal model, it seems better just to fit the model without weights. If the causal model is improperly specified, weighting is unlikely to help.

In applications where the statistical model is nearly correct, the Huber Sandwich Estimator makes little difference. On the other hand, if the model is seriously in error, the parameters being estimated are likely to be meaningless, except perhaps as descriptive statistics.

The usual Heckman two-step procedure should not be used for removing endogeneity bias in probit regression. From a theoretical perspective this procedure is unsatisfactory, and likelihood methods are superior. Unfortunately, standard software packages do a poor job of maximizing the biprobit likelihood function, even if the number of covariates is small.

Model diagnostics cannot have much power against omnibus alternatives. For instance, the hypothesis that observations are independent cannot be tested against the general alternative that they are dependent with power that exceeds the level of the test. Thus, the basic assumptions of regression cannot be validated from data.

Part IV
Shoe Leather Revisited

Causal inference can be strengthened in fields ranging from epidemiology to political science by linking statistical analysis to qualitative knowledge. Examples from epidemiology show that substantial progress can derive from informal reasoning, qualitative insights, and the creation of novel data sets that require deep substantive understanding and a great expenditure of effort and shoe leather. Scientific progress depends on refuting conventional ideas if they are wrong, developing new ideas that are better, and testing the new ideas as well as the old ones. Qualitative evidence can play a key role in all three tasks.

Preface

David A. Freedman presents in this book the foundations of statistical models and their limitations for causal inference. Examples, drawn from political science, public policy, law, and epidemiology, are real and important.

A statistical model is a set of equations that relate observable data to underlying parameters. The parameters are supposed to characterize the real world. Formulating a statistical model requires assumptions. Rarely are those assumptions tested. Indeed, some are untestable in principle, as Freedman shows in this volume. Assumptions are involved in choosing which parameters to include, the functional relationship between the data and the parameters, and how chance enters the model. It is common to assume that the data are a simple function of one or more parameters, plus random error. Linear regression is often used to estimate those parameters. More complicated models are increasingly common, but all models are limited by the validity of the assumptions on which they ride.

Freedman's observation that statistical models are fragile pervades this volume. Modeling assumptions—rarely examined or even enunciated —fail in ways that undermine model-based causal inference. Because of their unrealistic assumptions, many new techniques constitute not progress but regress. Freedman advocates instead "shoe leather" methods, which identify and exploit natural variation to mitigate confounding and which require intimate subject-matter knowledge to develop appropriate research designs and eliminate rival explanations.

Freedman assembled much of this book in the fall of 2008, shortly before his death. His goal was to offer an integrated presentation of his views on applied statistics, with case studies from the social and health sciences, and to encourage discussion of those views. We made some changes to Freedman's initial selection of topics to reduce length and broaden coverage. The text has been lightly edited; in a few cases chapter titles have been altered. The source is cited on the first page of each chapter and in the reference list, which has been consolidated at the end. When available, references to unpublished articles have been updated with the published versions. To alert the reader, chapter numbers have been added for citations to Freedman's works that appear in this book.

Many people deserve acknowledgment for their roles in bringing these ideas and this book to life, including the original co-authors and acknowledged reviewers. Colleagues at Berkeley and elsewhere contributed valuable suggestions, and Janet Macher provided astute assistance in editing the manuscript. Donald W. DeLand converted Chapters 3 and 8 into TeX. Josephine Marks also converted files and edited the references. Ed Parsons of Cambridge University Press helped shape the project and moved it to press with amazing speed. Above all, we admire David Freedman's tenacity and lucidity during his final days, and we are deeply grateful for his friendship, collaboration, and tutelage.

David Collier, Jasjeet S. Sekhon, and Philip B. Stark
Berkeley, California
July 2009

Companion website

http://statistics.berkeley.edu/~freedman/Dialogue.htm
Supplementary material, including errata, will be posted to the companion website.

Editors' Introduction:
Inference and Shoe Leather

David Collier, Jasjeet S. Sekhon, and Philip B. Stark

Drawing sound causal inferences from observational data is a central goal in social science. How to do so is controversial. Technical approaches based on statistical models—graphical models, non-parametric structural equation models, instrumental variable estimators, hierarchical Bayesian models, etc.—are proliferating. But David Freedman has long argued that these methods are not reliable. He demonstrated repeatedly that it can be better to rely on subject-matter expertise and to exploit natural variation to mitigate confounding and rule out competing explanations.

When Freedman first enunciated this position decades ago, many were skeptical. They found it hard to believe that a probabilist and mathematical statistician of his stature would favor "low-tech" approaches. But the tide is turning. An increasing number of social scientists now agree that statistical technique cannot substitute for good research design and subject-matter knowledge. This view is particularly common among those who understand the mathematics and have on-the-ground experience.

Historically, "shoe-leather epidemiology" is epitomized by intensive, door-to-door canvassing that wears out investigators' shoes. In contrast, advocates of statistical modeling sometimes claim that their methods can salvage poor research design or low-quality data. Some suggest that their algorithms are general-purpose inference engines: Put in data, turn the crank, out come quantitative causal relationships, no knowledge of the subject required.

This is tantamount to pulling a rabbit from a hat. Freedman's conservation of rabbits principle says "to pull a rabbit from a hat, a rabbit must first be placed in the hat."[1] In statistical modeling, assumptions put the rabbit in the hat.

Modeling assumptions are made primarily for mathematical convenience, not for verisimilitude. The assumptions can be true or false—usually false. When the assumptions are true, theorems about the methods hold. When the assumptions are false, the theorems do not apply. How well do the methods behave then? When the assumptions are "just a little wrong," are the results "just a little wrong"? Can the assumptions be tested empirically? Do they violate common sense?

Freedman asked and answered these questions, again and again. He showed that scientific problems cannot be solved by "one-size-fits-all" methods. Rather, they require shoe leather: careful empirical work tailored to the subject and the research question, informed both by subject-matter knowledge and statistical principles. Witness his mature perspective:

> Causal inferences can be drawn from nonexperimental data. However, no mechanical rules can be laid down for the activity. Since Hume, that is almost a truism. Instead, causal inference seems to require an enormous investment of skill, intelligence, and hard work. Many convergent lines of evidence must be developed. Natural variation needs to be identified and exploited. Data must be collected. Confounders need to be considered. Alternative explanations have to be exhaustively tested. Before anything else, the right question needs to be framed.

> Naturally, there is a desire to substitute intellectual capital for labor. That is why investigators try to base causal inference on statistical models. The technology is relatively easy to use, and promises to open a wide variety of questions to the research effort. However, the appearance of methodological rigor can be deceptive. The models themselves demand critical scrutiny. Mathematical equations are used to adjust for confounding and other sources of bias. These equations may appear formidably precise, but they typically derive from many somewhat arbitrary choices. Which variables to enter in the regression? What functional form to use? What assumptions to make about parameters and error terms? These choices are seldom dictated either by data or prior scientific knowledge. That is why judgment is so critical, the opportunity for error so large, and the number of successful applications so limited.[2]

Causal inference from randomized controlled experiments using the intention-to-treat principle is not controversial—provided the inference is based on the actual underlying probability model implicit in the randomization. But some scientists ignore the design and instead use regression to analyze data from randomized experiments. Chapters 12 and 13 show that the result is generally unsound.

Nonexperimental data range from "natural experiments," where Nature provides data as if from a randomized experiment, to observational studies where there is not even a comparison between groups. The epitome of a natural experiment is Snow's study of cholera, discussed in Chapters 3 and 20. Snow was able to show—by expending an enormous amount of shoe leather—that Nature had mixed subjects across "treatments" in a way that was tantamount to a randomized controlled experiment.

To assess how close an observational study is to an experiment requires hard work and subject-matter knowledge. Even without a real or natural experiment, a scientist with sufficient expertise and field experience may be able to combine case studies and other observational data to rule out possible confounders and make sound inferences.

Freedman was convinced by dozens of causal inferences from observational data—but not hundreds. Chapter 20 gives examples, primarily from epidemiology, and considers the implications for social science. In Freedman's view, the number of sound causal inferences from observational data in epidemiology and social sciences is limited by the difficulty of eliminating confounding. Only shoe leather and wisdom can tell good assumptions from bad ones or rule out confounders without deliberate randomization and intervention. These resources are scarce.

Researchers who rely on observational data need qualitative and quantitative evidence, including case studies. They also need to be mindful of statistical principles and alert to anomalies, which can suggest sharp research questions. No single tool is best: They must find a combination suited to the particulars of the problem.

Freedman taught students—and researchers—to evaluate the quality of information and the structure of empirical arguments. He emphasized critical thinking over technical wizardry. This focus shines through two influential textbooks. His widely acclaimed undergraduate text, *Statistics*,[3] transformed statistical pedagogy. *Statistical Models: Theory and Practice*,[4] written at the advanced undergraduate and graduate level, presents standard techniques in statistical modeling and explains their shortcomings. These texts illuminate the sometimes tenuous relationship between statistical theory and scientific applications by taking apart serious examples.

The present volume brings together twenty articles by David Freedman and co-authors on the foundations of statistics, statistical modeling, and causal inference in social science, public policy, law, and epidemiology. They show when, why, and by how much statistical modeling is likely to fail. They show that assumptions are not a good substitute for subject-matter knowledge and relevant data. They show when qualitative, shoe-leather approaches may well succeed where modeling will not. And they point out that in some situations, the only honest answer is, "we can't tell from the data available."

This book is the perfect companion to *Statistical Models*. It covers some of the same topics in greater depth and technical detail and provides more case studies and close analysis of newer and more sophisticated tools for causal inference. Like all of Freedman's writing, this compilation is engaging and a pleasure to read: vivid, clear, and dryly funny. He does not use mathematics when English will do. Two-thirds of the chapters are relatively non-mathematical, readily accessible to most readers. The entire book—except perhaps a few proofs—is within the reach of social science graduate students who have basic methods training.

Freedman sought to get to the bottom of statistical modeling. He showed that sanguine faith in statistical models is largely unfounded. Advocates of modeling have responded by inventing escape routes, attempts to rescue the models when the underlying assumptions fail. As Part III of this volume makes clear, there is no exit: The fixes ride on *other* assumptions that are often harder to think about, justify, and test than those they replace.

This volume will not end the modeling enterprise. As Freedman wrote, there will always be "a desire to substitute intellectual capital for labor" by using statistical models to avoid the hard work of examining problems in their full specificity and complexity. We hope, however, that readers will find themselves better informed, less credulous, and more alert to the moment the rabbit is placed in the hat.

Notes

1. See, e.g., Freedman and Humphreys (1999), p. 102.

2. Freedman (2003), p. 19. See also Freedman (1999), pp. 255–56.

3. David Freedman, Robert Pisani, and Roger Purves (2007). *Statistics*, 4th edn. New York: Norton.

4. David A. Freedman (2009). *Statistical Models: Theory and Practice*, rev. edn. New York: Cambridge.

Part I

Statistical Modeling:
Foundations and Limitations

1

Issues in the Foundations of Statistics:
Probability and Statistical Models

> "Son, no matter how far you travel, or how smart you get, always remember this: Someday, somewhere, a guy is going to show you a nice brand-new deck of cards on which the seal is never broken, and this guy is going to offer to bet you that the jack of spades will jump out of this deck and squirt cider in your ear. But, son, do not bet him, for as sure as you do you are going to get an ear full of cider."
>
> — Damon Runyon[1]

ABSTRACT. *After sketching the conflict between objectivists and subjectivists on the foundations of statistics, this chapter discusses an issue facing statisticians of both schools, namely, model validation. Statistical models originate in the study of games of chance and have been successfully applied in the physical and life sciences. However, there are basic problems in applying the models to social phenomena; some of the difficulties will be pointed out. Hooke's law will be contrasted with regression models for salary discrimination, the latter being a fairly typical application in the social sciences.*

Foundations of Science (1995) 1: 19–39. With kind permission from Springer Science+Business Media.

1.1 What is probability?

For a contemporary mathematician, probability is easy to define, as a countably additive set function on a σ-field, with a total mass of one. This definition, perhaps cryptic for non-mathematicians, was introduced by A. N. Kolmogorov around 1930, and has been extremely convenient for mathematical work; theorems can be stated with clarity, and proved with rigor.[2]

For applied workers, the definition is less useful; countable additivity and σ-fields are not observed in nature. The issue is of a familiar type—what objects in the world correspond to probabilities? This question divides statisticians into two camps:

   (i) the "objectivist" school, also called the "frequentists,"
   (ii) the "subjectivist" school, also called the "Bayesians," after the Reverend Thomas Bayes (England, c. 1701–61) (Bayes, 1764).

Other positions have now largely fallen into disfavor; for example, there were "fiducial" probabilities introduced by R. A. Fisher (England, 1890–1962). Fisher was one of the two great statisticians of the century; the other, Jerzy Neyman (b. Russia, 1894; d. U.S.A., 1981), turned to objectivism after a Bayesian start. Indeed, the objectivist position now seems to be the dominant one in the field, although the subjectivists are still a strong presence. Of course, the names are imperfect descriptors. Furthermore, statisticians agree amongst themselves about as well as philosophers; many shades of opinion will be represented in each school.

1.2 The objectivist position

Objectivists hold that probabilities are inherent properties of the systems being studied. For a simple example, like the toss of a coin, the idea seems quite clear at first. You toss the coin, it will land heads or tails, and the probability of heads is around 50%. A more exact value can be determined experimentally, by tossing the coin repeatedly and taking the long run relative frequency of heads. In one such experiment, John Kerrich (a South African mathematician interned by the Germans during World War II) tossed a coin 10,000 times and got 5067 heads: The relative frequency was $5067/10,000 = 50.67\%$. For an objectivist such as myself, the probability of Kerrich's coin landing heads has its own existence, separate from the data; the latter enable us to estimate the probability, or test hypotheses concerning it.

The objectivist position exposes one to certain difficulties. As Keynes said, "In the long run, we are all dead." Heraclitus (also out of context) is even more severe: "You can't step into the same river twice." Still, the tosses of a coin, like the throws of a die and the results of other such chance processes, do exhibit remarkable statistical regularities. These regularities can be described, predicted, and analyzed by technical probability theory. Using Kolmogorov's axioms (or more primitive definitions), we can construct statistical models that correspond to empirical phenomena; although verification of the correspondence is not the easiest of tasks.

1.3 The subjectivist position

For the subjectivist, probabilities describe "degrees of belief." There are two camps within the subjectivist school, the "classical" and the "radical." For a "classical" subjectivist, like Bayes himself or Laplace—although such historical readings are quite tricky—there are objective "parameters" which are unknown and to be estimated from the data. (A parameter is a numerical characteristic of a statistical model for data—for instance, the probability of a coin landing heads; other examples will be given below.) Even before data collection, the classical subjectivist has information about the parameters, expressed in the form of a "prior probability distribution."

The crucial distinction between a classical subjectivist and an objectivist: The former will make probability statements about parameters—for example, in a certain coin-tossing experiment, there is a 25% chance that the probability of heads exceeds .67. However, objectivists usually do not find that such statements are meaningful; they view the probability of heads as an unknown constant, which either is—or is not—bigger than .67. In replications of the experiment, the probability of heads will always exceed .67, or never; 25% cannot be relevant. As a technical matter, if the parameter has a probability distribution given the data, it must have a "marginal" distribution—that is, a prior. On this point, objectivists and subjectivists agree; the hold-out was R. A. Fisher, whose fiducial probabilities come into existence only after data collection.

"Radical" subjectivists, like Bruno de Finetti or Jimmie Savage, differ from classical subjectivists and objectivists; radical subjectivists deny the very existence of unknown parameters. For such statisticians, probabilities express degrees of belief about observables. You pull a coin out of your pocket, and—Damon Runyon notwithstanding—they can assign a probability to the event that it will land heads when you toss it. The braver ones can even assign a probability to the event that you really will toss the coin. (These are "prior" probabilities, or "opinions.") Subjectivists can also "update" opinions in the light of the data; for example, if the coin is tossed ten times, landing heads six times and tails four times, what is the chance that it will land heads on the eleventh toss? This involves computing a "conditional" probability using Kolmogorov's calculus, which applies whether the probabilities are subjective or objective.

Here is an example with a different flavor: What is the chance that a Republican will be president of the U.S. in the year 2025? For many subjectivists, this is a meaningful question, which can in principle be answered by introspection. For many objectivists, this question is beyond the scope of statistical theory. As best I can judge, however, complications will be found on both sides of the divide. Some subjectivists will not have quantifiable opinions about remote political events; likewise, there are objectivists who might develop statistical models for presidential elections, and compute probabilities on that basis.[3]

The difference between the radical and classical subjectivists rides on the distinction between parameters and observables; this distinction is made by objectivists too and is often quite helpful. (In some cases, of course, the issue may be rather subtle.) The radical subjectivist denial of parameters exposes members of this school to some rhetorical awkwardness; for example, they are required not to understand the idea of tossing a coin with an unknown probability of heads. Indeed, if they admit the coin, they will soon be stuck with all the unknown parameters that were previously banished.[4]

1.3.1 Probability and relative frequency

In ordinary language, "probabilities" are not distinguished at all sharply from empirical percentages—"relative frequencies." In statistics, the distinction may be more critical. With Kerrich's coin, the relative frequency of heads in 10,000 tosses, 50.67%, is unlikely to be the exact probability of heads; but it is unlikely to be very far off. For an example with a different texture, suppose you see the following sequence of ten heads and ten tails:

$$T\,H\,T\,H\,T\,H\,T\,H\,T\,H\,T\,H\,T\,H\,T\,H\,T\,H\,T\,H.$$

What is the probability that the next observation will be a head? In this case, relative frequency and probability are quite different.[5]

One more illustration along that line: United Airlines Flight 140 operates daily from San Francisco to Philadelphia. In 192 out of the last 365 days, Flight 140 landed on time. You are going to take this flight tomorrow. Is your probability of landing on time given by 192/365? For a radical subjectivist, the question is clear; not so for an objectivist or a classical subjectivist. Whatever the question really means, 192/365 is the wrong answer—if you are flying on the Friday before Christmas. This is Fisher's "relevant subset" issue; and he seems to have been anticipated by von Mises. Of course, if you pick a day at random from the data set, the chance of getting one with an on-time landing is indeed 192/365; that would not be controversial. The difficulties come with (i) extrapolation and (ii) judging the exchangeability of the data, in a useful Bayesian phrase. Probability is a subtler idea than relative frequency.[6]

1.3.2 Labels do not settle the issue

Objectivists sometimes argue that they have the advantage, because science is objective. This is not serious; "objectivist" statistical analysis must often rely on judgment and experience: Subjective elements come in. Likewise, subjectivists may tell you that objectivists (i) use "prior information," and (ii) are therefore closet Bayesians. Point (i) may be granted. The issue for (ii) is how prior information enters the analysis, and whether this information can be quantified or updated the way subjectivists insist it must be. The real questions are not to be settled on the basis of labels.

1.4 A critique of the subjectivist position

The subjectivist position seems to be internally consistent, and fairly immune to logical attack from the outside. Perhaps as a result, scholars of that school have been quite energetic in pointing out the flaws in the objectivist position. From an applied perspective, however, the subjectivist position is not free of difficulties either. What are subjective degrees of belief, where do they come from, and why can they be quantified? No convincing answers have been produced. At a more practical level, a Bayesian's opinion may be of great interest to himself, and he is surely free to develop it in any way that pleases him; but why should the results carry any weight for others?

To answer the last question, Bayesians often cite theorems showing "inter-subjective agreement." Under certain circumstances, as more and more data become available, two Bayesians will come to agree: The data swamp the prior. Of course, other theorems show that the prior swamps the data, even when the size of the data set grows without bounds— particularly in complex, high-dimensional situations. (For a review, see Diaconis and Freedman 1986.) Theorems do not settle the issue, especially for those who are not Bayesians to start with.

My own experience suggests that neither decision-makers nor their statisticians do in fact have prior probabilities. A large part of Bayesian statistics is about what you would do *if* you had a prior.[7] For the rest, statisticians make up priors that are mathematically convenient or attractive. Once used, priors become familiar; therefore, they come to be accepted as "natural" and are liable to be used again. Such priors may eventually generate their own technical literature.

1.4.1 Other arguments for the Bayesian position

Coherence. Well-known theorems, including one by Freedman and Purves (1969), show that stubborn non-Bayesian behavior has costs. Your opponents can make a "dutch book," and extract your last penny—if you are generous enough to cover all the bets needed to prove the results.[7] However, most of us don't bet at all; even the professionals bet on relatively few events. Thus, coherence has little practical relevance. (Its rhetorical power is undeniable—who wants to be incoherent?)

Rationality. It is often urged that to be rational is to be Bayesian. Indeed, there are elaborate axiom systems about preference orderings, acts, consequences, and states of nature, whose conclusion is—that you are a Bayesian. The empirical evidence shows, fairly clearly, that those axioms do not describe human behavior at all well. The theory is not descriptive; people do not have stable, coherent prior probabilities.

Now the argument shifts to the "normative": If you were rational, you would obey the axioms and be a Bayesian. This, however, assumes what must be proved. Why would a rational person obey those axioms? The axioms represent decision problems in schematic and highly stylized ways. Therefore, as I see it, the theory addresses only limited aspects of rationality. Some Bayesians have tried to win this argument on the cheap: To be rational is, by definition, to obey their axioms. (Objectivists do not always stay on the rhetorical high road either.)

Detailed examination of the flaws in the normative argument is a complicated task, beyond the scope of the present article. In brief, my position is this. Many of the axioms, on their own, have considerable normative force. For example, if I am found to be in violation of the "sure thing principle," I would probably reconsider.[9] On the other hand, taken as a whole, decision theory seems to have about the same connection to real decisions as war games do to real wars.

What are the main complications? For some events, I may have a rough idea of likelihood: One event is very likely, another is unlikely, a third is uncertain. However, I may not be able to quantify these likelihoods, even to one or two decimal places; and there will be many events whose probabilities are simply unknown—even if definable.[10] Likewise, there are some benefits that can be assessed with reasonable accuracy; others can be estimated only to rough orders of magnitude; in some cases, quantification may not be possible at all. Thus, utilities may be just as problematic as priors.

The theorems that derive probabilities and utilities from axioms push the difficulties back one step.[11] In real examples, the existence of many states of nature must remain unsuspected. Only some acts can be contemplated; others are not imaginable until the moment of truth arrives. Of the acts that can be imagined, the decision-maker will have preferences between some pairs but not others. Too, common knowledge suggests that consequences are often quite different in the foreseeing and in the experiencing.

Intransitivity would be an argument for revision, although not a decisive one; for example, a person choosing among several job offers might well have intransitive preferences, which it would be a mistake to ignore. By way of contrast, an arbitrageur who trades bonds intransitively is likely to lose a lot of money. (There is an active market in bonds, while the market in job offers—largely nontransferable—must be rather thin; the practical details make a difference.) The axioms do not capture the texture of real decision making. Therefore, the theory has little normative force.

The fallback defense. Some Bayesians will concede much of what I have said: The axioms are not binding; rational decision-makers may have neither priors nor utilities. Still, the following sorts of arguments can be heard. The decision-maker must have some ideas about relative likelihoods for a few events; a prior probability can be made up to capture such intuitions, at least in gross outline. The details (for instance, that distributions are normal) can be chosen on the basis of convenience. A utility function can be put together using similar logic: The decision-maker must perceive some consequences as very good, and big utility numbers can be assigned to these; he must perceive some other consequences as trivial, and small utilities can be assigned to those; and in between is in between. The Bayesian engine can now be put to work, using such approximate priors and utilities. Even with these fairly crude approximations, Bayesian analysis is held to dominate other forms of inference: That is the fallback defense.

Here is my reaction to such arguments. Approximate Bayesian analysis may in principle be useful. That this mode of analysis dominates other forms of inference, however, seems quite debatable. In a statistical decision problem, where the model and loss function are given, Bayes procedures are often hard to beat, as are objectivist likelihood procedures; with many of the familiar textbook models, objectivist and subjectivist procedures should give similar results if the data set is large. There are sharp mathematical theorems to back up such statements.[12] On the other hand, in real problems—where models and loss functions are mere approximations—the optimality of Bayes procedures cannot be a mathematical proposition. And empirical proof is conspicuously absent.

If we could quantify breakdowns in model assumptions, or degrees of error in approximate priors and loss functions, the balance of argument might shift considerably. The rhetoric of "robustness" may suggest that such error analyses are routine. This is hardly the case even for the models. For priors and utilities, the position is even worse, since the entities being approximated do not have any independent existence—outside the Bayesian framework that has been imposed on the problem.

De Finetti's theorem. Suppose you are a radical subjectivist, watching a sequence of 0's and 1's. In your prior opinion, this sequence is exchangeable: Permuting the order of the variables will not change your opinion about them. A beautiful theorem of de Finetti's asserts that your opinion can be represented as coin tossing, the probability of heads being selected at random from a suitable prior distribution. This theorem is often said to "explain" subjective or objective probabilities, or justify one system in terms of the other.[13]

Such claims cannot be right. What the theorem does is this: It enables the subjectivist to discover features of his prior by mathematical proof, rather than introspection. For example, suppose you have an exchangeable prior about those 0's and 1's. Before data collection starts, de Finetti will prove to you by pure mathematics that in your own opinion the relative frequency of 1's among the first n observations will almost surely converge to a limit as $n \to \infty$. (Of course, the theorem has other consequences too, but all have the same logical texture.)

This notion of "almost surely," and the limiting relative frequency, are features of your opinion not of any external reality. ("Almost surely" means with probability 1, and the probability in question is your prior.) Indeed, if you had not noticed these consequences of your prior by introspection, and now do not like them, you are free to revise your opinion— which will have no impact outside your head. What the theorem does is to show how various aspects of your prior opinion are related to each other. That is all the theorem can do, because the conditions of the theorem are conditions on the prior alone.

To illustrate the difficulty, I cite an old friend rather than making a new enemy. According to Jeffrey (1983, p. 199), de Finetti's result proves "your subjective probability measure [is] a certain mixture or weighted average of the various possible objective probability measures"—an unusually clear statement of the interpretation that I deny. Each of Jeffrey's "objective" probability measures governs the tosses of a p-coin, where p is your limiting relative frequency of 1's. (Of course, p has a probability distribution of its own, in your opinion.) Thus, p is a feature of your opinion, not of the real world: The mixands in de Finetti's theorem are "objective" only by terminological courtesy. In short, the "p-coins" that come out of de Finetti's theorem are just as subjective as the prior that went in.

1.4.2 To sum up

The theory—as developed by Ramsey, von Neumann and Morgenstern, de Finetti, and Savage, among others—is great work. They solved an important historical problem of interest to economists, mathematicians, statisticians, and philosophers alike. On a more practical level, the language of subjective probability is evocative. Some investigators find the consistency of Bayesian statistics to be a useful discipline; for some (including me), the Bayesian approach can suggest statistical procedures whose behavior is worth investigating. But the theory is not a complete account of rationality, or even close. Nor is it the prescribed solution for any large number of problems in applied statistics, at least as I see matters.

1.5 Statistical models

Of course, statistical models are applied not only to coin tossing but also to more complex systems. For example, "regression models" are widely used in the social sciences, as indicated below; such applications raise serious epistemological questions. (This idea will be developed from an objectivist perspective, but similar issues are felt in the other camp.)

The problem is not purely academic. The census suffers an undercount, more severe in some places than others; if certain statistical models are to be believed, the undercount can be corrected—moving seats in Congress and millions of dollars a year in entitlement funds (*Survey Methodology* (1992) 18(1); *Jurimetrics* (1993) 34(1); *Statistical Science* (1994) 9(4). If yet other statistical models are to be believed, the veil of secrecy can be lifted from the ballot box, enabling the experts to determine how racial or ethnic groups have voted—a crucial step in litigation to enforce minority voting rights (*Evaluation Review*, (1991) 1(6); Klein and Freedman, 1993).

1.5.1 Examples

Here, I begin with a noncontroversial example from physics, namely, Hooke's law: Strain is proportional to stress. We will have some number n of observations. For the ith observation, indicated by the subscript i, we hang weight$_i$ on a spring. The length of the spring is measured as length$_i$. The regression model says that[14]

(1) $$\text{length}_i = a + b \times \text{weight}_i + \epsilon_i.$$

The "error" term ϵ_i is needed because length$_i$ will not be exactly equal to $a + b \times$ weight$_i$. If nothing else, measurement error must be reckoned with. We model ϵ_i as a sequence of draws, made at random with replacement from a box of tickets; each ticket shows a potential error—the ϵ_i that will be realized if that ticket is the ith one drawn. The average of all the potential errors in the box is assumed to be 0. In more standard terminology, the ϵ_i are assumed to be "independent and identically distributed, with mean 0." Such assumptions can present difficult scientific issues, because error terms are not observable.

In equation (1), a and b are parameters, unknown constants of nature that characterize the spring: a is the length of the spring under no load, and b is stretchiness—the increase in length per unit increase in weight. These parameters are not observable, but they can be estimated by "the method of least squares," developed by Adrien-Marie Legendre (France, 1752–1833) and Carl Friedrich Gauss (Germany, 1777–1855) to fit astronomical orbits. Basically, you choose the values of $\hat{a}$ and $\hat{b}$ to minimize the sum of the squared "prediction errors," $\sum_i e_i{}^2$, where e_i is the prediction error for the ith observation:[15]

$$(2) \qquad e_i = \text{length}_i - \hat{a} - \hat{b} \times \text{weight}_i.$$

These prediction errors are often called "residuals": They measure the difference between the actual length and the predicted length, the latter being $\hat{a} - \hat{b} \times$ weight.

No one really imagines there to be a box of tickets hidden in the spring. However, the variability of physical measurements (under many but by no means all circumstances) does seem to be remarkably like the variability in draws from a box. This is Gauss' model for measurement error. In short, statistical models can be constructed that correspond rather closely to empirical phenomena.

I turn now to social-science applications. A case study would take us too far afield, but a stylized example—regression analysis used to demonstrate sex discrimination in salaries (adapted from Kaye and Freedman, 2000)—may give the idea. We use a regression model to predict salaries (dollars per year) of employees in a firm from:

- education (years of schooling completed),
- experience (years with the firm),
- the dummy variable "man," which takes the value 1 for men and 0 for women.

Employees are indexed by the subscript i; for example, salary$_i$ is the salary of the ith employee. The equation is[16]

$$(3) \quad \text{salary}_i = a + b \times \text{education}_i + c \times \text{experience}_i + d \times \text{man}_i + \epsilon_i.$$

Equation (3) is a statistical model for the data, with unknown parameters a, b, c, d; here, a is the "intercept" and the others are "regression coefficients"; ϵ_i is an unobservable error term. This is a formal analog of Hooke's law (1); the same assumptions are made about the errors. In other words, an employee's salary is determined as if by computing

(4) $a + b \times \text{education} + c \times \text{experience} + d \times \text{man}$,

then adding an error drawn at random from a box of tickets. The display (4) is the expected value for salary given the explanatory variables (education, experience, man); the error term in (3) represents deviations from the expected.

The parameters in (3) are estimated from the data using least squares. If the estimated coefficient d for the dummy variable turns out to be positive and "statistically significant" (by a "t-test"), that would be taken as evidence of disparate impact: Men earn more than women, even after adjusting for differences in background factors that might affect productivity. Education and experience are entered into equation (3) as "statistical controls," precisely in order to claim that adjustment has been made for differences in backgrounds.

Suppose the estimated equation turns out as follows:

(5)   predicted salary $= \$7100 + \$1300 \times \text{education}$
$+ \$2200 \times \text{experience} + \$700 \times \text{man}$.

That is, $\hat{a} = \$7100$, $\hat{b} = \$1300$, and so forth. According to equation (5), every extra year of education is worth on average \$1300; similarly, every extra year of experience is worth on average \$2200; and, most important, men get a premium of \$700 over women with the same education and experience, on average.

An example will illustrate (5). A male employee with twelve years of education (high school) and ten years of experience would have a predicted salary of

(6) $\$7100 + \$1300 \times 12 + \$2200 \times 10 + \$700 \times 1$
$= \$7100 + \$15,600 + \$22,000 + \$700$
$= \$45,400.$

A similarly situated female employee has a predicted salary of only

(7) $\$7100 + \$1300 \times 12 + \$2200 \times 10 + \$700 \times 0$
$= \$7100 + \$15,600 + \$22,000 + \$0$
$= \$44,700.$

Notice the impact of the dummy variable: $700 is added to (6), but not to (7).

A major step in the argument is establishing that the estimated coefficient of the dummy variable in (3) is "statistically significant." This step turns out to depend on the statistical assumptions built into the model. For instance, each extra year of education is assumed to be worth the same (on average) across all levels of experience, both for men and women. Similarly, each extra year of experience is worth the same across all levels of education, both for men and women. Furthermore, the premium paid to men does not depend systematically on education or experience. Ability, quality of education, or quality of experience are assumed not to make any systematic difference to the predictions of the model.

The story about the error term—that the ϵ's are independent and identically distributed from person to person in the data set—turns out to be critical for computing statistical significance. Discrimination cannot be proved by regression modeling unless statistical significance can be established, and statistical significance cannot be established unless conventional presuppositions are made about unobservable error terms.

Lurking behind the typical regression model will be found a host of such assumptions; without them, legitimate inferences cannot be drawn from the model. There are statistical procedures for testing some of these assumptions. However, the tests often lack the power to detect substantial failures. Furthermore, model testing may become circular; breakdowns in assumptions are detected, and the model is redefined to accommodate. In short, hiding the problems can become a major goal of model building.

Using models to make predictions of the future, or the results of interventions, would be a valuable corrective. Testing the model on a variety of data sets—rather than fitting refinements over and over again to the same data set—might be a good second-best (Ehrenberg and Bound 1993). With Hooke's law (1), the model makes predictions that are relatively easy to test experimentally. For the salary discrimination model (3), validation seems much more difficult. Thus, built into the equation is a model for nondiscriminatory behavior: The coefficient d vanishes. If the company discriminates, that part of the model cannot be validated at all.

Regression models like (3) are widely used by social scientists to make causal inferences; such models are now almost a routine way of demonstrating counterfactuals. However, the "demonstrations" generally turn out to depend on a series of untested, even unarticulated, technical assumptions. Under the circumstances, reliance on model outputs may be quite unjustified. Making the ideas of validation somewhat more precise is a serious problem in the philosophy of science. That models should correspond to reality is, after all, a useful but not totally straightforward idea—with some history to it. Developing appropriate models is a serious problem in statistics; testing the connection to the phenomena is even more serious.[17]

1.5.2 Standard errors, t-statistics, and statistical significance

The "standard error" of $\hat{d}$ measures the likely difference between $\hat{d}$ and d, due to the action of the error terms in equation (3). The "t-statistic" is $\hat{d}$ divided by its standard error. Under the "null hypothesis" that $d = 0$, there is only about a 5% chance that $|t| > 2$. Such a large value of t would demonstrate "statistical significance." Of course, the parameter d is only a construct in a model. If the model is wrong, the standard error, t-statistic, and significance level are rather difficult to interpret.

Even if the model is granted, there is a further issue: The 5% is a probability for the data given the model, namely, $P\{|t| > 2 \,||\, d = 0\}$. However, the 5% is often misinterpreted as $P\{d = 0|\text{data}\}$. Indeed, this misinterpretation is a commonplace in the social-science literature, and seems to have been picked up by the courts from expert testimony.[18] For an objectivist, $P\{d = 0|\text{data}\}$ makes no sense: Parameters do not exhibit chance variation. For a subjectivist, $P\{d = 0|\text{data}\}$ makes good sense, but its computation via the t-test is grossly wrong, because the prior probability that $d = 0$ has not been taken into account: The calculation exemplifies the "base rate fallacy."

The single vertical bar "|" is standard notation for conditional probability. The double vertical bar "||" is not standard; Bayesians might want to read this as a conditional probability; for an objectivist, "||" is intended to mean "computed on the assumption that . . . ."

1.5.3 Statistical models and the problem of induction

How do we learn from experience? What makes us think that the future will be like the past? With contemporary modeling techniques, such questions are easily answered—in form if not in substance.

- The objectivist invents a regression model for the data, and assumes the error terms to be independent and identically distributed; "IID" is the conventional abbreviation. It is this assumption of IID-ness that enables us to predict data we have not seen from a training sample— without doing the hard work of validating the model.

- The classical subjectivist invents a regression model for the data, assumes IID errors, and then makes up a prior for unknown parameters.

- The radical subjectivist adopts a prior that is exchangeable or partially exchangeable, and calls you irrational or incoherent (or both) for not following suit.

In our days, serious arguments have been made from data. Beautiful, delicate theorems have been proved, although the connection with data analysis often remains to be established. And an enormous amount of fiction has been produced, masquerading as rigorous science.

1.6 Conclusions

I have sketched two main positions in contemporary statistics, objectivist and subjectivist, and tried to indicate the difficulties. Some questions confront statisticians from both camps. How do statistical models connect with reality? What areas lend themselves to investigation by statistical modeling? When are such investigations likely to be sterile?

These questions have philosophical components as well as technical ones. I believe model validation to be a central issue. Of course, many of my colleagues will be found to disagree. For them, fitting models to data, computing standard errors, and performing significance tests is "informative," even though the basic statistical assumptions (linearity, independence of errors, etc.) cannot be validated. This position seems indefensible, nor are the consequences trivial. Perhaps it is time to reconsider.

Notes

1. From "The Idyll of Miss Sarah Brown," *Collier's Magazine*, 1933. Reprinted in *Guys and Dolls: The Stories of Damon Runyon*. Penguin Books, New York, 199 pp. 14–26. The quote is edited slightly, for continuity.

2. This note will give a compact statement of Kolmogorov's axioms. Let Ω be a set. By definition, a σ-field $\mathcal{F}$ is a collection of subsets of Ω, which has Ω itself as a member. Furthermore,

(i) $\mathcal{F}$ is closed under complementation (if $A \in \mathcal{F}$ then $A^c \in \mathcal{F}$), and

(ii) $\mathcal{F}$ is closed under the formation of countable unions (if $A_i \in \mathcal{F}$ for $i = 1, 2, \ldots$, then $\bigcup_i A_i \in \mathcal{F}$).

A probability P is a non-negative, real-valued function on $\mathcal{F}$ such that $P(\Omega) = 1$ and P is "countably additive": If $A_i \in \mathcal{F}$ for $i = 1, 2, \ldots$, and the sets are pairwise disjoint, in the sense that $A_i \cap A_j = 0$ for $i \neq j$, then $P(\bigcup_i A_i) = \sum_i P(A_i)$. A random variable X is an $\mathcal{F}$-measurable function on Ω. Informally, probabilists might say that Nature chooses

$\omega \in \Omega$ according to P, and shows you $X(\omega)$; the latter would be the "observed value" of X.

3. Models will be discussed in Section 1.5. Those for presidential elections may not be compelling. For genetics, however, chance models are well established; and many statistical calculations are therefore on a secure footing. Much controversy remains, for example, in the area of DNA identification (*Jurimetrics* (1993) 34(1)).

4. The distinction between classical and radical subjectivists made here is not often discussed in the statistical literature; the terminology is not standard. See, for instance, Diaconis and Freedman (1980a), Efron (1986), and Jeffrey (1983, section 12.6).

5. Some readers may say to themselves that here, probability is just the relative frequency of transitions. However, a similar but slightly more complicated example can be rigged up for transition counts. An infinite regress lies just ahead. My point is only this: Relative frequencies are not probabilities. Of course, if circumstances are favorable, the two are strongly connected—that is one reason why chance models are useful for applied work.

6. To illustrate the objectivist way of handling probabilities and relative frequencies, I consider repeated tosses of a fair coin: The probability of heads is 50%. In a sequence of 10,000 tosses, the chance of getting between 49% and 51% heads is about 95%. In replications of this (large) experiment, about 95% of the time, there will be between 49% and 51% heads. On each replication, however, the probability of heads stays the same—namely, 50%.

The strong law of large numbers provides another illustration. Consider n repeated tosses of a fair coin. With probability 1, as $n \rightarrow \infty$, the relative frequency of heads in the first n tosses eventually gets trapped inside the interval from 49% to 51%; ditto, for the interval from 49.9% to 50.1%; ditto, for the interval from 49.99% to 50.01%; and so forth. No matter what the relative frequency of heads happens to be at any given moment, the probability of heads stays the same—namely, 50%. Probability is not relative frequency.

7. Similarly, a large part of objectivist statistics is about what you would do *if* you had a model; and all of us spend enormous amounts of energy finding out what would happen if the data kept pouring in. I wish we could learn to look at the data more directly, without the fictional models and priors. On the same wish-list: We should stop pretending to fix bad designs and inadequate measurements by modeling.

8. A "dutch book" is a collection of bets on various events such that the bettor makes money, no matter what the outcome.

9. According to the "sure thing principle," if I prefer A to B given that C occurs, and I also prefer A to B given that C does not occur, I must prefer A to B when I am in doubt as to the occurrence of C.

10. Although one-sentence concessions in a book are not binding, Savage (1972 [1954], p. 59) does say that his theory "is a code of consistency for the person applying it, not a system of predictions about the world"; and personal probabilities can be known "only roughly." Another comment on this book may be in order. According to Savage (1972 [1954], pp. 61–62), "on no ordinary objectivistic view would it be meaningful, let alone true, to say that on the basis of the available evidence it is very improbable, though not impossible, that France will become a monarchy within the next decade." As anthropology of science, this seems wrong. I make qualitative statements about likelihoods and possibilities, and expect to be understood; I find such statements meaningful when others make them. Only the quantification seems problematic. What would it mean to say that P(France will become a monarchy) = .0032? Many objectivists of my acquaintance share such views, although caution is in order when extrapolating from such a sample of convenience.

11. The argument in the text is addressed to readers who have some familiarity with the axioms. This note gives a very brief review; Kreps (1988) has a chatty and sympathetic discussion (although some of the details are not quite in focus); Le Cam (1977) is more technical and critical; the arguments are crisp. In the axiomatic setup, there is a space of "states of nature," like the possible orders in which horses finish a race. There is another space of "consequences"; these can be pecuniary or non-pecuniary (win $1000, lose $5000, win a weekend in Philadelphia, etc.). Mathematically, an "act" is a function whose domain is the space of states of nature and whose values are consequences. You have to choose an act: That is the decision problem. Informally, if you choose the act f, and the state of nature happens to be s, you enjoy (or suffer) the consequence $f(s)$. For example, if you bet on those horses, the payoff depends on the order in which they finish: The bet is an act, and the consequence depends on the state of nature. The set of possible states of nature, the set of possible consequences, and the set of possible acts are all viewed as fixed and known. You are supposed to have a transitive preference ordering on the acts, not just the consequences. The sure thing principle is an axiom in Savage's setup.

12. Wald's idea of a statistical decision problem can be sketched as follows. There is an unobservable parameter θ. Corresponding to each

θ, there is a known probability distribution P_θ for an observable random quantity X. (This family of probability distributions is a "statistical model" for X, with parameter θ.) There is a set of possible "decisions"; there is a "loss function" $L(d, \theta)$ which tells you how much is lost by making the decision d when the parameter is really θ. (For example, d might be an estimate of θ, and loss might be squared error.) You have to choose a "decision rule," which is a mapping from observed values of X to decisions. Your objective is to minimize "risk," that is, expected loss.

A comparison with the setup in note 11 may be useful. The "state of nature" seems to consist of the observable value of X, together with the unobservable value θ of the parameter. The "consequences" are the decisions, and "acts" are decision rules. (The conflict in terminology is regrettable, but there is no going back.) The utility function is replaced by L, which is given but depends on θ as well as d.

The risk of a Bayes' procedure cannot be reduced for all values of θ; any "admissible" procedure is a limit of Bayes' procedures ("the complete class theorem"). The maximum-likelihood estimator is "efficient"; and its sampling distribution is close to the posterior distribution of θ by the "Bernstein–von Mises theorem," which is actually due to Laplace. More or less stringent regularity conditions must be imposed to prove any of these results, and some of the theorems must be read rather literally; Stein's paradox and Bahadur's example should at least be mentioned.

Standard monographs and texts include Berger (1985), Berger and Wolpert (1988), Bickel and Doksum (1977), Casella and Berger (1990), Ferguson (1967), Le Cam (1986), Lehmann and Casella (2003), Lehmann and Romano (2005), and Rao (1973). The Bernstein–von Mises theorem is discussed in Le Cam and Yang (1990) and Prakasa Rao (1987). Of course, in many contexts, Bayes procedures and frequentist procedures will go in opposite directions; for a review, see Diaconis and Freedman (1986). These references are all fairly technical.

13. Diaconis and Freedman (1980a,b; 1981) review the issues and the mathematics. The first-cited paper is relatively informal; the second gives a version of de Finetti's theorem applicable to a finite number of observations, with bounds; the last gives a fairly general mathematical treatment of partial exchangeability, with numerous examples, and is more technical. More recent work is described in Diaconis and Freedman (1988, 1990).

The usual hyperbole can be sampled in Kreps (1988, p. 145): de Finetti's theorem is "the fundamental theorem of statistical inference— the theorem that from a subjectivist point of view makes sense out of most statistical procedures." This interpretation of the theorem fails to distinguish between what is assumed and what is proved. It is the assumption of exchangeability that enables you to predict the future from the past, at least to your own satisfaction—not the conclusions of the theorem or the elegance of the proof. Also see Section 1.5. If you pretend to have an exchangeable prior, the statistical world is your oyster, de Finetti or no de Finetti.

14. The equation holds for quite a large range of weights. With large enough weights, a quadratic term will be needed in equation (1). Moreover, beyond some point, the spring passes its "elastic limit" and snaps. The law is named after Robert Hooke, England, 1653–1703.

15. The residual e_i is observable, but is only an approximation to the disturbance term ϵ_i in (1); that is because the estimates $\hat{a}$ and $\hat{b}$ are only approximations to the parameters a and b.

16. Such equations are suggested, somewhat loosely, by "human capital theory." However, there remains considerable uncertainty about which variables to put into the equation, what functional form to assume, and how error terms are supposed to behave. Adding more variables is no panacea: Freedman (1983) and Clogg and Haritou (1997).

17. For more discussion in the context of real examples, with citations to the literature of model validation, see Freedman (1985, 1987, 1991 [Chapter 3], 1997). Many recent issues of *Sociological Methodology* have essays on this topic. Also see Oakes (1990), who discusses modeling issues, significance tests, and the objectivist-subjectivist divide.

18. Some legal citations may be of interest (Kaye and Freedman 2000): *Waisome v. Port Authority*, 948 F.2d 1370, 1376 (2d Cir. 1991) ("Social scientists consider a finding of two standard deviations significant, meaning there is about 1 chance in 20 that the explanation for a deviation could be random"); *Rivera v. City of Wichita Falls*, 665 F.2d 531, 545 n.22 (5th Cir. 1982) ("A variation of two standard deviations would indicate that the probability of the observed outcome occurring purely by chance would be approximately five out of 100; that is, it could be said with a 95% certainty that the outcome was not merely a fluke."); *Vuyanich v. Republic Nat'l Bank*, 505 F. Supp. 22 271 (N.D. Tex. 1980), vacated and remanded, 723 F.2d 1195 (5th Cir. 1984) ("if a 5% level of significance is used, a sufficiently large t-statistic for the coefficient indicates that the chances are less than one in 20 that the true coefficient is actually zero.").

An example from the underlying technical literature may also be of interest. According to (Fisher 1980, p. 717), "in large samples, a t-statistic of approximately two means that the chances are less than one in twenty that the true coefficient is actually zero and that we are observing a larger coefficient just by chance.... A t-statistic of approximately two and one half means the chances are only one in one hundred that the true coefficient is zero...." No. If the true coefficient is zero, there is only one chance in one hundred that $|t| > 2.5$. (Frank Fisher is a well-known econometrician who often testifies as an expert witness, although I do not believe he figures in any of the cases cited above.)

Acknowledgments

I would like to thank Dick Berk, Cliff Clogg, Persi Diaconis, Joe Eaton, Neil Henry, Paul Humphreys, Lucien Le Cam, Diana Petitti, Brian Skyrms, Terry Speed, Steve Turner, Amos Tversky, Ken Wachter, and Don Ylvisaker for many helpful suggestions—some of which I could implement.

2

Statistical Assumptions
as Empirical Commitments

With Richard A. Berk

ABSTRACT. *Statistical inference with convenience samples is a risky business. Technical issues and substantive issues overlap. No amount of statistical maneuvering can get very far without deep understanding of how the data were generated. Empirical generalizations from a single data set should be viewed with suspicion. Rather than ask what would happen in principle if the study were repeated, it is better to repeat the study— as is standard in physical science. Indeed, it is generally impossible to predict variability across replications of an experiment without replicating the experiment, just as it is generally impossible to predict the effect of intervention without actually intervening.*

2.1 Introduction

Researchers who study punishment and social control, like those who study other social phenomena, typically seek to generalize their findings from the data they have to some larger context: In statistical jargon, they

Law, Punishment, and Social Control: Essays in Honor of Sheldon Messinger (2005) 2nd edn. T. G. Blomberg and S. Cohen, eds. Aldine de Gruyter, pp. 235–54. Copyright © 2003 by Aldine Publishers. Reprinted by permission of AldineTransaction, a division of Transaction Publishers.

generalize from a sample to a population. Generalizations are one important product of empirical inquiry. Of course, the process by which the data are selected introduces uncertainty. Indeed, any given data set is but one of many that could have been studied. If the data set had been different, the statistical summaries would have been different, and so would the conclusions, at least by a little.

How do we calibrate the uncertainty introduced by data collection? Nowadays, this question has become quite salient, and it is routinely answered using well-known methods of statistical inference, with standard errors, t-tests, and P-values, culminating in the "tabular asterisks" of Meehl (1978). These conventional answers, however, turn out to depend critically on certain rather restrictive assumptions, for instance, random sampling.[1]

When the data are generated by random sampling from a clearly defined population, and when the goal is to estimate population parameters from sample statistics, statistical inference can be relatively straightforward. The usual textbook formulas apply; tests of statistical significance and confidence intervals follow.

If the random-sampling assumptions do not apply, or the parameters are not clearly defined, or the inferences are to a population that is only vaguely defined, the calibration of uncertainty offered by contemporary statistical technique is in turn rather questionable.[2]

Thus, investigators who use conventional statistical technique turn out to be making, explicitly or implicitly, quite restrictive behavioral assumptions about their data collection process. By using apparently familiar arithmetic, they have made substantial empirical commitments; the research enterprise may be distorted by statistical technique, not helped. At least, that is our thesis, which we will develop in the pages that follow.

Random sampling is hardly universal in contemporary studies of punishment and social control. More typically, perhaps, the data in hand are simply the data most readily available (e.g., Gross and Mauro 1989; MacKenzie 1991; Nagin and Paternoster 1993; Berk and Campbell 1993; Phillips and Grattet 2000; White 2000). For instance, information on the use of prison "good time" may come from one prison in a certain state. Records on police use of force may be available only for encounters in which a suspect requires medical attention. Prosecutors' charging decisions may be documented only after the resolution of a lawsuit.

"Convenience samples" of this sort are not random samples. Still, researchers may quite properly be worried about replicability. The generic concern is the same as for random sampling: If the study were repeated, the results would be different. What, then, can be said about the results obtained? For example, if the study of police use of force were repeated, it is almost certain that the sample statistics would change. What can be concluded, therefore, from the statistics?

These questions are natural, but may be answerable only in certain contexts. The moment that conventional statistical inferences are made from convenience samples, substantive assumptions are made about how the social world operates. Conventional statistical inferences (e.g., formulas for the standard error of the mean, t-tests, etc.) depend on the assumption of random sampling. This is not a matter of debate or opinion; it is a matter of mathematical necessity.[3] When applied to convenience samples, the random-sampling assumption is not a mere technicality or a minor revision on the periphery; the assumption becomes an integral part of the theory.

In the pages ahead, we will try to show how statistical and empirical concerns interact. The basic question will be this: What kinds of social processes are assumed by the application of conventional statistical techniques to convenience samples? Our answer will be that the assumptions are quite unrealistic. If so, probability calculations that depend on the assumptions must be viewed as unrealistic too.[4]

2.2 Treating the data as a population

Suppose that one has data from spouse abuse victims currently residing in a particular shelter. A summary statistic of interest is the proportion of women who want to obtain restraining orders. How should potential uncertainty be considered?

One strategy is to treat the women currently residing in the shelter as a population; the issue of what would happen if the study were repeated does not arise. All the investigator cares about are the data now in hand. The summary statistics describe the women in the data set. No statistical inference is needed since there is no sampling uncertainty to worry about.

Treating the data as a population and discarding statistical inference might well make sense if the summary statistics are used to plan for current shelter residents. A conclusion that "most" want to obtain restraining orders is one thing; a conclusion that a "few" want to obtain such orders has different implications. But there are no inferences about women who might use the shelter in the future, or women residing in other shelters. In short, the ability to generalize has been severely restricted.

2.3 Assuming a real population
and an imaginary sampling mechanism

Another way to treat uncertainty is to define a real population and assume that the data can be treated as a random sample from that population. Thus, current shelter residents could perhaps be treated as a random sample drawn from the population of residents in all shelters in the area during the previous twelve months. This "as-if" strategy would seem to set the stage for statistical business as usual.

An explicit goal of the "as-if" strategy is generalizing to a specific population. And one issue is this: Are the data representative? For example, did each member of the specified population have the same probability of coming into the sample? If not, and the investigator fails to weight the data, inferences from the sample to the population will likely be wrong.[5]

More subtle are the implications for estimates of standard errors.[6] The usual formulas require the investigator to believe that the women are sampled independently of one another. Even small departures from independence may have serious consequences, as we demonstrate later. Furthermore, the investigator is required to assume constant probabilities across occasions. This assumption of constant probabilities is almost certainly false.

Family violence has seasonal patterns. (Christmas is a particularly bad time.) The probabilities of admission therefore vary over the course of the year. In addition, shelters vary in catchment areas, referral patterns, interpersonal networks, and admissions policies. Thus, women with children may have low probability of admission to one shelter, but a high probability of admission to other shelters. Selection probabilities depend on a host of personal characteristics; such probabilities must vary across geography and over time.

The independence assumption seems even more unrealistic. Admissions policies evolve in response to daily life in the shelter. For example, some shelter residents may insist on keeping contact with their abusers. Experience may make the staff reluctant to admit similar women in the future. Likewise, shelter staff may eventually decide to exclude victims with drug or alcohol problems.

To summarize, the random-sampling assumption is required for statistical inference. But this assumption has substantive implications that are unrealistic. The consequences of failures in the assumptions will be discussed below.

2.4 An imaginary population
and imaginary sampling mechanism

Another way to treat uncertainty is to create an imaginary population from which the data are assumed to be a random sample. Consider the shelter story. The population might be taken as the set of all shelter residents that could have been produced by the social processes creating victims who seek shelter. These processes might include family violence, as well as more particular factors affecting possible victims, and external forces shaping the availability of shelter space.

With this approach, the investigator does not explicitly define a population that could in principle be studied, with unlimited resources of time and money. The investigator merely *assumes* that such a population exists in some ill-defined sense. And there is a further assumption, that the data set being analyzed can be treated *as if* it were based on a random sample from the assumed population. These are convenient fictions. Convenience will not be denied; the source of the fiction is twofold: (i) the population does not have any empirical existence of its own; and (ii) the sample was not in fact drawn at random.

In order to use the imaginary-population approach, it would seem necessary for investigators to demonstrate that the data can be treated as a random sample. It would be necessary to specify the social processes that are involved, how they work, and why they would produce the statistical equivalent of a random sample. Handwaving is inadequate. We doubt the case could be made for the shelter example or any similar illustration. Nevertheless, reliance on imaginary populations is widespread. Indeed, regression models are commonly used to analyze convenience samples: As we show later, such analyses are often predicated on random sampling from imaginary populations. The rhetoric of imaginary populations is seductive precisely because it seems to free the investigator from the necessity of understanding how data were generated.

2.5 When the statistical issues are substantive

Statistical calculations are often a technical sideshow; the primary interest is in some substantive question. Even so, the methodological issues need careful attention, as we have argued. However, in many cases the substantive issues are very close to the statistical ones. For example, in litigation involving claims of racial discrimination, the substantive research question is usually operationalized as a statistical hypothesis: Certain data are like a random sample from a specified population.

Suppose, for example, that in a certain jurisdiction there are 1084 probationers under federal supervision: 369 are black. Over a six-month period, 119 probationers are cited for technical violations: 54 are black. This is disparate impact, as one sees by computing the percents: In the total pool of probationers, 34% are black; however, among those cited, 45% are black.

A t-test for "statistical significance" would probably follow. The standard error on the 45% is $\sqrt{.45 \times .55/119} = .046$, or 4.6%. So, $t = (.45 - .34)/.046 = 2.41$, and the one-sided P is .01. (A more sophisticated analyst might use the hypergeometric distribution, but that would not change the outlines of the problem.) The null hypothesis is rejected, and there are at least two competing explanations: Either blacks are more prone to violate probation, or supervisors are racist. It is up to the probation office to demonstrate the former; the t-test shifts the burden of argument.

However, there is a crucial (and widely ignored) step in applying the t-test: translating the idea of a race-neutral citation process into a statistical null hypothesis. In a race-neutral world, the argument must go, the citation process would be like citing 119 people drawn at random from a pool consisting of 34% blacks. This random-sampling assumption is the critical one for computing the standard error.

In more detail, the t-statistic may be large for two reasons: (i) too many blacks are cited, so the numerator in the t-statistic is too big; or (ii) the standard error in the denominator is too small. The first explanation may be the salient one, but we think the second explanation needs to be considered as well. In a race-neutral world, it is plausible that blacks and whites should have the same overall citation probabilities. However, in any world, these probabilities seem likely to vary from person to person and time to time. Furthermore, dependence from occasion to occasion would seem to be the rule rather than the exception. As will be seen below, even fairly modest amounts of dependence can create substantial bias in estimated standard errors.

In the real world of the 1990's, the proportion of federal probationers convicted for drug offenses increased dramatically. Such probationers were often subjected to drug testing and required to participate in drug treatment programs. The mix of offenders and supervision policies changed dramatically. The assumption of probabilities constant over time is, therefore, highly suspect. Likewise, an assumption that all probationers faced the same risks of citation must be false. Even in a race-neutral world, the intensity of supervision must be in part determined by the nature of the offender's crime and background; the intensity of supervision obviously affects the likelihood of detecting probation violations.

The assumption of independence is even more problematic. Probation officers are likely to change their supervision policies, depending on past performance of the probationers. For example, violations of probation seem likely to lead to closer and more demanding supervision, with higher probabilities of detecting future violations. Similarly, behavior of the probationers is likely to depend on the supervision policies.

In short, the translation of race neutrality into a statistical hypothesis of random sampling is not innocuous. The statistical formulation seems inconsistent with the social processes on which it has been imposed. If so, the results of the statistical manipulations—the P-values—are of questionable utility.

This example is not special. For most convenience samples, the social processes responsible for the data likely will be inconsistent with what needs to be assumed to justify conventional formulas for standard errors. If so, translating research questions into statistical hypotheses may be quite problematic: Much can be lost in translation.

2.6 Does the random-sampling assumption make any difference?

For criminal justice research, we have tried to indicate the problems with making statistical inferences based on convenience samples. The assumption of independence is critical, and we believe this assumption will always be difficult to justify (Kruskal 1988). The next question is whether failures of the independence assumption matter. There is no definitive answer to this question; much depends on context. However, we will show that relatively modest violations of independence can lead to substantial bias in estimated standard errors. In turn, the confidence levels and significance probabilities will be biased too.

2.6.1 Violations of independence

Suppose the citation process violates the independence assumption in the following manner. Probation officers make contact with probationers on a regular basis. If contact leads to a citation, the probability of a subsequent citation goes up, because the law enforcement perspective is reinforced. If contact does not lead to a citation, the probability of a subsequent citation goes down (the law enforcement perspective is not reinforced). This does not seem to be an unreasonable model; indeed, it may be far more reasonable than independence.

More specifically, suppose the citation process is a "stationary Markov chain." If contact leads to a citation, the chance that the next case will be cited is .50. On the other hand, if contact does not lead to a citation, the chance of a citation on the next contact is only .10. To get started, we assume the chance of a citation on the first contact is .30; the starting probability makes little difference for this demonstration.

Suppose an investigator has a sample of 100 cases, and observes seventeen citations. The probability of citation would be estimated as $17/100 = .17$, with a standard error of $\sqrt{.17 \times .83/100} = .038$. Implicitly, this calculation assumes independence. However, Markov chains do not obey the independence assumption. The right standard error, computed by simulation, turns out to be .058. This is about 50% larger than the standard error computed by the usual formula. As a result, the conventional t-statistic is about 50% too large. For example, a researcher who might ordinarily use a critical value of 2.0 for statistical significance at the .05 level should really be using a critical value of about 3.0.

Alert investigators might notice the breakdown of the independence assumption: The first-order serial correlation for our Markov process is about .40. This is not large, but it is detectable with the right test. However, the dependencies could easily be more complicated and harder to find, as the next example shows.

Consider a "four-step Markov chain." The probation officer judges an offender in the light of recent experience with similar offenders. The officer thinks back over the past four cases and finds the case most like the current case. If this "reference" case was cited, the probability that the current case will be cited is .50. If the reference case was not cited, the probability that the current case will be cited is .10. In our example, the reference case is chosen at random from the four prior cases. Again, suppose an investigator has a sample of 100 cases, and observes seventeen citations. The probability of citation would still be estimated as $17/100 = .17$, with a standard error of $\sqrt{.17 \times .83/100} = .038$. Now, the right standard error, computed by simulation, turns out to be .062. This is about 60% larger than the standard error computed by the usual formula.

Conclusions are much the same as for the first simulation. However, the four-step Markov chain spreads out the dependence so that it is hard to detect: The first-order serial correlation is only about .12.[7] Similar problems come about if the Markov chain produces negative serial correlations rather than positive ones. Negative dependence can be just as hard to detect, and the estimated standard errors will still be biased. Now the bias is upward so the null hypothesis is not rejected when it should be: Significant findings are missed.

Of course, small correlations are easier to detect with large samples. Yet probation officers may use more than four previous cases to find a reference case; they may draw on their whole current case load, and on salient cases from past case loads. Furthermore, transition probabilities (here, .50 and .10) are likely to vary over time in response to changing penal codes, administrative procedures, and mix of offenders. As a result of such complications, even very large samples may not save the day.

The independence assumption is fragile. It is fragile as an empirical matter because real world criminal justice processes are unlikely to produce data for which independence can be reasonably assumed. (Indeed, if independence were the rule, criminal justice researchers would have little to study.) The assumption is fragile as a statistical matter, because modest violations of independence may have major consequences while being nearly impossible to detect. The Markov chain examples are not worst case scenarios, and they show what can happen when independence breaks down. The main point: Even modest violations of independence can introduce substantial biases into conventional procedures.

2.7 Dependence in other settings

2.7.1 Spatial dependence

In the probation example, dependence was generated by social processes that unfolded over time. Dependence can also result from spatial relationships rather than temporal ones. Spatial dependence may be even more difficult to handle than temporal dependence.

For example, if a researcher is studying crime rates across census tracts in a particular city, it may seem natural to assume that the correlation between tracts depends on the distance between them. However, the right measure of distance is by no means obvious. Barriers such as freeways, parks, and industrial concentrations may break up dependence irrespective of physical distance.

"Closeness" might be better defined by travel time. Perhaps tracts connected by major thoroughfares are more likely to violate the assumption of independence than tracts between which travel is inconvenient. Ethnic mix and demographic profiles matter too, since crimes tend to be committed within ethnic and income groups. Social distance rather than geographical distance may be the key. Our point is that spatial dependence matters. Its measurement will be difficult, and may depend on how distance itself is measured. Whatever measures are used, spatial dependence produces the same kinds of problems for statistical inference as temporal dependence.

2.7.2 Regression models

In research on punishment and social control, investigators often use complex models. In particular, regression and its elaborations (e.g., structural equation modeling) are now standard tools of the trade. Although rarely discussed, statistical assumptions have major impacts on analytic results obtained by such methods.

Consider the usual textbook exposition of least squares regression. We have n observational units, indexed by $i = 1, \ldots, n$. There is a response variable y_i, conceptualized as $\mu_i + \epsilon_i$, where μ_i is the theoretical mean of y_i while the disturbances or errors ϵ_i represent the impact of random variation (sometimes of omitted variables). The errors are assumed to be drawn independently from a common (Gaussian) distribution with mean 0 and finite variance.

Generally, the error distribution is not empirically identifiable outside the model, so it cannot be studied directly—even in principle—without the model. The error distribution is an imaginary population and the errors ϵ_i are treated as if they were a random sample from this imaginary population—a research strategy whose frailty was discussed earlier.

Usually, explanatory variables are introduced and μ_i is hypothesized to be a linear combination of such variables. The assumptions about the μ_i and ϵ_i are seldom justified or even made explicit—although minor correlations in the ϵ_i can create major bias in estimated standard errors for coefficients. For one representative textbook exposition, see Weisberg (1985). Conventional econometric expositions are for all practical purposes identical (e.g., Johnston 1984).

Structural equation models introduce further complications (Freedman, 1987, 1991 [Chapter 3], 1995 [Chapter 1], 1997, 1999; Berk, 1988, 1991). Although the models seem sophisticated, the same old problems have been swept under the carpet, because random variation is represented in the same old way. Why do μ_i and ϵ_i behave as assumed? To answer this question, investigators would have to consider, much more closely than is commonly done, the connection between social processes and statistical assumptions.

2.7.3 Time series models

Similar issues arise in time series work. Typically, the data are highly aggregated; each observation characterizes a time period rather than a case; rates and averages are frequently used. There may be T time periods indexed by $t = 1, 2, \ldots, T$. The response variable y_t is taken to be

$\mu_t + \epsilon_t$ where the ϵ_t are assumed to have been drawn independently from a common distribution with mean 0 and finite variance. Then, μ_t will be assumed to depend linearly on values of the response variable for preceding time periods and on values of the explanatory variables. Why such assumptions should hold is a question that is seldom asked let alone answered.

Serial correlation in residuals may be too obvious to ignore. The common fix is to assume a specific form of dependence between the ϵ_t. For example, a researcher might assert that $\epsilon_t = \alpha \epsilon_{t-1} + \delta_t$, where now δ_t satisfy the familiar assumptions: The δ_t are drawn independently from a common distribution with mean 0 and finite variance. Clearly, the game has not changed except for additional layers of technical complexity.

2.7.4 Meta-analysis

Literature reviews are a staple of scientific work. Over the past twenty-five years, a new kind of review has emerged, claiming to be more systematic, more quantitative, more scientific: This is "meta-analysis." The initial step is to extract "the statistical results of numerous studies, perhaps hundreds, and assemble them in a database along with coded information about the important features of the studies producing these results. Analysis of this database can then yield generalizations about the body of research represented and relationships within it" (Lipsey 1997, p. 15). Attention is commonly focused on the key outcomes of each study, with the hope that by combining the results, one can learn what works. For example, Lipsey (1992) assesses the efficacy of a large number of juvenile delinquency treatment programs, while Sherman and his colleagues (1997) consider in a similar fashion a wide variety of other criminal justice interventions. Meta-analysis is discussed in any number of accessible texts (e.g., Lipsey and Wilson 2001). Statistical inference is usually a central feature of the exposition.

A meta-analysis identifies a set of studies, each of which provides one or more estimates of the effect of some intervention. For example, one might be interested in the impact of job training programs on prisoner behavior after release. For some studies, the outcome of interest might be earnings: Do inmates who participate in job training programs have higher earnings after release than those who do not? For other studies, the outcome might be the number of weeks employed during the first year after release. For a third set of studies, the outcome might be the time between release and getting a job. For each outcome, there would likely be several research reports with varying estimates of the treatment effect. The meta-analysis seeks to provide a summary estimate over all of the studies.

We turn to a brief description of how summary estimates are computed. We follow Hedges and Olkin (1985, Secs. 4AB), but relax some of their assumptions slightly. Outcomes for treated subjects ("experimentals") are denoted Y_{ij}^E, while the outcomes for the controls are denoted Y_{ij}^C. Here, i indexes the study and j indexes subject within study. Thus, Y_{ij}^E is the response of the jth experimental subject in the ith study. There are k studies in all, with n_i^E experimentals and n_i^C controls in the ith study. Although we use the "treatment-control" language, it should be clear that meta-analysis is commonly applied to observational studies in which the "treatments" can be virtually any variable that differs across subjects. In Archer's (2000) meta-analysis of sex differences in domestic violence, for example, the "treatment" is the sex of the perpetrator.

One key assumption is that for each $i = 1, \ldots, k$,

(A) Y_{ij}^E are independent and identically distributed for $j = 1, \ldots, n_i^E$; these variables have common expectation μ_i^E and variance σ_i^2.

Similarly,

(B) Y_{ij}^C are independent and identically distributed for $j = 1, \ldots, n_i^C$; these variables have common expectation μ_i^C and variance σ_i^2.

Notice that μ_i^E, μ_i^C, and σ_i^2 are parameters—population-level quantities that are unobservable. Notice too that the variances in (A) and (B) are assumed to be equal. Next, it is assumed that

(C) The responses of the experimentals and controls are independent.

Assumptions (A) and (B) specified within-group independence; (C) adds the assumption of between-group independence. Finally, it is assumed that

(D) studies are independent of one another.

Let $\bar{Y}_i^E$ be the average response for the experimentals in study i, and let $\bar{Y}_i^C$ be the average response for the controls. These averages are statistics, computable from study data. It follows from (A) and (B) that, to a reasonable approximation,

$$(1) \qquad \bar{Y}_i^E \sim N(\mu_i^E, \sigma_i^2/n_i^E) \quad \text{for } i = 1, \ldots, k$$

and

$$(2) \qquad \bar{Y}_i^C \sim N(\mu_i^C, \sigma_i^2/n_i^C) \quad \text{for } i = 1, \ldots, k.$$

For the ith study, the "effect size" is

$$(3) \qquad \eta_i = \frac{\mu_i^E - \mu_i^C}{\sigma_i}.$$

It is assumed that

$$(4) \qquad \eta_1 = \eta_2 = \ldots = \eta_k = \eta.$$

The goal is to estimate the value of η. For instance, if $\eta = .20$, the interpretation would be this: Treatment shifts the distribution of responses to the right by 20% of a standard deviation.[8]

There are a number of moves here. Assumptions (A), (B), and (C) mean that treatment and control subjects for each study are drawn as independent random samples from two different populations with a common standard deviation. The standardization in (3) eliminates differences in scale across studies.[9] After that, (4) requires that there is but a single parameter value for the effect size over all of the studies: There is only one true treatment effect, which all of the studies are attempting to measure.

Now the common effect can be estimated by taking a weighted average

$$(5) \qquad \hat{\eta} = w_1 \hat{\eta}_1 + \ldots + w_k \hat{\eta}_k,$$

where

$$(6) \qquad \hat{\eta}_i = (\bar{Y}_i^E - \bar{Y}_i^C)/\hat{\sigma}_i.$$

In (6), the statistic $\hat{\sigma}_i$ estimates the common standard deviation from the sample; the weights w_i adjust for differences in sample size across studies. (To minimize variance, w_i should be inversely proportional to $1/n_i^E + 1/n_i^C$; other weights are sometimes used.) Moreover, we can compute standard errors for $\hat{\eta}$, because this estimator is the product of a convenient and well-defined chance process. For details, see Hedges and Olkin (1985, chapter 6).

The outcome is both pleasing and illusory. The subjects in treatment and control (even in a randomized controlled experiment, as discussed below) are not drawn at random from populations with a common variance; with an observational study, there is no randomization at all. It is gratuitous to assume that *standardized* effects are constant across studies: It could be, for instance, that the average effects themselves are approximately constant but standard deviations vary widely. If we seek to combine studies with different kinds of outcome measures (earnings, weeks worked, time to first job), standardization seems helpful. And yet, *why* are standardized effects constant across these different measures? Is there really one underlying construct being measured, constant across studies, except for scale? We find no satisfactory answers to these critical questions.

The assumed independence of studies is worth a little more attention. Investigators are trained in similar ways, read the same papers, talk to one another, write proposals for funding to the same agencies, and publish the findings after peer review. Earlier studies beget later studies, just as each generation of Ph.D. students trains the next. After the first few million dollars are committed, granting agencies develop agendas of their own, which investigators learn to accommodate. Meta-analytic summaries of past work further channel the effort. There is, in short, a web of social dependence inherent in all scientific research. Does social dependence compromise statistical independence? Only if you think that investigators' expectations, attitudes, preferences, and motivations affect the written word—and never forget those peer reviewers.[10]

The basic model represented in equations (1–4) can be—and often is—extended in one way or another, although not in any way that makes the model substantially more believable. Perhaps the most common change is to allow for the possibility of different effect sizes. That is, equation (4) no longer applies; there is no longer an η characterizing all of the studies. Under a "random-effects model," the η_i's are assumed to be drawn as a random sample from some population of η's. Now the goal is to estimate the grand mean μ of this population of η's. However, insofar as meta-analysis rests on a convenience sample of studies, if not a whole population, the random-effects model is at a considerable distance from the facts.[11]

But wait. Perhaps the random-effects model can be reformulated: The ith study measures η_i, with an intrinsic error whose size is governed by equations (1), (2), and (3). Then, in turn, η_i differs from the sought-for grand mean μ by some random error; this error (i) has a mean value of 0 across all potential studies, and (ii) a variance that is constant across studies. This second formulation (a "components of variance" model) is equally phantasmagorical. Why would these new assumptions be true? Which potential studies are we talking about,[12] and what parameter are we estimating? Even if we could agree on answers to those questions, it seems likely—particularly with nonexperimental data—that each study deviates from truth by some intrinsic bias, whose size varies from one study to another. If so, the meta-analytic machine grinds to a halt.

There are further variations on the meta-analytic model, with biases related to study characteristics through some form of regression analysis. The unit of analysis is the study, and the response variable is the estimated effect size. Statistical inference is driven by the sort of random-sampling assumptions discussed earlier, when regression analysis was initially considered. However, with research studies as the unit of analysis, the random-sampling assumption becomes especially puzzling. The interesting question is why the technique is so widely used. One possible answer is this. Meta-analysis would be a wonderful method *if* the assumptions held. However, the assumptions are so esoteric as to be unfathomable and hence immune from rational consideration: The rest is history. For other commentaries, see Oakes (1990) or Petitti (1999).

2.7.5 Observational studies and experiments

We return to the basic assumptions (A–C) above. How are these to be understood? Meta-analysis is on its most secure footing with experiments, so we begin there. By way of example, consider an experiment with 1000 subjects. Each subject has two possible responses. One response will be manifest if the subject is put into the treatment condition; the other, in the control condition. For any particular subject, of course, one and only one of the two responses can be measured: The subject can be put into treatment or control, but not both.

Suppose 500 out of our 1000 subjects are chosen at random, and put into treatment; the other 500 are put in the control condition; the treatment and control averages will be compared. This is the cleanest of study designs. Do assumptions (A-B-C) hold? No, they do not—as a moment's reflection will show. There are two samples of size 500 each, but these are dependent, precisely because a subject assigned to treatment cannot be assigned to control, and vice versa. Thus, (C) fails. Similarly, the treatment group is drawn at random without replacement, so there is dependence between observations within each group: The first subject drawn cannot appear also as the second subject, and so forth. So the independence assumption in (A) fails, as does the corresponding assumption in (B).

To secure assumptions (A-B-C) in an experimental setting, we need an extremely large pool of subjects, most of whom will not be used. Suppose, for instance, we have 10,000 subjects: 500 will be chosen at random and put into treatment; another 500 will be chosen at random for the controls; and the remaining 9000 will be ignored. In this unusual design, we have the independence required by (A-B-C), at least to a first approximation. But we're not there yet. Assumptions (A) and (B) require that the variance be the same in treatment and control. In effect, treatment is only allowed to add one number—the same for all subjects—to the control response. If different subjects show different responses to treatment, then the constant-variance assumption is likely to be wrong.

To sum up, (A-B-C) hold—to a good approximation—for an experiment with a large pool of subjects, where a relatively small number are chosen at random for treatment, another small number are chosen at random for controls, and the only effect of treatment is to add a constant to all responses. Few experiments satisfy these conditions.[13]

Typically, of course, a meta-analysis starts not from a set of experiments, but from a set of observational studies. Then what? The basic conceit is that each observational study can be treated *as if* it were an experiment; not only that, but a very special kind of experiment, with the sampling structure described above. This is exactly the sort of unwarranted assumption whose consequences we have explored earlier in this essay. In brief, standard errors and P-values are liable to be quite misleading.

The assumptions underlying meta-analysis can be shown to give reasonable results in one situation; namely, combining a series of properly designed randomized controlled experiments, run with a common protocol, to test the global null hypothesis (treatment has no effect in any of the experiments).[14] Of course, even if the global null hypothesis is rejected, so the treatment has some effects on some subjects in some studies, the model underlying meta-analysis is far from demonstrated: The treatment may have different effects on different people, depending on context and circumstance. Indeed, that seems more plausible a priori than the hypothesis of a constant additive effect.[15]

2.8 Recommendations for practice

Convenience samples are a fact of scientific life in criminal justice research; so is uncertainty. However, the conventional techniques designed to measure uncertainty assume that the data are generated by the equivalent of random sampling, or probability sampling more generally.[16]

Real probability samples have two great benefits: (i) they allow unbiased extrapolation from the sample; and (ii) with data internal to the sample, it is possible to estimate how much results are likely to change if another sample is taken. These benefits, of course, have a price: Drawing probability samples is hard work. An investigator who assumes that a convenience sample is like a random sample seeks to obtain the benefits without the costs—just on the basis of assumptions.

If scrutinized, few convenience samples would pass muster as the equivalent of probability samples. Indeed, probability sampling is a technique whose use is justified because it is so unlikely that social processes will generate representative samples. Decades of survey research have demonstrated that when a probability sample is desired, probability sampling must be done. Assumptions do not suffice. Hence, our first recommendation for research practice: Whenever possible, use probability sampling.

If the data-generation mechanism is unexamined, statistical inference with convenience samples risks substantial error. Bias is to be expected and independence is problematic. When independence is lacking, the P-values produced by conventional formulas can be grossly misleading. In general, we think that reported P-values will be too small; in the social world, proximity seems to breed similarity. Thus, many research results are held to be statistically significant when they are the mere product of chance variation.

We are skeptical about conventional statistical adjustments for dependent data. These adjustments will be successful only under restrictive assumptions whose relevance to the social world is dubious. Moreover, adjustments require new layers of technical complexity, which tend to distance the researcher from the data. Very soon, the model rather than the data will be driving the research. Hence another recommendation: Do not rely on post hoc statistical adjustments to remove dependence.

No doubt, many researchers working with convenience samples will continue to attach standard errors to sample statistics. In such cases, sensitivity analyses may be helpful. Partial knowledge of how the data were generated might be used to construct simulations. It may be possible to determine which findings are robust against violations of independence. However, sensitivity analysis will be instructive only if it captures important features of the data-generation mechanism. Fictional sensitivity analysis will produce fictional results.

We recommend better focus on the questions that statistical inference is supposed to answer. If the object is to evaluate what would happen were the study repeated, real replication is an excellent strategy (Freedman 1991 [Chapter 3]; Berk 1991; Ehrenberg and Bound 1993). Empirical results from one study can be used to forecast what should be found in another study. Forecasts about particular summary statistics, such as means or regression coefficients, can be instructive. For example, an average rate of offending estimated for teenagers in one neighborhood could be used as a forecast for teenagers in another similar neighborhood. Using data from one prison, a researcher might predict which inmates in another prison will be cited for rule infractions. Correct forecasts would be strong evidence for the model.

Cross validation is an easier alternative. Investigators can divide a large sample into two parts. One part of the data can be used to construct forecasting models which are then evaluated against the rest of the data. This offers some degree of protection against bias due to over-fitting or chance capitalization. But cross validation does not really address the issue of replicability. It cannot, because the data come from only one study.

Finally, with respect to meta-analysis, our recommendation is simple: Just say no. The suggested alternative is equally simple: Read the papers, think about them, and summarize them.[17] Try our alternative. Trust us: You will like it. And if you can't sort the papers into meaningful categories, neither can the meta-analysts. In the present state of our science, invoking a formal relationship between random samples and populations is more likely to obscure than to clarify.

2.9 Conclusions

We have tried to demonstrate that statistical inference with convenience samples is a risky business. While there are better and worse ways to proceed with the data at hand, real progress depends on deeper understanding of the data-generation mechanism. In practice, statistical issues and substantive issues overlap. No amount of statistical maneuvering will get very far without some understanding of how the data were produced.

More generally, we are highly suspicious of efforts to develop empirical generalizations from any single data set. Rather than ask what would happen in principle if the study were repeated, it makes sense to actually repeat the study. Indeed, it is probably impossible to predict the changes attendant on replication without doing replications. Similarly, it may be impossible to predict changes resulting from interventions without actually intervening.

Notes

1. "Random sampling" has a precise, technical meaning: Sample units are drawn independently, and each unit in the population has an equal chance to be drawn at each stage. Drawing a random sample of the U.S. population, in this technical sense, would cost several billion dollars (since it requires a census as a preliminary matter) and would probably require the suspension of major constitutional guarantees. Random sampling is not an idea to be lightly invoked.

2. As we shall explain, researchers may find themselves assuming that their sample is a random sample from an imaginary population. Such a population has no empirical existence, but is defined in an essentially

circular way—as that population from which the sample may be assumed to be randomly drawn. At the risk of the obvious, inferences to imaginary populations are also imaginary.

3. Of course, somewhat weaker assumptions may be sufficient for some purposes. However, as we discuss below, the outlines of the problem stay the same.

4. We use the term "parameter" for a characteristic of the population. A "sample statistic" or "estimate" is computed from the sample to estimate the value of a parameter. As indicated above, we use "random sampling" to mean sampling with replacement from a finite population: Each unit in the population is selected independently (with replacement) and with the same probability of selection. Sampling without replacement (i.e., simple random sampling) may be more familiar. In many practical situations, sampling without replacement is very close to sampling with replacement. Stratified cluster samples are often more cost effective than purely random samples, but estimates and standard errors then need to be computed taking the sample design into account. Convenience samples are often treated as if they were random samples, and sometimes as if they were stratified random samples—that is, random samples drawn within subgroups of some poorly defined super-population. Our analysis is framed in terms of the first model, but applies equally well to the second.

5. Weighting requires that the investigator know the probability of selection for each member of the population. It is hard to imagine that such precise knowledge will be available for convenience samples. Without reweighting, estimates will be biased, perhaps severely.

6. The standard error measures sampling variability; it does not take bias into account. Our basic model is random sampling. In the time-honored way, suppose we draw women into the sample one after another (with replacement). The conventional formula for the standard error assumes that the selection probabilities stay the same from draw to draw; on any given draw, the selection probabilities do not have to be identical across women.

7. The standard error is affected not only by first-order correlations, but also by higher-order correlations. Without a priori knowledge that the data were generated by a four-step Markov chain, a researcher is unlikely to identify the dependence.

8. We are not quite following the notation in Hedges and Olkin (1985): Our standardized effect size is η rather than δ, corresponding to d in Cohen (1988).

9. Temperature can measured in degrees Celsius or degrees Fahrenheit. The two temperature scales are different, but they are linearly related: $F° = \frac{9}{5}C° + 32°$. The Hedges-Olkin model for meta-analysis described above does not account for transformations more complicated than the linear one. In short, units do not matter; but anything more substantive than a difference in units between studies is beyond the scope of the model.

10. Meta-analysts deal with publication bias by making the "file-drawer" calculation: How many studies would have to be withheld from publication to change the outcome of the meta-analysis from significant to insignificant? Typically, the number is astronomical. This is because of a crucial assumption in the procedure—that the missing estimates are centered on zero. The calculation ignores the possibility that studies with contrarian findings—significant or insignificant—are the ones that have been withheld. There is still another possibility, which is ignored by the calculation: Study designs may get changed in midstream if results are going the wrong way. See Rosenthal (1979), Oakes (1990, p. 158), or Petitti (1999, p. 134).

11. The model now requires two kinds of random sampling: A random sample of studies and then a random sample of study subjects.

12. If the answer is "all possible studies," then the next question might be, with what assumptions about government spending in fiscal 2025? or for that matter, in 1975? What about the respective penal codes and inmate populations? The point is that hypothetical super-populations don't generate real statistics.

13. With a binary response variable—"success" or "failure"—there does seem to be a logical contradiction in the model: Changing the probability p of success automatically changes the variance $p(1 - p)$. Naturally, other models can then be used, with different definitions for η. But then, combining binary and continuous responses in the same meta-analysis almost seems to be a logical contradiction, because the two kinds of studies are measuring incommensurable parameters.

For example, in Lipsey (1992), half the studies use a binary response variable (item 87, p. 111). Following Cohen (1988), Lipsey (p. 91) handles these binary responses by making the "arcsine transformation" $f(x) = 2 \arcsin \sqrt{x}$. In more detail, suppose we have n independent trials, each leading to success with probability p and failure with the remaining probability $1 - p$. We would estimate p by $\hat{p}$, the proportion of successes in the sample. The sampling variance of $\hat{p}$ is $p(1 - p)/n$, which depends on the parameter p and the sample size n. The charm of the arcsine transformation—which is considerable—is that the asymptotic variance of $f(\hat{p})$ is $1/n$, and does not depend on the unknown p.

If now $\hat{p}^T$ is the proportion of successes in the treatment group, while $\hat{p}^C$ is the proportion of successes in the control group, $f(\hat{p}^T) - f(\hat{p}^C) = f(p^T) - f(p^C)$, up to an additive random error that is asymptotically normal with mean 0 and variance $1/n^T + 1/n^C$. Lipsey—like many others who follow Cohen—would define the effect size as $f(\hat{p}^T) - f(\hat{p}^C)$. But why is a reduction of 0.20 standard deviations in time to rearrest—for instance—comparable to a reduction of 0.20 in twice the arcsine of the square root of the recidivism rate, i.e., a reduction of 0.10 in the arcsine itself. We see no rationale for combining studies this way, and Lipsey does not address such questions, although he does provide a numerical example on pp. 97–98 to illustrate the claimed equivalence.

14. Although (A-B-C) are false, as shown above, the statistic $\hat{\eta}_i$ in (6) should be essentially normal. Under the global null hypothesis that all the η_i are zero, the expected value of $\hat{\eta}_i$ is approximately zero, and the variance of $\hat{\eta}_i / \sqrt{1/n_i^E + 1/n_i^C}$ is approximately 1, by a combinatorial argument. Other tests are available, too. For example, the χ^2-test is a more standard, and more powerful, test of the global null. Similar calculations can be made if the treatment effect is any additive constant—the same for all subjects in the study. If the treatment effect varies from subject to subject, the situation is more complicated; still, conventional procedures often provide useful approximations to the (correct) permutation distributions—just as the χ^2 is a good approximation to Fisher's exact test.

15. Some readers will, no doubt, reach for Occam's razor. But this is a two-edged sword. (i) Isn't it simpler to have one number than 100? (ii) Isn't it simpler to drop the assumption that all the numbers are the same? Finally, if the numbers are different, Occam's razor can even cut away the next assumption—that the studies are a random sample from a hypothetical super-population of studies. Occam's razor is to be unsheathed only with great caution.

16. A probability sample starts from a well-defined population; units are drawn into the sample by some objective chance mechanism, so the probability that any particular set of units falls into the sample is computable. Each sample unit can be weighted by the inverse of the selection probability to get unbiased estimates.

17. Descriptive statistics can be very helpful in the last-mentioned activity. For one lovely example out of many, see Grace, Muench, and Chalmers (1966).

3

Statistical Models and Shoe Leather

ABSTRACT. *Regression models have been used in the social sciences at least since 1899, when Yule published a paper on the causes of pauperism. Regression models are now used to make causal arguments in a wide variety of applications, and it is perhaps time to evaluate the results. No definitive answers can be given, but this chapter takes a rather negative view. Snow's work on cholera is presented as a success story for scientific reasoning based on nonexperimental data. Failure stories are also discussed, and comparisons may provide some insight. In particular, this chapter suggests that statistical technique can seldom be an adequate substitute for good design, relevant data, and testing predictions against reality in a variety of settings.*

3.1 Introduction

Regression models have been used in social sciences at least since 1899, when Yule published his paper on changes in "out-relief" as a cause of pauperism: He argued that providing income support outside the poorhouse increased the number of people on relief. At present, regression models are used to make causal arguments in a wide variety of social science applications, and it is perhaps time to evaluate the results.

Sociological Methodology (1991) 21: 291–313.

A crude four-point scale may be useful:

1. Regression usually works, although it is (like anything else) imperfect and may sometimes go wrong.
2. Regression sometimes works in the hands of skillful practitioners, but it isn't suitable for routine use.
3. Regression might work, but it hasn't yet.
4. Regression can't work.

Textbooks, courtroom testimony, and newspaper interviews seem to put regression into category 1. Category 4 seems too pessimistic. My own view is bracketed by categories 2 and 3, although good examples are quite hard to find.

Regression modeling is a dominant paradigm, and many investigators seem to consider that any piece of empirical research has to be equivalent to a regression model. Questioning the value of regression is then tantamount to denying the value of data. Some declarations of faith may therefore be necessary. Social science is possible, and sound conclusions can be drawn from nonexperimental data. (Experimental confirmation is always welcome, although some experiments have problems of their own.) Statistics can play a useful role. With multi-dimensional data sets, regression may provide helpful summaries of the data.

However, I do not think that regression can carry much of the burden in a causal argument. Nor do regression equations, by themselves, give much help in controlling for confounding variables. Arguments based on statistical significance of coefficients seem generally suspect; so do causal interpretations of coefficients. More recent developments, like two-stage least squares, latent-variable modeling, and specification tests, may be quite interesting. However, technical fixes do not solve the problems, which are at a deeper level. In the end, I see many illustrations of technique but few real examples with validation of the modeling assumptions.

Indeed, causal arguments based on significance tests and regression are almost necessarily circular. To derive a regression model, we need an elaborate theory that specifies the variables in the system, their causal interconnections, the functional form of the relationships, and the statistical properties of the error terms—independence, exogeneity, etc. (The stochastics may not matter for descriptive purposes, but they are crucial for significance tests.) *Given the model*, least squares and its variants can be used to estimate parameters and to decide whether or not these are zero. However, the model cannot in general be regarded as given, because current social science theory does not provide the requisite level of technical detail for deriving specifications.

There is an alternative validation strategy, which is less dependent on prior theory: Take the model as a black box and test it against empirical reality. Does the model predict new phenomena? Does it predict the results of interventions? Are the predictions right? The usual statistical tests are poor substitutes because they rely on strong maintained hypotheses. Without the right kind of theory, or reasonable empirical validation, the conclusions drawn from the models must be quite suspect.

At this point, it may be natural to ask for some real examples of good empirical work and strategies for research that do not involve regression. Illustrations from epidemiology may be useful. The problems in that field are quite similar to those faced by contemporary workers in the social sciences. Snow's work on cholera will be reviewed as an example of real science based on observational data. Regression is not involved.

A comparison will be made with some current regression studies in epidemiology and social science. This may give some insight into the weaknesses of regression methods. The possibility of technical fixes for the models will be discussed, other literature will be reviewed, and then some tentative conclusions will be drawn.

3.2 Some examples from Epidemiology

Quantitative methods in the study of disease precede Yule and regression. In 1835, Pierre Louis published a landmark study on bleeding as a cure for pneumonia. He compared outcomes for groups of pneumonia patients who had been bled at different times and found

> that bloodletting has a happy effect on the progress of pneumonitis; that is it shortens its duration; and this effect, however, is much less than has been commonly believed. (Louis 1986 [1835], p. 48)

The finding and the statistical method were roundly denounced by contemporary physicians:

> By invoking the inflexibility of arithmetic in order to escape the encroachments of the imagination, one commits an outrage upon good sense. (Louis 1986 [1835], p. 63)

Louis may have started a revolution in our thinking about empirical research in medicine, or his book may only provide a convenient line of demarcation. But there is no doubt that within a few decades, the "inflexibility of arithmetic" had helped identify the causes of some major diseases and the means for their prevention. Statistical modeling played almost no role in these developments.

In the 1850's, John Snow demonstrated that cholera was a water-borne infectious disease (Snow 1965 [1855]). A few years later, Ignaz Semmelweis discovered how to prevent puerperal fever (Semmelweis 1981 [1861]). Around 1914, Joseph Goldberger found the cause of pellagra (Carpenter 1981; Terris 1964). Later epidemiologists have shown, at least on balance of argument, that most lung cancer is caused by smoking (Lombard and Doering 1928; Mueller 1939; Cornfield et al. 1959; U.S. Public Health Service 1964). In epidemiology, careful reasoning on observational data has led to considerable progress. (For failure stories on that subject, see below.)

An explicit definition of good research methodology seems elusive; but an implicit definition is possible, by pointing to examples. In that spirit, I give a brief account of Snow's work. To see his achievement, I ask you to go back in time and forget that germs cause disease. Microscopes are available but their resolution is poor. Most human pathogens cannot be seen. The isolation of such microorganisms lies decades into the future. The infection theory has some supporters, but the dominant idea is that disease results from "miasmas": minute, inanimate poison particles in the air. (Belief that disease-causing poisons are in the ground comes later.)

Snow was studying cholera, which had arrived in Europe in the early 1800's. Cholera came in epidemic waves, attacked its victims suddenly, and was often fatal. Early symptoms were vomiting and acute diarrhea. Based on the clinical course of the disease, Snow conjectured that the active agent was a living organism that got into the alimentary canal with food or drink, multiplied in the body, and generated some poison that caused the body to expel water. The organism passed out of the body with these evacuations, got back into the water supply, and infected new victims.

Snow marshaled a series of persuasive arguments for this conjecture. For example, cholera spreads along the tracks of human commerce. If a ship goes from a cholera-free country to a cholera-stricken port, the sailors get the disease only after they land or take on supplies. The disease strikes hardest at the poor, who live in the most crowded housing with the worst hygiene. These facts are consistent with the infection theory and hard to explain with the miasma theory.

Snow also did a lot of scientific detective work. In one of the earliest epidemics in England, he was able to identify the first case, "a seaman named John Harnold, who had newly arrived by the *Elbe* steamer from Hamburgh, where the disease was prevailing" (p. 3). Snow also found the second case, a man who had taken the room in which Harnold had stayed. More evidence for the infection theory.

Snow found even better evidence in later epidemics. For example, he studied two adjacent apartment buildings, one heavily hit by cholera, the other not. He found that the water supply in the first building was contaminated by runoff from privies and that the water supply in the second building was much cleaner. He also made several "ecological" studies to demonstrate the influence of water supply on the incidence of cholera. In the London of the 1800's, there were many different water companies serving different areas of the city, and some areas were served by more than one company. Several companies took their water from the Thames, which was heavily polluted by sewage. The service areas of such companies had much higher rates of cholera. The Chelsea water company was an exception, but it had an exceptionally good filtration system.

In the epidemic of 1853–54, Snow made a spot map showing where the cases occurred and found that they clustered around the Broad Street pump. He identified the pump as a source of contaminated water and persuaded the public authorities to remove the handle. As the story goes, removing the handle stopped the epidemic and proved Snow's theory. In fact, he did get the handle removed and the epidemic did stop. However, as he demonstrated with some clarity, the epidemic was stopping anyway, and he attached little weight to the episode.

For our purposes, what Snow actually did in 1853–54 is even more interesting than the fable. For example, there was a large poorhouse in the Broad Street area with few cholera cases. Why? Snow found that the poorhouse had its own well and that the inmates did not take water from the pump. There was also a large brewery with no cases. The reason is obvious: The workers drank beer, not water. (But if any wanted water, there was a well on these premises.)

To set up Snow's main argument, I have to back up just a bit. In 1849, the Lambeth water company had moved its intake point upstream along the Thames, above the main sewage discharge points, so that its water was fairly pure. The Southwark and Vauxhall water company, however, left its intake point downstream from the sewage discharges. An ecological analysis of the data for the epidemic of 1853–54 showed that cholera hit harder in the Southwark and Vauxhall service areas and largely spared the Lambeth areas. Now let Snow finish in his own words.

> Although the facts show in the above table [the ecological data; Table 3.1, p. 51] afford very strong evidence of the powerful influence which the drinking of water containing the sewage of a town exerts over the spread of cholera, when that disease is present, yet the question does not end here; for the intermixing of the water supply of the Southwark and Vauxhall Company

with that of the Lambeth Company, over an extensive part of
London, admitted of the subject being sifted in such a way as to
yield the most incontrovertible proof on one side or the other. In
the subdistricts enumerated in the above table [Table 3.1, p. 51]
as being supplied by both Companies, the mixing of the supply
is of the most intimate kind. The pipes of each Company go
down all the streets, and into nearly all the courts and alleys.
A few houses are supplied by one Company and a few by the
other, according to the decision of the owner or occupier at that
time when the Water Companies were in active competition. In
many cases a single house has a supply different from that on
either side. Each company supplies both rich and poor, both
large houses and small; there is no difference either in the condition or occupation of the persons receiving the water of the
different Companies. Now it must be evident that, if the diminution of cholera, in the districts partly supplied with improved
water, depended on this supply, the houses receiving it would
be the houses enjoying the whole benefit of the diminution of
the malady, whilst the houses supplied with the water from the
Battersea Fields would suffer the same mortality as they would
if the improved supply did not exist at all. As there is no difference whatever in the houses or the people receiving the supply
of the two Water Companies, or in any of the physical conditions
with which they are surrounded, it is obvious that no experiment
could have been devised which would more thoroughly test the
effect of water supply on the progress of cholera than this, which
circumstances placed ready made before the observer.

The experiment, too, was on the grandest scale. No fewer than
three hundred thousand people of both sexes, of every age and
occupation, and of every rank and station, from gentlefolks
down to the very poor, were divided into two groups without
their choice, and in most cases, without their knowledge; one
group being supplied with water containing the sewage of London, and amongst it, whatever might have come from the cholera patients, the other group having water quite free from such
impurity.

To turn this grand experiment to account, all that was required
was to learn the supply of water to each individual house where
a fatal attack of cholera might occur. (pp. 74–75)

Table 3.1 Snow's Table IX

	Number of houses	Deaths from cholera	Deaths per 10,000 houses
Southwark and Vauxhall	40,046	1263	315
Lambeth	26,107	98	37
Rest of London	256,423	1422	59

Snow identified the companies supplying water to the houses of cholera victims in his study area. This gave him the numerators in Table 3.1. (The denominators were taken from parliamentary records.)

Snow concluded that *if* the Southwark and Vauxhall company had moved their intake point as Lambeth did, about 1000 lives would have been saved. He was very clear about quasi randomization as the control for potential confounding variables. He was equally clear about the differences between ecological correlations and individual correlations. And his counterfactual inference is compelling.

As a piece of statistical technology, Table 3.1 is by no means remarkable. But the story it tells is very persuasive. The force of the argument results from the clarity of the prior reasoning, the bringing together of many different lines of evidence, and the amount of shoe leather Snow was willing to use to get the data.

Later, there was to be more confirmation of Snow's conclusions. For example, the cholera epidemics of 1832 and 1849 in New York were handled by traditional methods: exhorting the population to temperance, bringing in pure water to wash the streets, treating the sick by bleeding and mercury. After the publication of Snow's book, the epidemic of 1866 was dealt with using the methods suggested by his theory: boiling the drinking water, isolating the sick individuals, and disinfecting their evacuations. The death rate was cut by a factor of 10 or more (Rosenberg 1962).

In 1892, there was an epidemic in Hamburg. The leaders of Hamburg rejected Snow's arguments. They followed Max von Pettenkofer, who taught the miasma theory: Contamination of the ground caused cholera. Thus, Hamburg paid little attention to its water supply but spent a great deal of effort digging up and carting away carcasses buried by slaughterhouses. The results were disastrous (Evans 1987).

What about evidence from microbiology? In 1880, Pasteur created a sensation by showing that the cause of rabies was a microorganism. In 1884, Koch isolated the cholera vibrio [*Vibrio cholerae*], confirming all the essential features of Snow's account; Filipo Pacini may have discovered this organism even earlier (see Howard-Jones 1975). The vibrio is a water-borne bacterium that invades the human gut and causes cholera. Today, the molecular biology of cholera is reasonably well understood (Finlay, Heffron, and Falkow 1989; Miller, Mekalanos, and Falkow 1989). The vibrio makes a protein enterotoxin, which affects the metabolism of human cells and causes them to expel water. The interaction of the enterotoxin with the cell has been worked out, and so has the genetic mechanism used by the vibrio to manufacture this protein.

Snow did some brilliant detective work on nonexperimental data. What is impressive is not the statistical technique but the handling of the scientific issues. He made steady progress from shrewd observation through case studies to analysis of ecological data. In the end, he found and analyzed a natural experiment. Of course, he also made his share of mistakes: For example, based on rather flimsy analogies, he concluded that plague and yellow fever were also propagated through the water (Snow 1965 [1855], pp. 125–27).

The next example is from modern epidemiology, which had adopted regression methods. The example shows how modeling can go off the rails. In 1980, Kanarek et al. published an article in the *American Journal of Epidemiology*—perhaps the leading journal in the field—which argued that asbestos fibers in the drinking water caused lung cancer. The study was based on 722 census tracts in the San Francisco Bay Area. There were huge variations in fiber concentrations from one tract to another; factors of ten or more were commonplace.

Kanarek et al. examined cancer rates at 35 sites, for blacks and whites, men and women. They controlled for age by standardization and for sex and race by cross-tabulation. But the main tool was log-linear regression, to control for other covariates (marital status, education, income, occupation). Causation was inferred, as usual, if a coefficient was statistically significant after controlling for covariates.

Kanarek et al. did not discuss their stochastic assumptions, that outcomes are independent and identically distributed given covariates. The argument for the functional form was only that "theoretical construction of the probability of developing cancer by a certain time yields a function of the log form" (1980, p. 62). However, this model of cancer causation is open to serious objections (Freedman and Navidi 1989).

For lung cancer in white males, the asbestos fiber coefficient was highly significant ($P < .001$), so the effect was described as strong. Actually, the model predicts a risk multiplier of only about 1.05 for a 100-fold increase in fiber concentrations. There was no effect in women or blacks. Moreover, Kanarek et al. had no data on cigarette smoking, which affects lung cancer rates by factors of ten or more. Thus, imperfect control over smoking could easily account for the observed effect, as could even minor errors in functional form. Finally, Kanarek et al. ran upwards of 200 equations; only one of the P values was below .001. So the real significance level may be closer to $200 \times .001 = .20$. The model-based argument is not a good one.

What is the difference between Kanarek et al.'s study and Snow's? Kanarek et al. ignored the ecological fallacy. Snow dealt with it. Kanarek et al. tried to control for covariates by modeling, using socioeconomic status as a proxy for smoking. Snow found a natural experiment and collected the data as he needed. Kanarek et al.'s argument for causation rides on the statistical significance of a coefficient. Snow's argument used logic and shoe leather. Regression models make it all too easy to substitute technique for work.

3.3 Some examples from the Social Sciences

If regression is a successful methodology, the routine paper in a good journal should be a modest success story. However, the situation is quite otherwise. I recently spent some time looking through leading American journals in quantitative social science: *American Journal of Sociology*, *American Sociological Review*, and *American Political Science Review*. These refereed journals accept perhaps ten percent of their submissions. For analysis, I selected papers that were published in 1987–88, that posed reasonably clear research questions, and that used regression to answer them. I will discuss three of these papers. These papers may not be the best of their kind, but they are far from the worst. Indeed, one was later awarded a prize for the best article published in *American Political Science Review* in 1988. In sum, I believe these papers are quite typical of good current research practice.

Example 1. Bahry and Silver (1987) hypothesized that in Russia perception of the KGB as efficient deterred political activism. Their study was based on questionnaires filled out by Russian emigres in New York. There was a lot of missing data and perhaps some confusion between response variables and control variables. Leave all that aside. In the end, the argument was that after adjustment for covariates, subjects who viewed the KGB as efficient were less likely to describe themselves as activists. And this negative correlation was statistically significant.

Of course, that could be evidence to support the research hypothesis of the paper: If you think the KGB is efficient, you don't demonstrate. Or the line of causality could run the other way: If you're an activist, you find out that the KGB is inefficient. Or the association could be driven by a third variable: People of certain personality types are more likely to describe themselves as activists and also more likely to describe the KGB as inefficient. Correlation is not the same as causation; statistical technique, alone, does not make the connection. The familiarity of this point should not be allowed to obscure its force.

Example 2. Erikson, McIver, and Wright (1987) argued that in the U.S., different states really do have different political cultures. After controlling for demographics and geographical region, adding state dummy variables increased R^2 for predicting party identification from .0898 to .0953. The F to enter the state dummies was about eight. The data base consisted of 55,000 questionnaires from CBS/*New York Times* opinion surveys. With forty degrees of freedom in the numerator and 55,000 in the denominator, P is spectacular.

On the other hand, the R^2's are trivial—never mind the increase. The authors argued that the state dummies are not proxies for omitted variables. As proof, they put in trade union membership and found that the estimated state effects did not change much. This argument does not support the specification, but it is weak.

Example 3. Gibson (1988) asked whether the political intolerance during the McCarthy era was driven by mass opinion or elite opinion. The unit of analysis was the state. Legislation was coded on a tolerance/intolerance scale; there were questionnaire surveys of elite opinion and mass

Figure 3.1 Path model of political intolerance. Adapted by permission from Gibson (1988).

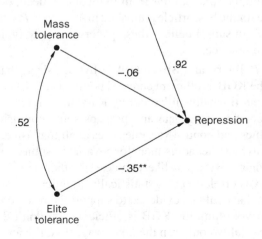

opinion. Then comes a path model (Figure 3.1); one coefficient is significant, one is not. Gibson concluded: "Generally it seems that elites, not masses, were responsible for the repression of the era" (p. 511).

Of the three papers, I thought Gibson's had the clearest question and the best summary data. However, the path diagram seems to be an extremely weak causal model. Moreover, even granting the model, the difference between the two path coefficients is not significant. The paper's conclusion does not follow from the data.

3.4 Summary of the position

In this set of papers, and in many papers outside the set, the adjustment for covariates is by regressions; the argument for causality rides on the significance of a coefficient. But significance levels depend on specifications, especially of error structure. For example, if the errors are correlated or heteroscedastic, the conventional formulas will give the wrong answers. And the stochastic specification is never argued in any detail. (Nor does modeling the covariances fix the problem, unless the model for the covariances can be validated; more about technical fixes below.)

To sum up, each of the examples has these characteristics:

1. There is an interesting research question, which may or may not be sharp enough to be empirically testable.

2. Relevant data are collected, although there may be considerable difficulty in quantifying some of the concepts, and important data may be missing.

3. The research hypothesis is quickly translated into a regression equation, more specifically, into an assertion that certain coefficients are (or are not) statistically significant.

4. Some attention is paid to getting the right variables into the equation, although the choice of covariates is usually not compelling.

5. Little attention is paid to functional form or stochastic specification; textbook linear models are just taken for granted.

Clearly, evaluating the use of regression models in a whole field is a difficult business; there are no well-beaten paths to follow. Here, I have selected for review three papers that, in my opinion, are good of their kind and that fairly represent a large (but poorly delineated) class. These papers illustrate some basic obstacles in applying regression technology to make causal inferences.

In Freedman (1987), I took a different approach and reviewed a modern version of the classic model for status attainment. I tried to state the technical assumptions needed for drawing causal inferences from path diagrams—assumptions that seem to be very difficult to validate in applications. I also summarized previous work on these issues. Modelers had an extended opportunity to answer. The technical analysis was not in dispute, and serious examples were not forthcoming.

If the assumptions of a model are not derived from theory, and if predictions are not tested against reality, then deductions from the model must be quite shaky. However, without the model, the data cannot be used to answer the research question. Indeed, the research hypothesis may not be really translatable into an empirical claim except as a statement about nominal significance levels of coefficients in a model.

Two authorities may be worth quoting in this regard. Of course, both of them have said other things in other places.

> The aim . . . is to provide a clear and rigorous basis for determining when a causal ordering can be said to hold between two variables or groups of variables in a model . . . . *The concepts . . . all refer to a model—a system of equations—and not to the "real" world the model purports to describe.* (Simon 1957, p. 12 [emphasis added])

> If . . . we choose a group of social phenomena with no antecedent knowledge of the causation or absence of causation among them, then the calculation of correlation coefficients, total or partial, will not advance us a step toward evaluating the importance of the causes at work. (Fisher 1958, p. 190)

In my view, regression models are not a particularly good way of doing empirical work in the social sciences today, because the technique depends on knowledge that we do not have. Investigators who use the technique are not paying adequate attention to the connection—if any—between the models and the phenomena they are studying. Their conclusions may be valid for the computer code they have created, but the claims are hard to transfer from that microcosm to the larger world.

For me, Snow's work exemplifies one point on a continuum of research styles; the regression examples mark another. My judgment on the relative merits of the two styles will be clear—and with it, some implicit recommendations. Comparisons may be invidious, but I think Snow's research stayed much closer to reality than the modeling exercises. He was not interested in the properties of systems of equations but in ways of preventing a real disease. He formulated sharp, empirical questions that could be answered using data that could, with effort, be collected. At every turn, he anchored his argument in stubborn fact. And he exposed his theory to harsh tests in a variety of settings. That may explain how he discovered something extraordinarily important about cholera, and why his book is still worth reading more than a century later.

3.5 Can technical fixes rescue the models?

Regression models often seem to be used to compensate for problems in measurement, data collection, and study design. By the time the models are deployed, the scientific position is nearly hopeless. Reliance on models in such cases is Panglossian. At any rate, that is my view. By contrast, some readers may be concerned to defend the technique of regression modeling: According to them, the technique is sound and only the applications are flawed. Other readers may think that the criticisms of regression modeling are merely technical, so that the technical fixes—e.g., robust estimators, generalized least squares, and specification tests—will make the problems go away.

The mathematical basis for regression is well established. My question is whether the technique applies to present-day social science problems. In other words, are the assumptions valid? Moreover, technical fixes become relevant only when models are nearly right. For instance, robust estimators may be useful if the error terms are independent, identically distributed, and symmetric but long-tailed. If the error terms are neither independent nor identically distributed and there is no way to find out whether they are symmetric, robust estimators probably distract from the real issues.

This point is so uncongenial that another illustration may be in order. Suppose $y_i = \alpha + \epsilon_i$, the ϵ_i have mean 0, and the ϵ_i are *either* independent and identically distributed *or* autoregressive of order 1. Then the well-oiled statistics machine springs into action. However, if the ϵ_i are just a sequence of random variables, the situation is nearly hopeless—with respect to standard errors and hypothesis testing. So much the worse if the y_i have no stochastic pedigree. The last possibility seems to me the most realistic. Then formal statistical procedures are irrelevant, and we are reduced (or should be) to old-fashioned thinking.

A well-known discussion of technical fixes starts from the evaluation of manpower-training programs using nonexperimental data. LaLonde (1986) and Fraker and Maynard (1987) compare evaluation results from modeling with results from experiments. The idea is to see whether regression models fitted to observational data can predict the results of experimental interventions. Fraker and Maynard conclude:

The results indicate that nonexperimental designs cannot be relied on to estimate the effectiveness of employment programs. Impact estimates tend to be sensitive both to the comparison group construction methodology and to the analytic model used. There is currently no way a priori to ensure that the results of comparison group studies will be valid indicators of the program impacts. (p. 194)

Heckman and Hotz (1989, pp. 862, 874) reply that specification tests can be used to rule out models that give wrong predictions:

A simple testing procedure eliminates the range of nonexperimental estimators at variance with the experimental estimates of program impact.... Thus, while not definitive, our results are certainly encouraging for the use of nonexperimental methods in social-program evaluation.

Heckman and Hotz have in hand (i) the experimental data, (ii) the nonexperimental data, and (iii) LaLonde's results as well as Fraker and Maynard's. Heckman and Hotz proceed by modeling the selection bias in the nonexperimental comparison groups. There are three types of models, each with two main variants. These are fitted to several different time periods, with several sets of control variables. Averages of different models are allowed, and there is a "slight extension" of one model.

By my count, twenty-four models are fitted to the nonexperimental data on female AFDC recipients and thirty-two to the data on high school dropouts. *Ex post facto*, models that pass certain specification tests can more or less reproduce the experimental results (up to very large standard errors). However, the real question is what can be done *ex ante*, before the right estimate is known. Heckman and Hotz may have an argument, but it is not a strong one. It may even point us in the wrong direction. Testing one model on twenty-four different data sets could open a serious inquiry: Have we identified an empirical regularity that has some degree of invariance? Testing twenty-four models on one data set is less serious.

Generally, replication and prediction of new results provide a harsher and more useful validating regime than statistical testing of many models on one data set. Fewer assumptions are needed, there is less chance of artifact, more kinds of variation can be explored, and alternative explanations can be ruled out. Indeed, taken to the extreme, developing a model by specification tests just comes back to curve fitting—with a complicated set of constraints on the residuals.

Given the limits to present knowledge, I doubt that models can be rescued by technical fixes. Arguments about the theoretical merit of regression or the asymptotic behavior of specification tests for picking one version of a model over another seem like the arguments about how to build desalination plants with cold fusion as the energy source. The concept may be admirable, the technical details may be fascinating, but thirsty people should look elsewhere.

3.6 Other literature

The issues raised here are hardly new, and this section reviews some recent literature. No brief summary can do justice to Lieberson (1985), who presents a complicated and subtle critique of current empirical work in the social sciences. I offer a crude paraphrase of one important message: When there are significant differences between comparison groups in an observational study, it is extraordinarily difficult if not impossible to achieve balance by statistical adjustments. Arminger and Bohrnstedt (1987, p. 366) respond by describing this as a special case of "misspecification of the mean structure caused by the omission of relevant causal variables" and cite literature on that topic.

This trivializes the problem and almost endorses the idea of fixing misspecification by elaborating the model. However, that idea is unlikely to work. Current specification tests need independent, identically distributed observations, and lots of them; the relevant variables must be identified; some variables must be taken as exogenous; additive errors are needed; and a parametric or semiparametric form for the mean function is required. These ingredients are rarely found in the social sciences, except by assumption. To model a bias, we need to know what causes it, and how. In practice, this may be even more difficult than the original research question. Some empirical evidence is provided by the discussion of manpower-training program evaluations above (also see Stolzenberg and Relles 1990).

As Arminger and Bohrnstedt concede (1987, p. 370),

There is no doubt that experimental data are to be preferred over nonexperimental data, which practically demand that one knows the mean structure except for the parameters to be estimated.

In the physical or life sciences, there are some situations in which the mean function is known, and regression models are correspondingly useful. In the social sciences, I do not see this precondition for regression modeling as being met, even to a first approximation.

In commenting on Lieberson (1985), Singer and Marini (1987) emphasize two points:

1. "It requires rather yeoman assumptions or unusual phenomena to conduct a comparative analysis of an observational study as though it represented the conclusions (inferences) from an experiment." (p. 376)

2. "There seems to be an implicit view in much of social science that any question that might be asked about a society is answerable in principle." (p. 382)

In my view, point 1 says that in the current state of knowledge in the social sciences, regression models are seldom if ever reliable for causal inference. With respect to point 2, it is exactly the reliance on models that makes all questions seem "answerable in principle"—a great obstacle to the development of the subject. It is the beginning of scientific wisdom to recognize that not all questions have answers. For some discussion along these lines, see Lieberson (1988).

Marini and Singer (1988) continue the argument:

Few would question that the use of "causal" models has improved our knowledge of causes and is likely to do so increasingly as the models are refined and become more attuned to the phenomena under investigation. (p. 394)

However, much of the analysis in Marini and Singer contradicts this presumed majority view:

Causal analysis ... is not a way of deducing causation but of quantifying already hypothesized relationships.... Information external to the model is needed to warrant the use of one specific representation as truly "structural." The information must come from the existing body of knowledge relevant to the domain under consideration. (pp. 388, 391)

As I read the current empirical research literature, causal arguments depend mainly on the statistical significance of regression coefficients. If so, Marini and Singer are pointing to the fundamental circularity in the regression strategy: The information needed for building regression models comes only from such models. Indeed, Marini and Singer continue:

The relevance of causal models to empirical phenomena is often open to question because assumptions made for the purpose of model identification are arbitrary or patently false. The models take on an importance of their own, and convenience or elegance in the model building overrides faithfulness to the phenomena. (p. 392)

Holland (1988) raises similar points. Causal inferences from nonexperimental data using path models require assumptions that are quite close to the conclusions; so the analysis is driven by the model, not the data. In effect, given a set of covariates, the mean response over the "treatment group" minus the mean over the "controls" must be assumed to equal the causal effect being estimated (1988, p. 481).

> The effect ... cannot be estimated by the usual regression methods of path analysis without making untestable assumptions about the counterfactual regression function. (p. 470)

Berk (1988, p. 161) discusses causal inferences based on path diagrams, including "unobservable disturbances meeting the usual (and sometimes heroic) assumptions." He considers the oft-recited arguments that biases will be small, or if large will tend to cancel, and concludes, "Unfortunately, it is difficult to find any evidence for these beliefs" (p. 163). He recommends quasi-experimental designs, which

> are terribly underutilized by sociologists despite their considerable potential. While they are certainly no substitute for random assignment, the stronger quasi-experimental designs can usually produce far more compelling causal inferences than conventional cross-sectional data sets. (p. 163)

He comments on model development by testing, including the use of the specification tests:

> The results may well be misleading if there are any other statistical assumptions that are substantially violated. (p. 165)

I found little to disagree with in Berk's essay. Casual observation suggests that no dramatic change in research practice took place following publication of his essay; further discussion of the issues may be needed.

Of course, Meehl (1978) already said most of what needs saying in 1978, in his article, "Theoretical Risks and Tabular Asterisks: Sir Karl, Sir Ronald, and the Slow Progress of Soft Psychology." In paraphrase, the good knight is Karl Popper, whose motto calls for subjecting scientific theories to grave danger of refutation. The bad knight is Ronald Fisher, whose significance tests are trampled in the dust:

> The almost universal reliance on merely refuting the null hypothesis as the standard method for corroborating substantive theories in the soft areas is ... basically unsound. (p. 817)

Paul Meehl is an eminent psychologist, and he has one of the best data sets available for demonstrating the predictive power of regression models. His judgment deserves some consideration.

3.7 Conclusion

One fairly common way to attack a problem involves collecting data and then making a set of statistical assumptions about the process that generated the data—for example, linear regression with normal errors, conditional independence of categorical data given covariates, random censoring of observations, independence of competing hazards.

Once the assumptions are in place, the model is fitted to the data, and quite intricate statistical calculations may come into play: three-stage least squares, penalized maximum likelihood, second-order efficiency, and so on. The statistical inferences sometimes lead to rather strong empirical claims about structure and causality.

Typically, the assumptions in a statistical model are quite hard to prove or disprove, and little effort is spent in that direction. The strength of empirical claims made on the basis of such modeling therefore does not derive from the solidity of the assumptions. Equally, these beliefs cannot be justified by the complexity of the calculations. Success in controlling observable phenomena is a relevant argument, but one that is seldom made.

These observations lead to uncomfortable questions. Are the models helpful? Is it possible to differentiate between successful and unsuccessful uses of the models? How can the models be tested and evaluated? Regression models have been used on social science data since Yule (1899), so it may be time ask these questions—although definitive answers cannot be expected.

Acknowledgments

This research was partially supported by NSF grant DMS 86-01634 and by the Miller Institute for Basic Research in Science. Much help was provided by Richard Berk, John Cairns, David Collier, Persi Diaconis, Sander Greenland, Steve Klein, Jan de Leeuw, Thomas Rothenberg, and Amos Tversky. Special thanks go to Peter Marsden.

Part II

Studies in Political Science, Public Policy, and Epidemiology

4

Methods for Census 2000 and Statistical Adjustments

With Kenneth W. Wachter

ABSTRACT. *The U.S. Census is a sophisticated, complex undertaking, carried out on a vast scale. It is remarkably accurate. Statistical adjustment is unlikely to improve on the census, because adjustment can easily introduce more error than it takes out. The data suggest a strong geographical pattern to such errors even after controlling for demographic variables, which contradicts basic premises of adjustment. In fact, the complex demographic controls built into the adjustment process seem on whole to have been counter-productive.*

4.1 Introduction

The census has been taken every ten years since 1790, and provides a wealth of demographic information for researchers and policy-makers. Beyond that, counts are used to apportion Congress and redistrict states. Moreover, census data are the basis for allocating federal tax money to cities and other local governments. For such purposes, the geographical distribution of the population matters more than counts for the nation as a whole. Data from 1990 and previous censuses suggested there would

Handbook of Social Science Methodology. (2007) S. Turner and W. Outhwaite, eds. Sage Publications, pp. 232–45.

be a net undercount in 2000. Furthermore, the undercount would depend on age, race, ethnicity, gender, and—most importantly—geography. This differential undercount, with its implications for sharing power and money, attracted considerable attention in the media and the courthouse.

There were proposals to adjust the census by statistical methods, but this is advisable only if the adjustment gives a truer picture of the population and its geographical distribution. The census turned out to be remarkably good, despite much critical commentary. Statistical adjustment was unlikely to improve the accuracy, because adjustment can easily put in more error than it takes out.

We sketch procedures for taking the census, making adjustments, and evaluating results. (Detailed descriptions cover thousands of pages; summaries are a necessity.) Data are presented on errors in the census, in the adjustment, and on geographical variation in error rates. Alternative adjustments are discussed, as are methods for comparing the accuracy of the census and the adjustments. There are pointers to the literature, including citations to the main arguments for and against adjustment. The present article is based on Freedman and Wachter (2003), which may be consulted for additional detail and bibliographic information.

4.2 The census

The census is a sophisticated enterprise whose scale is remarkable. In round numbers, there are 10,000 permanent staff at the Bureau of the Census. Between October 1999 and September 2000, the staff opened 500 field offices, where they hired and trained 500,000 temporary employees. In spring 2000, a media campaign encouraged people to cooperate with the census, and community outreach efforts were targeted at hard-to-count groups.

The population of the United States is about 280 million persons in 120 million housing units, distributed across seven million "blocks," the smallest pieces of census geography. (In Boston or San Francisco, a block is usually a block; in rural Wyoming, a "block" may cover a lot of rangeland.) Statistics for larger areas like cities, counties, or states are obtained by adding up data for component blocks.

From the perspective of a census-taker, there are three types of areas to consider. In city delivery areas (high-density urban housing with good addresses), the Bureau develops a Master Address File. Questionnaires are mailed to each address in the file. About seventy percent of these questionnaires are filled out and returned by the respondents. Then "Non-Response Follow-Up" procedures go into effect: For instance, census enumerators go out several times and attempt to contact non-responding households, by knocking on doors and working the telephone. City delivery areas include roughly 100 million housing units.

Update/leave areas, comprising less than twenty million households, are mainly suburban and have lower population densities; address lists are more difficult to construct. In such areas, the Bureau leaves the census questionnaire with the household while updating the Master Address File. Beyond that, procedures are similar to those in the city delivery areas.

In update/enumerate areas, the Bureau tries to enumerate respondents by interviewing them as it updates the Master Address File. These areas are mainly rural, and post office addresses are poorly defined, so address lists are problematic. (A typical address might be something like Smith, Rural Route #1, south of Willacoochee, GA.) Perhaps a million housing units fall into such areas. There are also special populations that need to be enumerated—institutional (prisons and the military), as well as non-institutional "group quarters." (For instance, twelve nuns sharing a house in New Orleans are living in group quarters.) About eight million persons fall into these special populations.

4.3 Demographic analysis

DA (Demographic Analysis) estimates the population using birth certificates, death certificates, and other administrative record systems. The estimates are made for national demographic groups defined by age, gender, and race (Black and non-Black). Estimates for subnational geographic areas like states are currently not available. According to DA, the undercount in 1970 was about three percent nationally. In 1980, it was one to two percent, and the result for 1990 was similar. DA reported the undercount for Blacks at about five percentage points above non-Blacks, in all three censuses.

DA starts from an accounting identity:

$$\text{Population} = \text{Births} - \text{Deaths} + \text{Immigration} - \text{Emigration}.$$

However, data on emigration are incomplete. And there is substantial illegal immigration, which cannot be measured directly. Thus, estimates need to be made for illegals, but these are (necessarily) somewhat speculative.

Evidence on differential undercounts depends on racial classifications, which may be problematic. Procedures vary widely from one data collection system to another. For the census, race of all household members is reported by the person who fills out the form. In Census 2000, respondents were allowed for the first time to classify themselves into multiple racial categories. This is a good idea from many perspectives, but creates a discontinuity with past data. On death certificates, race of decedent is often determined by the undertaker. Birth certificates show the race of the mother and (usually) the race of father; procedures for ascertaining race differ from hospital to hospital. A computer algorithm is used to determine race of infant from race of parents.

Prior to 1935, many states did not collect birth certificate data at all; and the further back in time, the less complete is the system. This makes it harder to estimate the population aged sixty-five and over. In 2000, DA estimates the number of such persons starting from Medicare records. Despite its flaws, DA has generally been considered to be the best yardstick for measuring census undercounts. Recently, however, another procedure has come to the fore, the DSE ("Dual System Estimator").

4.4 DSE—Dual System Estimator

The DSE is based on a special sample survey done after the census—a PES ("Post Enumeration Survey"). The PES of 2000 was renamed ACE ("Accuracy and Coverage Evaluation Survey"). The ACE sample covers 25,000 blocks, containing 300,000 housing units and 700,000 people. An independent listing is made of the housing units in the sample blocks, and persons in these units are interviewed after the census is complete. This process yields the "P-sample."

The "E-sample" comprises the census records in the same blocks, and the two samples are then matched up against each other. In most cases, a match validates both the census record and the PES record. A P-sample record that does not match to the census may be a gross omission, that is, a person who should have been counted in the census but was missed. Conversely, a census record that does not match to the P-sample may be an erroneous enumeration, in other words, a person who got into the census by mistake. For instance, a person can be counted twice in the census—because he sent in two forms. Another person can be counted correctly but assigned to the wrong unit of geography: She is a gross omission in one place and an erroneous enumeration in the other.

Of course, an unmatched P-sample record may just reflect an error in ACE; likewise, an unmatched census record could just mean that the corresponding person was found by the census and missed by ACE. Fieldwork is done to resolve the status of some unmatched cases, deciding whether the error should be charged against the census or ACE. Other cases are resolved using computer algorithms. However, even after fieldwork is complete and the computer shuts down, some cases remain unresolved. Such cases are handled by statistical models that fill in the missing data.

The number of unresolved cases is relatively small, but it is large enough to have an appreciable influence on the final results (Section 4.9).

Movers—people who change address between census day and ACE interview—represent another complication. Unless persons can be correctly identified as movers or non-movers, they cannot be matched correctly. Identification depends on getting accurate information from respondents as to where they were living at the time of the census. Again, the number of movers is relatively small, but they are a large factor in the adjustment equation. More generally, matching records between the ACE and the census becomes problematic if respondents give inaccurate information to the ACE, or the census, or both. Thus, even cases that are resolved though ACE fieldwork and computer operations may be resolved incorrectly. We refer to such errors as "processing errors."

The statistical power of the DSE comes from matching, not from counting better. In fact, the E-sample counts came out a bit higher than the P-sample counts, in 1990 and in 2000: The census found more people than the post enumeration survey in the sample blocks. As the discussion of processing error shows, however, matching is easier said than done.

Some persons are missed both by the census and by ACE. Their number is estimated using a statistical model, assuming that ACE is as likely to find people missed by the census as people counted in the census—"the independence assumption." Following this assumption, a gross omission rate estimated from the people found by ACE can be extrapolated to people in the census who were missed by ACE, although the true gross omission rate for that group may well be different. Failures in the independence assumption lead to "correlation bias." Data on processing error and correlation bias will be presented later.

4.5 Small-area estimation

The Bureau divides the population into post strata defined by demographic and geographic characteristics. For Census 2000, there were 448 post strata. One post stratum, for example, consisted of Asian male renters age thirty to forty-nine, living anywhere in the United States. Another post stratum consisted of Blacks age zero to seventeen (male or female) living in owner-occupied housing in big or medium-size cities with high mail return rates across the whole country. Persons in the P-sample are assigned to post strata on the basis of information collected during the ACE interview. (For the E-sample, assignment is based on the census return.)

Each sample person gets a weight. If one person in 500 were sampled, each person in the sample would stand for 500 in the population and be given a weight of 500. The actual sampling plan for ACE is more complex, so different people are given different weights. To estimate the total number of gross omissions in a post stratum, one simply adds the weights of all ACE respondents who were identified as (i) gross omissions and (ii) being in the relevant post stratum.

To a first approximation, the estimated undercount in a post stratum is the difference between the estimated numbers of gross omissions and erroneous enumerations. In more detail, ACE data are used to compute an adjustment factor for each post stratum. When multiplied by this factor, the census count for a post stratum equals the estimated true count from the DSE. About two-thirds of the adjustment factors exceed one. These post strata are estimated to have undercounts. The remaining post strata are estimated to have been overcounted by the census; their adjustment factors are less than one.

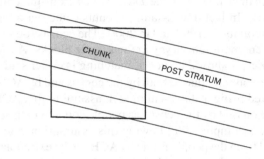

How to adjust small areas like blocks, cities, or states? Take any particular area. As the sketch indicates, this area will be carved up into "chunks" by post strata. Each chunk has some number of persons counted by the census in that area. (The number may be zero.) This census number is multiplied by the adjustment factor for the post stratum. The process is repeated for all post strata, and the adjusted count is obtained by adding the products; complications due to rounding are ignored here. The adjustment process makes the "homogeneity assumption" that undercount rates are constant within each post stratum across all geographical units. This is not plausible, and was strongly contradicted by census data on variables related to the undercount. Failures in the homogeneity assumption are termed "heterogeneity." Ordinarily, samples are used to extrapolate upwards, from the part to the whole. In census adjustment, samples are used to extrapolate sideways, from 25,000 sample blocks to each and every one of the seven million blocks in the United States. That is where the homogeneity assumption comes into play.

Heterogeneity is endemic. Undercount rates differ from place to place within population groups treated as homogeneous by adjustment.

Heterogeneity puts limits on the accuracy of adjustments for areas like states, counties, or legislative districts. Studies of the 1990 data, along with more recent work discussed in Section 4.11 below, show that heterogeneity is a serious concern.

The adjustment issue was often framed in terms of sampling: "Sampling is scientific." However, from a technical perspective, sampling is not the point. The crucial questions are about the size of processing errors, and the validity of statistical models for missing data, correlation bias, and homogeneity—in a context where the margin of allowable error is relatively small.

4.6 State shares

All states would gain population from adjustment. Some, however, gain more than others. In terms of population share, the gains and losses must balance. This point was often overlooked in the political debate. In 2000, even more so than in 1990, share changes were tiny. According to Census 2000, for example, Texas had 7.4094 percent of the population. Adjustment would have given it 7.4524 percent, an increase of

$$7.4524 - 7.4094 = 0.0430 \text{ percent},$$

or 430 parts per million. The next biggest winner was California, at 409 parts per million; third was Georgia, at 88 parts per million.

Ohio would have been the biggest loser, at 241 parts per million; then Michigan, at 162 parts per million. Minnesota came third in this sorry competition at 152 parts per million. The median change (up or down) is about twenty-eight parts per million. These changes are tiny, and most are easily explained as the result of sampling error in ACE. "Sampling error" means random error introduced by the luck of the draw in choosing blocks for the ACE sample: You get a few too many blocks of one kind or not quite enough of another. The contrast is with "systematic" or "non-sampling" error like processing error.

The map (Figure 4.1) shows share changes that exceed fifty parts per million. Share increases are marked "+"; share decreases, "−". The size of the mark corresponds to the size of the change. As the map indicates, adjustment would have moved population share from the Northeast and Midwest to the South and West. This is paradoxical, given the heavy concentrations of minorities in the big cities of the Northeast and Midwest, and political rhetoric contending that the census shortchanges such areas ("statistical grand larceny," according to New York's ex-Mayor Dinkins). One explanation for the paradox is correlation bias. The older urban centers of the Northeast and Midwest may be harder to reach, both for census and for ACE.

Figure 4.1 ACE adjustment: State share changes exceeding fifty parts per million.

4.7 The 1990 adjustment decision

A brief look at the 1990 adjustment decision provides some context for discussions of Census 2000. In July 1991, the Secretary of Commerce declined to adjust Census 1990. At the time, the undercount was estimated as 5.3 million persons. Of this, 1.7 million persons were thought by the Bureau to reflect processing errors in the post enumeration survey, rather than census errors. Later research has shown the 1.7 million to be a serious underestimate. Current estimates range from 3.0 million to 4.2 million, with a central value of 3.6 million. (These figures are all nationwide, and net; given the data that are available, parceling the figures down to local areas would require heroic assumptions.)

The bulk of the 1990 adjustment resulted from errors not in the census but in the PES. Processing errors generally inflate estimated undercounts, and subtracting them leaves a corrected adjustment of 1.7 million. (There is an irritating numerical coincidence here as 1.7 million enters the discussion with two different meanings.) Correlation bias, estimated at 3.0 million, works in the opposite direction, and brings the undercount estimate up to the Demographic Analysis figure of 4.7 million (Table 4.1). On the scale of interest, most of the estimated undercount is noise.

4.8 Census 2000

Census 2000 succeeded in reducing differential undercounts from their 1990 levels. That sharpened questions about the accuracy of pro-

Table 4.1 Errors in the adjustment of 1990

The adjustment	+5.3	
Processing error	−3.6	
Corrected adjustment		+1.7
Correlation bias		+3.0
Demographic Analysis		+4.7

posed statistical adjustments. Errors in statistical adjustments are not new. Studies of the 1980 and 1990 data have quantified, at least to some degree, the three main kinds of error: processing error, correlation bias, and heterogeneity. In the face of these errors, it is hard for adjustment to improve on the accuracy of census numbers for states, counties, legislative districts, and smaller areas.

Errors in the ACE statistical operations may from some perspectives have been under better control than they were in 1990. But error rates may have been worse in other respects. There is continuing research, both inside the Bureau and outside, on the nature of the difficulties. Troubles occurred with a new treatment of movers (discussed in the next section) and duplicates. Some twenty-five million duplicate persons were detected in various stages of the census process and removed. But how many slipped through? And how many of those were missed by ACE?

Besides processing error, correlation bias is an endemic problem that makes it difficult for adjustment to improve on the census. Correlation bias is the tendency for people missed in the census to be missed by ACE as well. Correlation bias in 2000 probably amounted, as it did in 1990, to millions of persons. Surely these people are unevenly distributed across the country ("differential correlation bias"). The more uneven is the distribution, the more distorted a picture of census undercounts is created by the DSE.

4.9 The adjustment decision for Census 2000

In March 2001, the Secretary of Commerce—on the advice of the Census Bureau—decided to certify the census counts rather than the adjusted counts for use in redistricting (drawing congressional districts within state). The principal reason was that, according to DA, the census had overcounted the population by perhaps two million people. Proposed adjustments would have added another three million people, making the overcounts even worse. Thus, DA and ACE pointed in opposite directions. The three population totals are shown in Table 4.2.

Table 4.2 The population of the United States

Demographic Analysis	279.6 million
Census 2000	281.4 million
ACE	284.7 million

If DA is right, there is a census overcount of 0.7 percent. If ACE is right, there is a census undercount of 1.2 percent. DA is a particularly valuable benchmark because it is independent (at least in principle) of both the census and the post enumeration survey that underlies proposed adjustments. While DA is hardly perfect, it was a stretch to blame DA for the whole of the discrepancy with ACE. Instead, the discrepancy pointed to undiscovered error in ACE. When the Secretary made his decision, there was some information on missing data and on the influence of movers, summarized in Table 4.3.

These figures are weighted to national totals, and should be compared to (i) a total census population around 280 million, and (ii) errors in the census that may amount to a few million persons. For some three million P-sample persons, a usable interview could not be completed; for six million, a household roster as of census day could not be obtained (lines 1 and 2 in the table). Another three million persons in the P-sample and seven million in the E-sample had unresolved match status after field-work: Were they gross omissions, erroneous enumerations, or what? For six million, residence status was indeterminate—where *were* they living on census day? (National totals are obtained by adding up the weights for the corresponding sample people; non-interviews are weighted out of the sample and ignored in the DSE, but we use average weights.) If the idea is to correct an undercount of a few million in the census, these are serious gaps. Much of the statistical adjustment therefore depends on models used to fill in missing data. Efforts to validate such models remain unconvincing.

The 2000 adjustment tried to identify both inmovers and outmovers, a departure from past practice. Gross omission rates were computed for the outmovers and applied to the inmovers, although it is not clear why rates are equal within local areas. For outmovers, information must have been obtained largely from neighbors. Such "proxy responses" are usually thought to be of poor quality, inevitably creating false non-matches and inflating the estimated undercount. As the table shows, movers contribute about three million gross omissions (a significant number on the scale of interest) and ACE failed to detect a significant number of outmovers. That is why the number of outmovers is so much less than the number

Table 4.3 Missing data in ACE and results for movers

Non-interviews	
P-sample	3 million
E-sample	6 million
Imputed match status	
P-sample	3 million
E-sample	7 million
Inmovers and outmovers	
Imputed residence status	6 million
Outmovers	9 million
Inmovers	13 million
Mover gross omissions	3 million

of inmovers. Again, the amount of missing data is small relative to the total population, but large relative to errors that need fixing. The conflict between these two sorts of comparisons is the central difficulty of census adjustment. ACE may have been a great success by the ordinary standards of survey research, but not nearly good enough for adjusting the census.

4.10 Gross or net?

Errors can reported either gross or net, and there are many possible ways to refine the distinction. (Net error allows overcounts to balance undercounts; gross error does not.) Some commentary suggests that the argument for adjustment may be stronger if gross error is the yardstick. Certain places may have an excess number of census omissions while other places will have an excess number of erroneous enumerations. Such imbalances could be masked by net error rates, when errors of one kind in one place offset error of another kind in another place. In this section, we consider gross error rates.

Some number of persons were left out of Census 2000 and some were counted in error. There is no easy way to estimate the size of these two errors separately. Many people were counted a few blocks away from where they should have been counted: They are both gross omissions and erroneous enumerations. Many other people were classified as erroneous enumerations because they were counted with insufficient information for matching; they should also come back as gross omissions in the ACE fieldwork. With some rough-and-ready allowances for this sort of double-counting, the Bureau estimated that six to eight million people were left out of the census while three to four million were wrongly included, for a gross error in the census of nine to twelve million; the Bureau's preferred values are 6.4 and 3.1, for a gross error of 9.5 million in Census 2000.

Before presenting comparable numbers for ACE, we mention some institutional history. The census is used as a base for post-censal population estimates. This may sound even drier than redistricting, but $200 billion a year of tax money is allocated using post-censal estimates. In October 2001, the Bureau revisited the adjustment issue: Should the census be adjusted as a base for the post-censals? The decision against adjustment was made after further analysis of the data. Some 2.2 million persons were added to the Demographic Analysis. Estimates for processing error in ACE were sharply increased. Among other things, ACE had failed to detect large numbers of duplicate enumerations in the census because interviewers did not get accurate census-day addresses from respondents. That is why ACE had over-estimated the population. The Bureau's work confirmed that gross errors in ACE were well above ten million, with another fifteen million cases whose status remains to be resolved. Error rates in ACE are hard to determine with precision, but are quite large relative to error rates in the census.

4.11 Heterogeneity in 2000

This section demonstrates that substantial heterogeneity remains in the data despite elaborate post stratification. In fact, post stratification seems on the whole to be counter-productive. Heterogeneity is measured as in Freedman and Wachter (1994, 2003), with SUB ("whole-person substitutions") and LA ("late census adds") as proxies—surrogates—for the undercount: see the notes to Table 4.4. For example, 0.0210 of the census count (just over 2%) came from whole-person substitutions. This figure is in the first line of the table, under the column headed "Level." Substitution rates are computed not only for the whole country, but for each of the 435 congressional districts: The standard deviation of the 435 rates is 0.0114, in the "Across CD" column. The rate is also computed for each post stratum: Across the 448 post strata, the standard deviation of the substitution rates is 0.0136, in the "Across P-S" column. The post strata exhibit more variation than the geographical districts, which is one hallmark of a successful post stratification.

To compute the last column of the table, we think of each post stratum as being divided into chunks by the congressional districts. We compute the substitution rate for each chunk with a non-zero census count, then take the standard deviation across chunks within post stratum, and finally the root-mean-square over post strata. The result is 0.0727, in the last column of Table 4.4. If rates were constant across geography within post strata,

Table 4.4 Measuring heterogeneity across Congressional Districts (CD). In the first column, post stratification is either (i) by 448 post strata; or (ii) by the sixty-four post-stratum groups, collapsing age and sex; or (iii) by the sixteen evaluation post strata. "SUB" means whole-person substitutions, and "LA" is late census adds. In the last two columns, "P-S" stands for post strata; there are three different kinds, labeled according to row.

Proxy & post stratification	Level	Standard deviation		
		Across CD	Across P-S	Within P-S across CD
SUB 448	0.0210	0.0114	0.0136	0.0727
SUB 64	0.0210	0.0114	0.0133	0.0731
SUB 16	0.0210	0.0114	0.0135	0.0750
LA 448	0.0085	0.0054	0.0070	0.0360
LA 64	0.0085	0.0054	0.0069	0.0363
LA 16	0.0085	0.0054	0.0056	0.0341

Note: The level of a proxy does not depend on the post stratification, and neither does the standard deviation across CDs. These two statistics do depend on the proxy. A "substitution" is a person counted in the census with no personal information, which is later imputed. A "late add" is a person originally thought to be a duplicate, but later put back into the census production process. Substitutions include late adds that are not "data defined," i.e., do not have enough information for matching. Substitutions and late adds have poor data quality, which is why they may be good proxies for undercount. Table 5 in Freedman and Wachter (2003) uses slightly different conventions and includes the District of Columbia.

the homogeneity assumption requires, this standard deviation should be zero. Instead, it is much larger than the variability across congressional districts. This points to a serious failure in the post stratification. If the proxies are good, there is a lot of heterogeneity within post strata across geography.

Similar calculations can be made for two coarser post stratifications. (i) The Bureau considers its 448 post strata as coming from sixty-four PSG's. (Each PSG, or "post-stratum group," divides into seven age-sex groups, giving back $64 \times 7 = 448$ post strata.) The sixty-four PSG's are used as post strata in the second line of Table 4.4. (ii) The Bureau groups

PSG's into sixteen EPS, or "evaluation post strata." These are the post strata in the third line of Table 4.4. Variability across post strata or within post strata across geography is not much affected by the coarseness of the post stratification, which is surprising. Results for late census adds (LA) are similar, in lines 4–6 of the table. Refining the post stratification is not productive. There are similar results for states in Freedman and Wachter (2003).

The Bureau computed "direct DSEs" for the sixteen evaluation post strata, by pooling the data in each. From these, an adjustment factor can be constructed, as the direct DSE divided by the census count. We adjusted the United States using these sixteen factors rather than the 448. For states and congressional districts, there is hardly any difference. The scatter diagram in Figure 4.2 shows results for congressional districts. There are 435 dots, one for each congressional district. The horizontal axis shows the change in population count that would have resulted from adjustment with 448 post strata; the vertical, from adjustment with sixteen post strata.

For example, take CD 1 in Alabama, with a 2000 census population of 646,181. Adjustment with 448 post strata would have increased this figure by 7630; with sixteen post strata, the increase would have been

Figure 4.2 Changes to congressional district populations. The production adjustment, with 448 post strata, is plotted on the horizontal. An alternative, based only on the sixteen evaluation post strata (EPS), is plotted on the vertical.

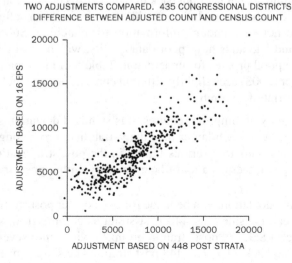

TWO ADJUSTMENTS COMPARED. 435 CONGRESSIONAL DISTRICTS
DIFFERENCE BETWEEN ADJUSTED COUNT AND CENSUS COUNT

Table 4.5 Comparing the production adjustment based on 448 post strata to one based on sixteen evaluation post strata. Correlation coefficients for changes due to adjustments.

Changes in state population counts	.99
Changes in state population shares	.90
Changes in congressional district counts	.87
Changes in congressional district shares	.85

7486. The corresponding point is (7630, 7486). The correlation between the 435 pairs of changes is .87, as shown in the third line of Table 4.5. For two out of the 435 districts, adjustment by 448 post strata would have reduced the population count: Their points are plotted just outside the axes, at the lower left. On this basis, and on the basis of Table 4.4, we suggest that 448 post strata are no better than sixteen. (For some geographical areas with populations below 100,000, however, the two adjustments are likely to be different.)

Tables 4.4–4.5 and Figure 4.2 show that an elaborate post stratification does not remove much heterogeneity. We doubt that heterogeneity can be removed by the sort of post stratification—no matter how elaborate—that can be constructed in real census conditions. The impact of heterogeneity on errors in adjustment is discussed by Freedman and Wachter (1994, pp. 479–81). Heterogeneity is more of a problem than sampling error.

Within a state, districts are by case law almost exactly equal in size—when redistricting is done shortly after census counts are released. Over the decade, people move from one district to another. Variation in population sizes at the end of the decade is therefore of policy interest. In California, for one example, fifty-two districts were drawn to have equal populations according to Census 1990. According to Census 2000, the range in their populations is 583,000 to 773,000. Exact equality at the beginning of the decade does not seem like a compelling goal.

4.12 Loss function analysis

A statistical technique called "loss function analysis" has been used to justify adjustment. In effect, this technique attempts to make summary estimates of the error levels in the census and the adjustment. However, the apparent gains in accuracy—like the gains from adjustment—tend to be concentrated in a few geographical areas, and heavily influenced by the vagaries of chance. At a deeper level, loss function analysis turns out to depend more on assumptions than on data.

For example, loss function analysis depends on models for correlation bias, and the model used in 2000 assumes there is no correlation bias for women. The idea that only men are hard to reach—for the census and the post enumeration survey—is unlikely on its face. It is also at loggerheads with the data from 1990: see Wachter and Freedman (2000). A second example: Loss function analysis depends on having precise estimates of error rates in ACE. But there is considerable uncertainty about these error rates, even at the national level (Sections 4.9–4.10). A last example: Adjustment makes the homogeneity assumption—census errors occur at a uniform rate within post strata across wide stretches of geography. Loss function analysis assumes that and more: Error rates in the census are uniform, and so are error rates in ACE. That is how processing errors and correlation bias in ACE can be parceled out to local areas without creating unmanageably large variances. But these homogeneity assumptions are not tenable (Section 4.11).

4.13 Pointers to the literature

Reviews and discussions of the 1980 and 1990 adjustments can be found in *Survey Methodology* (1992) 18: 1–74, *Journal of the American Statistical Association* (1993) 88: 1044–1166, and *Statistical Science* (1994) 9: 458–537. Other exchanges worth noting include *Jurimetrics* (1993) 34: 59–115 and *Society* (2001) 39: 3–53. These are easy to read, and informative. Pro-adjustment arguments are made by Anderson and Fienberg (1999), but see Stark (2001) and Ylvisaker (2001). Prewitt (2000) may be a better source, and Zaslavsky (1993) is often cited. Cohen, White, and Rust (1999) try to answer arguments on the 1990 adjustment; but see Freedman and Wachter (2003). Skerry (2000) has an accessible summary of the issues. Darga (2000) is a critic. Freedman, Stark, and Wachter (2001) have a probability model for census adjustment, which may help to clarify some of the issues.

The decision against adjustment for 1990 is explained in U.S. Department of Commerce (1991). On the 2000 adjustment decision, see U.S. Bureau of the Census (2001a,b), U.S. Census Bureau (2003). For another perspective on Census 2000, see Citro, Cork, and Norwood (2004). Problems with the PES, especially with respect to detecting duplicates, are discussed at pp. 214 ff and 240 ff. However, there is residual enthusiasm for a PES in 2010 and a corresponding lack of enthusiasm for Demographic Analysis (p. 8). Cork, Cohen, and King (2004) reach different conclusions (p. 11).

4.14 Litigation

The Commerce Department's decision not to adjust the 1980 census was upheld after trial. *Cuomo v Baldrige*, 674 F.Supp. 1089 (S.D.N.Y. 1987). The Department's decision not to adjust the 1990 census was also upheld after trial and appeal to the Supreme Court. 517 U.S. 1 (1996). Later in the decade, the Court found that use of adjustment for reapportionment, that is, allocating congressional seats among the states, violated the Census Act. 525 U.S. 316 (1999). The administration had at the time planned to adjust, so the Court's decision necessitated a substantial revision to the design of ACE (Brown et al. 1999).

Efforts by Los Angeles and the Bronx among others to compel adjustment of Census 2000 were rejected by the courts (*City of Los Angeles et al. v. Evans et al.*, Central District, California); the decision was upheld on appeal to the Ninth Circuit. 307 F.3d 859 (9th Cir. 2002). There was a similar outcome in an unpublished case, *Cameron County et al. v. Evans et al.*, Southern District, Texas. Utah sued to preclude the use of imputations but the suit was denied by the Supreme Court. *Utah et al. v. Evans et al.*, 536 U.S. 452 (2002).

The Commerce Department did not wish to release block-level adjusted counts, but was compelled to do so as a result of several lawsuits. The lead case was *Carter v. U.S. Dept. of Commerce* in Oregon. The decision was upheld on appeal to the Ninth Circuit. 307 F.3d 1084 (9th Cir. 2002).

4.15 Other countries

For context, this section gives a bird's-eye view of the census process in a few other countries. In Canada, the census is taken every five years (1996, 2001, 2006, ...). Unadjusted census counts are published. Coverage errors are estimated, using variations on the PES (including a "reverse record check") and other resources. A couple of years later, when the work is complete, post-censal population estimates are made for provinces and many subprovincial areas. These estimates are based on adjusted census counts. The process in Australia is similar; the PES there is like a scaled-down version of the one in the U.S.

The U.K. takes its census every ten years (1991, 2001, 2011, ...). Coverage errors are estimated using a PES. Only the adjusted census counts are published. The official acronym is ONC, for One-Number Census. Failure to release the original counts cannot enhance the possibility of informed discussion. Moreover, results dating back to 1982 are adjusted to agree with current estimates. "Superseded" data sets seem to be withdrawn from the official U.K. web page (http://www.statistics.gov.uk).

Anomalies are found in the demographic structure of the estimated population (not enough males age twenty to twenty-four). See Redfern (2004) for discussion; also see http://www.statistics.gov.uk/downloads/theme_population/PT113.pdf, pp. 17 and 48.

In Scandinavian countries, the census is based on administrative records and population registries. In Sweden, for example, virtually every resident has a PIN [Personal Identification Number]; the authorities try to track down movers—even persons who leave the country. Norway conducted a census by mail in 2001, to complete its registry of housing, but is switching to an administrative census in the future. The accuracy of a registry census may not be so easy to determine.

4.16 Summary and conclusion

The idea behind the census is simple: You try to count everybody in the population, once and only once, at their place of residence rather than somewhere else. The U.S. Bureau of the Census does this sort of thing about as well as it can be done. Of course, the details are complicated, the expense is huge, compromises must be made, and mistakes are inevitable. The idea behind adjustment is to supplement imperfect data collection in the census with imperfect data collection in a post enumeration survey, and with modeling. It turns out, however, that the imperfections in the adjustment process are substantial, relative to the imperfections in the census. Moreover, the arguments for adjustment turn out to be based on hopeful assumptions rather than on data.

The lesson extends beyond the census context. Models look objective and scientific. If they are complicated, they appear to take into account many factors of interest. Furthermore, complexity is by itself a good first line of defense against criticism. Finally, modelers can try to buttress their results with another layer of models, designed to show that outcomes are insensitive to assumptions, or that different approaches lead to similar findings. Thus, modeling has considerable appeal. Moreover, technique is seductive, and seems to offer badly needed answers. However, conclusions may be driven by assumptions rather than data. Indeed, that is likely to be so. Otherwise, a model with unsupported assumptions would hardly be needed in the first place.

Note

Freedman and Wachter testified against adjustment in *Cuomo v. Baldrige* (1980 census) and *New York v. Department of Commerce* (1990 census). They consulted for the Department of Commerce on the 2000 census.

5

On "Solutions" to the Ecological Inference Problem

With Stephen P. Klein, Michael Ostland, and Michael R. Roberts

ABSTRACT. *In his 1997 book, King announced 'A Solution to the Ecological Inference Problem."King's method may be tested with data where truth is known. In the test data, his method produces results that are far from truth, and diagnostics are unreliable. Ecological regression makes estimates that are similar to King's, while the neighborhood model is more accurate. His announcement is premature.*

5.1 Introduction

Before discussing King (1997), we explain the problem of "eco-logical inference." Suppose, for instance, that in a certain precinct there are 500 registered voters of whom 100 are Hispanic and 400 are non-Hispanic. Suppose too that a Hispanic candidate gets ninety votes in this precinct. (Such data would be available from public records.) We would like to know how many of the votes for the Hispanic candidate came from

Journal of the American Statistical Association (1998) 93: 1518–22.

the Hispanics. That is a typical ecological-inference problem. The secrecy of the ballot box prevents a direct solution, so indirect methods are used.

This review will compare three methods for making ecological inferences. First and easiest is the "neighborhood model." This model makes its estimates by assuming that, within a precinct, ethnicity has no influence on voting behavior: In the example, of the ninety votes for the Hispanic candidate, $90 \times 100/(100 + 400) = 18$ are estimated to come from the Hispanic voters. The second method to consider is "ecological regression," which requires data on many precincts (indexed by i). Let n_i^h be the number of Hispanics in precinct i, and n_i^a the number of non-Hispanics; let v_i be the number of votes for the Hispanic candidate. (The superscript a is for "anglo"; this is only a mnemonic.) If our example precinct is indexed by $i = 1$, say, then $n_1^h = 100$, $n_1^a = 400$, and $v_1 = 90$. Ecological regression is based on the "constancy assumption": There is a fixed propensity p for Hispanics to vote for the Hispanic candidate and another fixed propensity q for non-Hispanics to vote for that candidate. These propensities are fixed in the sense of being constant across precincts. On this basis, the expected number of votes for the Hispanic candidate in precinct i is $pn_i^h + qn_i^a$. Then p and q can be estimated by doing some kind of regression of v on n^h and n^a.

More recently, King published "a solution to the ecological inference problem." His method will be sketched now, with a more detailed treatment below. In precinct i, the Hispanics have propensity p_i to vote for the Hispanic candidate, while the non-Hispanics have propensity q_i: The number of votes for the Hispanic candidate is then $v_i = p_i n_i^h + q_i n_i^a$. The precinct-specific propensities p_i and q_i are assumed to vary independently from precinct to precinct, being drawn at random from a fixed bivariate distribution—fixed in the sense that the same distribution is used for every precinct. (That replaces the "constancy assumption" of ecological regression.) The bivariate distribution is assumed to belong to a family of similar distributions, characterized by a few unknown parameters. These parameters are estimated by maximum likelihood, and then the precinct-level propensities p_i and q_i can be estimated too.

According to King, his "basic model is robust to aggregation bias" and "offers realistic estimates of the uncertainty of ecological estimates." Moreover, "all components of the proposed model are in large part verifiable in aggregate data" using "diagnostic tests to evaluate the appropriateness of the model to each application" (pp. 19–20). The model is validated on two main data sets, in chapters 10 and 11:

- registration by race in 275 southern counties, and
- poverty status by sex in 3187 block groups in South Carolina.

In the South Carolina data, "there are high levels of aggregation bias" (p. 219), but "even in this data set, chosen for its difficulty in making ecological inferences, the inferences are accurate" (p. 225). Chapter 13 considers two additional data sets: voter turnout in successive years in Fulton County, Georgia, and literacy by race and county in the U.S. in 1910. Apparently, the model succeeds in the latter example if two thirds of the counties are eliminated (p. 243). A fifth data set, voter turnout by race in Louisiana, is considered briefly on pp. 22–33.

King contends that (i) his method works even if the assumptions are violated, and (ii) his diagnostics will detect the cases where assumptions are violated. With respect to claim (i), the method should of course work when its assumptions are satisfied. Furthermore, the method may work when assumptions are violated—but it may also fail, as we show by example. With respect to claim (ii), the diagnostics do not reliably identify cases where assumptions are problematic. Indeed, we give examples where the data satisfy the diagnostics but the estimates are seriously in error. In other examples, data are generated according to the model but the diagnostics indicate trouble.

We apply King's method, and three of his main diagnostics, to several data sets where truth is known:

- an exit poll in Stockton where the unit of analysis is the precinct,
- demographic data from the 1980 census in Los Angeles County where the unit of analysis is the tract, and
- Registration data from the 1988 general election in Los Angeles County, aggregated to the tract level.

In these cases, as in King's examples discussed above, truth is known. We aggregate the data, deliberately losing (for the moment) information about individuals or subgroups, and then use three methods to make ecological inferences:

(i) the neighborhood model,
(ii) ecological regression, and
(iii) King's method.

The inferences having been made, they can be compared to truth. Moreover, King's method can be compared to other methods for ecological inference. King's method (estimation, calculation of standard errors, and diagnostic plots) is implemented in the software package EZIDOS—version 1.31 dated 8/22/97—which we downloaded in Fall 1997 from his Web page after publication of the book. We used this software for Tables 5.1 and 5.2 below.

Table 5.1 Comparison of three methods for making ecological inferences, in situations where the truth is known. King's method gives an estimate and a standard error, reported in the format "estimate ± standard error," and

$$Z = (\text{King's estimate} - \text{Truth})/\text{standard error.}$$

	Nbd	ER	King	Truth	Z
Stockton					
Exit poll	46%	109%	61% ± 18%	35%	+1.4
Artificial data	39%	36%	40% ± 15%	56%	−1.1
Los Angeles					
Education	65.1%	30.7%	30.1% ± 1.1%	55.6%	−23.2
High Hispanic	55.8%	38.9%	40.4% ± 1.2%	48.5%	−6.8
Income	48.5%	31.5%	32.9% ± 1.2%	48.8%	−13.2
Ownership	56.7%	51.7%	49.0% ± 1.5%	53.6%	−3.1
Party affiliation	65.0%	85.7%	90.8% ± 0.5%	73.5%	+34.6
Artificial data	67.2%	90.3%	90.3% ± 0.5%	89.5%	+1.6
High Hispanic	73.4%	90.1%	90.3% ± 0.5%	81.0%	+18.6

Note: "Nbd" is the neighborhood model; "ER" is ecological regression.

Table 5.2 Which estimation procedure comes closest to truth?

Data Set	Nbd	ER	King	King's diagnostics
Stockton				
Exit poll	x			Fails bias plot
Artificial data			x	Warning messages
Los Angeles				
Education	x			Marginal bias plot
High Hispanic	x			Passes
Income	x			Passes
Ownership	x			Passes
Party affiliation	x			Fails $E\{t|x\}$ plot
Artificial data		x	x	Fails $E\{t|x\}$ plot
High Hispanic	x			Passes
Number of wins	7	1	2	

5.2 The test data

The exit poll was done in Stockton during the 1988 presidential primary; the outcome measure is Hispanic support for Jackson: Data were collected on 1867 voters in thirty-nine sample precincts. The data set differs slightly from the one used in Freedman, Klein, Sacks et al. (1991) or Klein, Sacks, and Freedman (1991). For our purposes, "truth" is defined by the exit poll data at the level of individuals. (As it happens, the poll tracked the election results; but that does enter into the calculations here.)

The other data sets are based on 1409 census tracts in Los Angeles County, using demographic data from the 1980 census and registration data from the 1988 general election. Tracts that were small, or had inconsistent data, were eliminated; again, the data differ slightly from those in Freedman et al. (1991). The "high Hispanic" tracts have more than 25% Hispanics. The outcome measures on the demographic side are percent with high school degrees, percent with household incomes of $20,000 a year or more, and percent living in owner-occupied housing units. We also consider registration in the Democratic party. For demographic data, the base is citizen voting age population, and there are 314 high-Hispanic tracts. For registration data, the base is registered voters, and there are 271 high-Hispanic tracts.

Two artificial data sets were generated using King's model in order to assess the quality of the diagnostics when the model is correct. In Stockton, for instance, King's software was used to fit his model to the real exit poll data, and estimated parameters were used to generate an artificial data set. In these data, King's assumptions hold by construction. The artificial data were aggregated and run through the three estimation procedures. A similar procedure was followed for the registration data in Los Angeles (all 1409 tracts).

5.3 Empirical results

In Stockton, ecological regression gives impossible estimates: 109% of the Hispanics supported Jesse Jackson for president in 1988. King's method gives estimates that are far from the truth, but the standard error is large too (Table 5.1). In the Los Angeles data, King's method gives essentially the same estimates as ecological regression. These estimates are seriously wrong, and the standard errors are much too small. For example, 55.6% of Hispanics in Los Angeles are high school graduates. King's model estimates 30.1%, with a standard error of 1.1%: The model is off by 23.2 standard errors. The ecological regression estimate of 30.7 is virtually the same as King's, while the neighborhood model does noticeably better at 65.1%. As discussed below, the diagnostics are mildly suggestive of model failure, with indications that the high-Hispanic tracts are different from others. So we looked at tracts that are more than 25% Hispanic (compare King, pp. 241ff). The diagnostic plots for the restricted data were unremarkable, but King's estimates were off by 8.1 percentage points, or 6.8 standard errors. For these tracts, ecological regression does a little worse than King, while the neighborhood model is a bit better. Other lines in the table can be interpreted in the same way.

5.4 Diagnostics

We examined plots of $E\{t|x\}$ vs x as in King (p. 206) and "bias plots" of the estimated p or q vs x as in King (p. 183). We also examined "tomography plots" as in King (p. 176); these were generally unrevealing. The diagnostics will be defined more carefully below, and some examples will be given. In brief, x is the fraction of Hispanics in each area and t is the response: The $E\{t|x\}$ plot, for instance, shows the data and confidence bands derived from the model. In the Stockton exit poll data set, the $E\{t|x\}$ plot looks fine. The estimated p vs x plot has a significant slope of about 0.6. To calibrate the diagnostics, we used artificial data generated from King's model as fitted to the exit poll. Diagnostic plots indicated no problems, but the software generated numerous error messages. For instance,

Warning: Some bounds are very far from distribution mean.
Forcing 36 simulations to their closest bound.

(Similar warning messages were generated for the real data.)

We turn to Los Angeles. In the education data, there is a slight nonlinearity in the $E\{t|x\}$ figure—the data are too high at the right. Furthermore, there is a small but significant slope in the bias plot of estimated p vs x. In the high-Hispanic tracts, by contrast, the diagnostic plots are fine. For income and ownership, the diagnostics are unremarkable; there is a small but significant slope in the plot of estimated p vs x, for instance, $.05 \pm .02$ for ownership. For party affiliation, heterogeneity is visible in the scatter plot, with a cluster of tracts that have a low proportion of Hispanics but are highly democratic in registration. (These tracts are in South-Central Los Angeles, with a high concentration of black voters.) Heterogeneity is barely detectable in the tomography plot. The plot of $E\{t|x\}$ is problematic: Most of the tracts are above their expected responses. An artificial data set was constructed to satisfy King's assumptions, but the $E\{t|x\}$ plot looked as problematic as the one for the real data. In the high-Hispanic tracts, the diagnostic plots are unrevealing. Our overall judgments on the diagnostics for the various data sets are shown in Table 5.2.

5.5 Summary on diagnostics

The diagnostics are quite subjective, with no clear guidelines as to when King's model should *not* be used. Of course, some degree of subjectivity may be inescapable. In several data sets where estimates are far from truth, diagnostics are passed. On the other hand, the diagnostics indicate problems where none exist, in artificial data generated according to the assumptions of the model. Finally, when diagnostics are passed, standard errors produced by the model do not reliably indicate the magnitude of the actual errors (Tables 5.1 and 5.2).

5.6 Summary of empirical findings

Table 5.2 shows for each data set which method comes closest to truth. For the artificial registration data in Los Angeles, generated to satisfy the assumptions of King's model, his method ties with ecological regression and beats the neighborhood model. Likewise, his model wins on the artificial data set generated from the Stockton exit poll. Paradoxically, his diagnostics suggest trouble in these two data sets. In all the real data sets, even those selected to pass the diagnostics, the neighborhood model prevails. The neighborhood model was introduced to demonstrate the power of assumptions in determining statistical estimates from aggregate data, not as a substantive model for group behavior (Freedman et al., 1991, pp. 682, 806; compare King, pp. 43–44). Still, the neighborhood model handily outperforms the other methods, at least in our collection of data sets.

There is some possibility of error in EZIDOS. In the Los Angeles party affiliation data (1409 tracts), the mean non-Hispanic propensity to register democratic is estimated by King's software as 37%, while 56% is suggested by our calculations based on his model. Such an error might explain paradoxical results obtained from the diagnostics. There is a further numerical issue: Although the diagnostics that we consulted do not pick up the problem, the covariance matrix for the parameter estimates is nearly singular.

5.7 Counting success

King (p. xvii) claims that his method has been validated in a "myriad" comparisons between estimates and truth; on p. 19, the number of comparisons is said to be "over sixteen thousand." However, as far as we can see, King tests the model only on five data sets. Apparently, the figure of sixteen thousand is obtained by considering each geographical area in each data set. For instance, "the first application [to Louisiana data on turnout by race] provides 3262 evaluations of the ecological inference model presented in [the] book—sixty-seven times as many comparisons between estimates from an aggregate model and truth as exist in the entire history of ecological inference research" (p. 22). The Louisiana data may indeed cover 3262 precincts. However, if our arithmetic is correct, to arrive at sixteen thousand comparisons, King must count each area twice—once for each of the two groups about whom inferences are being made.

We do not believe that King's counting procedure is a good one, but let us see how it would apply to Table 5.1. In the education data, for instance, the neighborhood model is more accurate than King's model in 1133 out of 1409 tracts. That represents 1133 failures for King's model. Moreover, King provides 80% confidence intervals for tract-level truth. But these intervals cover the parameters only 20% of the time—another 844 failures, since $(0.80 - 0.20) \times 1409 = 844$. In the education data alone, King's approach fails two thousand times for the Hispanics, never mind the non-Hispanics. On this basis, Table 5.1 provides thousands of counterexamples to the theory. Evidently, King's way of summarizing comparisons is not a good one. What seems fair to say is only this: His model works on some data sets but not others, nor do the diagnostics indicate which are which.

5.8 A checklist

In chapter 16, King has "a concluding checklist." However, this checklist does not offer any very specific guidance in thinking about when or how to use the model. For instance, the first point advises the reader to "begin by deciding what you would do with the ecological inferences once they were made." The last point is that "it may also be desirable to use the methods described in . . . chapter 15," but that chapter only "generalize[s] the model to tables of any size and complexity." See pp. 263, 277, and 291.

5.9 Other literature

Robinson (1950) documented the bias in ecological correlations. Goodman (1953, 1959) showed that with the constancy assumption, ecological inference was possible: Otherwise, misleading results could easily be obtained. For current perspectives from the social sciences, see Achen and Shively (1995); Cho (1998) gives a number of empirical results like the ones described here. The validity of the constancy assumption for Hispanics is addressed, albeit indirectly, by Massey (1981), Massey and Denton (1985), and Lieberson and Waters (1988), among others.

Skerry (1995) discusses recent developments. For more background and pointers to the extensive literature, see Klein and Freedman (1993).

5.10 Some details

Let i index the units to be analyzed (precincts, tracts, and so forth). Let n_i^h be the number of Hispanics in area i, and n_i^a the number of non-Hispanics. These quantities are known. The total population in area i is then $n_i = n_i^h + n_i^a$. The population may be restricted to those interviewed in an exit poll, or to citizens of voting age as reported on census questionnaires, among other possibilities. Let v_i be the number of responses in area i, for instance, the number of persons who voted for a certain candidate, or the number who graduated from high school. Then $v_i = v_i^h + v_i^a$, where v_i^h is the number of Hispanics with the response in question, and v_i^a is the corresponding number of non-Hispanics. Although v_i is observable, its components v_i^h and v_i^a are generally unobservable. The main issue is to estimate

$$(1) \qquad P^h = \sum_i v_i^h / \sum_i n_i^h.$$

Generally, the denominator of P^h is known but the numerator is not. In the Stockton exit poll, P^h is the percentage of Hispanics who support Jackson; in the Los Angeles education data, P^h is the percentage of Hispanics with high school degrees, for two examples. Estimating P^h from $\{v_i, n_i^h, n_i^a\}$ is an "ecological inference." In Table 5.1, $\{v_i^h, v_i^a\}$ are known, so the quality of the ecological estimates can be checked; likewise for the test data used by King.

Let $x_i = n_i^h / n_i$, the fraction of the population in area i that is Hispanic; and let $t_i = v_i / n_i$, which is the ratio of response to population in area i. The three methods for ecological inference will be described in terms of (t_i, x_i, n_i), which are observable. The neighborhood model assumes that ethnicity has no impact within an area, so P^h can be estimated as $\sum t_i x_i n_i / \sum x_i n_i$. The ecological regression model, in its simplest form, assumes that Hispanics have a propensity p to respond, constant across areas; likewise, non-Hispanics have propensity q. This leads to a regression equation

$$(2) \qquad t_i = px_i + q(1 - x_i) + \epsilon_i,$$

so that p and q can be estimated by least squares. Call these estimates $\hat{p}$ and $\hat{q}$, respectively. Then P^h is estimated as $\hat{p}$. The error terms ϵ_i

in (2) are not convincingly explained by the model. It is usual to assume $E\{\epsilon_i\} = 0$ and the ϵ_i are independent as i varies. Some authors assume constant variance, others assume variance inversely proportional to n_i, and so forth.

King's model is more complex. In area i, the Hispanics have propensity p_i to respond and the non-Hispanics have propensity q_i, so that by definition

$$(3) \qquad\qquad t_i = p_i x_i + q_i (1 - x_i).$$

It is assumed that the pairs (p_i, q_i) are independent and identically distributed across i. The distribution is taken to be conditioned bivariate normal. More specifically, the model begins with a bivariate normal distribution covering the plane. This distribution is characterized by five parameters: two means, two standard deviations, and the correlation coefficient. The propensities (p_i, q_i) that govern behavior in area i are drawn from this distribution, but are conditioned to fall in the unit square. The five parameters are estimated by maximum likelihood. Then p_i can be estimated as $E\{p_i | t_i\}$, the expectation being computed using estimated values for the parameters. Finally, P^h in (1) can be estimated as $\sum_i \hat{p}_i x_i n_i / \sum_i x_i n_i$. King seems to use average values generated by Monte Carlo rather than conditional means. There also seems to be a fiducial twist to his procedure, which resamples parameter values as it goes along (chapter 8).

With King's method, $(\hat{p}_i, \hat{q}_i)$ falls on the line defined by (3), so that bounds are respected. Of course, the neighborhood model also makes estimates falling on these tomography lines. Ecological regression does not obey the constraints, and therefore gives impossible estimates on occasion.

As a minor technical point, there may be a slip in King's value of the normalizing constant for the density of the truncated normal. One factor in this constant is the probability that a normal variate falls in an interval, given that it falls along a line. The conditional mean is incorrectly reported on pp. 109, 135, 307. In these formulas, $\omega_i \epsilon_i / \sigma_i$ should probably be $\omega_i \epsilon_i / \sigma_i^2$, as on pp. 108 and 304.

We turn now to King's diagnostic plots, illustrated on the Los Angeles education data. Data for every fifth tract are shown; with more tracts, the figures would be unreadable. The tomography plot (Figure 5.1) has one line per tract, representing the possible combinations of the propensities (p_i, q_i) in the unit square that satisfy equation (3). The Hispanic propensity p_i is on the horizontal axis and q_i on the vertical. The plot seems uninformative.

The "bias plot" (Figure 5.2) graphs $(x_i, \hat{p}_i)$. There is one dot per tract, with the fraction x_i of Hispanics on the horizontal axis and the esti-

Figure 5.1 Tomography plot.   Figure 5.2 Bias plot.

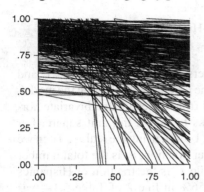
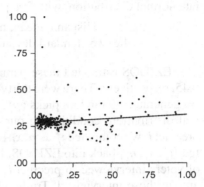

mated Hispanic propensity $\hat{p}_i$ on the vertical. The regression line should be flat. As it turns out, the slope is small but significant, indicating some breakdown in the constancy assumption.

Figure 5.3 plots (x_i, t_i). There is one dot per tract: x_i is on the horizontal axis and t_i, the fraction of persons in the tract with a high school education, is on the vertical. Also shown are 80% confidence bands derived from the model; the middle line is the estimated $E\{t|x\}$. The dots may be too high at the far right, hinting at nonlinearity. The $E\{t|x\}$ plot superimposes the data (x_i, t_i) on the graphs of three functions of x: (i) the lower 10%-point, (ii) the mean, and (iii) the upper 10% of the distribution of $px + q(1 - x)$, with (p, q) drawn from the conditioned normal with estimated values of the parameters.

We turn now to the artificial data for Stockton, mentioned above. To generate the data, we fitted King's model to the exit poll data using

Figure 5.3 The $E\{t|x\}$ plot.

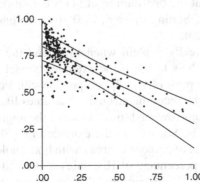

EZIDOS. As explained after equation (3), the key to the model is a bivariate normal distribution, with five parameters:

Hispanic mean, non-Hispanic mean,
the two standard deviations, and the correlation.

EZIDOS estimated these parameters as 0.68, 0.37, 0.43, 0.21, and 0.45, respectively. There were thirty-nine precincts. Following the model, we generated 39 random picks (p_i^*, q_i^*) from the estimated bivariate normal distribution, conditioning our picks to fall in the unit square. For precinct i, we computed t_i^* as $p_i^* x_i + q_i^* (1 - x_i)$, using the real x_i. Then we fed $\{t_i^*, x_i, n_i\}$ back into EZIDOS. In our notation, n_i is the total number of voters interviewed in precinct i, while x_i is the fraction of Hispanics among those interviewed. Truth—the 56% in line 2 of Table 5.1—was computed as $\sum p_i^* x_i n_i / \sum x_i n_i$. The procedure for the registration data in Los Angeles was similar.

5.11 The extended model

The discussion so far covers the "basic model." In principle, the model can be modified so the distribution of (p_i, q_i) depends on covariates (chapter 9), although we found no real examples in the book. The specification seems to be the following. Let u_i and w_i be covariates for area i. Then (p_i, q_i) is modeled as a random draw from the distribution of

$$(4) \qquad \alpha_0 + \alpha_1 u_i + \delta_i, \quad \beta_0 + \beta_1 w_i + \epsilon_i.$$

Here $\alpha_0, \alpha_1, \beta_0, \beta_1$ are parameters, constant across areas. The disturbances (δ_i, ϵ_i) are independent across areas, with a common bivariate normal distribution, having mean 0 and a covariance matrix Σ that is constant across areas; but the distribution of (4) is conditioned for each i to lie in the unit square. Setting $\alpha_1 = \beta_1 = 0$ gives the basic model—only the notation is different.

King does not really explain when to extend the model, when to stop extending it, or how to tell if the extended model fits the data. He does advise putting a prior on α_1, β_1: cf. pp. 288–89. For the Los Angeles registration data, he recommends using variables like "education, income, and rates of home ownership . . . to solve the aggregation problem in these data" (p. 171). So, we ran the extended model with u_i and w_i equal to the percentage of persons in area i with household incomes above $20,000 a year. The percentage of Hispanics registered as democrats is 73.5%; see Table 5.1. The basic model gives an estimate of 90.8% $\pm$

0.5%. The extended model gives 91.3% $\pm$ 0.5%. The change is tiny, and in the wrong direction. With education as the covariate, the extended model does very well: The estimate is 76.0% $\pm$ 1.5%. With housing as the covariate, the extended model goes back up to 91.0% $\pm$ 0.6%. In practice, of course, truth would be unknown and it would not be at all clear which model to use, if any. The diagnostics cannot help very much. In our example, all the models fail diagnostics: The scatter diagram is noticeably higher than the confidence bands in the $E\{t|x\}$ plots. There is also a "non-parametric" model (pp. 191–96); no real examples are given, and we made no computations of our own.

5.12 Identifiability and other *a priori* arguments

King's basic model constrains the observables:

(5)   the t_i are independent across areas.

Moreover, the expected value for t_i in area i is a linear function of x_i, namely,

(6) $E\{t_i|x_i\} = ax_i + b(1 - x_i),$

where a is the mean of p and b is the mean of q, with (p, q) being drawn at random from the conditioned normal distribution. Finally, the variance of t_i for area i is a quadratic function of x_i:

(7) $\text{var}(t_i|x_i) = c^2 x_i^2 + d^2 (1 - x_i)^2 + 2rcdx_i(1 - x_i),$

where c^2 is the variance of p, d^2 is the variance of q, and r is the correlation between p and q.

One difference between King's method and the ecological regression equation (2) is the heteroscedasticity expressed in (7). Another difference—perhaps more critical—is that King's estimate for area i falls on the tomography line (3). When ecological regression makes impossible estimates, as in Stockton, this second feature has some impact. When ecological regression makes sensible-looking (if highly erroneous) estimates, as in Los Angeles, there is little difference between estimates made by ecological regression and estimates made by King's method: The heteroscedasticity does not seem to matter very much. See Table 5.1.

In principle, the constraints (5), (6), and (7) are testable. On the other hand, assumptions about unobservable area-specific propensities are—obviously—not testable. Failure of such assumptions may have radical implications for the reliability of the estimates. For instance, suppose that Hispanics and non-Hispanics alike have propensity π_i to respond in area i: The π_i are assumed to be independent across areas, with a mean that depends linearly on x_i as in (6) and a variance that is a quadratic function of x_i as in (7). Indeed, we can choose (p_i, q_i) from King's distribution and set $\pi_i = p_i x_i + q_i (1 - x_i)$. This "equal-propensity" model cannot on the basis of aggregate data be distinguished from King's model but leads to very different imputations. Of course, the construction applies not only to the basic model but also to the extended model, a point King seems to overlook on pp. 175–83. No doubt, the specification of the equal-propensity model may seem a bit artificial. On the other hand, King's specifications cannot be viewed as entirely natural. Among other questions: Why are the propensities independent across areas? Why the bivariate normal?

According to King (p. 43), the neighborhood model "can be ruled out on theoretical grounds alone, even without data, since the assumptions are not invariant to the districting plan." This argument applies with equal force to his own model. If, for example, the model holds for a set of geographical areas, it will not hold when two adjacent areas are combined—even if the two areas have exactly the same size and demographic makeup. Equation (7) must be violated, because averaging reduces variance.

5.13 Summary and conclusions

King does not really verify conditions (5), (6), and (7) in any of his examples, although he compares estimated propensities to actual values. Nor does he say at all clearly how the diagnostics would be used to decide *against* using his methods. The critical behavioral assumption in his model cannot be validated on the basis of aggregate data. Empirically, his method does no better than ecological regression or the neighborhood model, and the standard errors are far too small. The diagnostics cannot distinguish between cases where estimates are accurate, and cases where estimates are far off the mark. In short, King's method is not a solution to the ecological inference problem.

6

Rejoinder to King

With Stephen P. Klein, Michael Ostland,
and Michael R. Roberts

ABSTRACT. *King's "solution" works with some data sets and fails
with others. As a theoretical matter, inferring the behavior of subgroups
from aggregate data is generally impossible: The relevant parameters are
not identifiable. Unfortunately, King's diagnostics do not discriminate
between probable successes and probable failures. Caution would seem
to be in order.*

6.1 Introduction

King (1997) proposed a method for ecological inference and made
sweeping claims about its validity. According to King, his method provided realistic estimates of uncertainty, with diagnostics capable of detecting failures in assumptions. He also claimed that his method was robust,
giving correct inferences even when the model is wrong.

Journal of the American Statistical Association (1999) 94: 355–57.

Our review (Freedman, Klein, Ostland, and Roberts 1998 [Chapter 5]) showed that the claims were exaggerated. King's method works if its assumptions hold. If assumptions fail, estimates are unreliable: so are internally-generated estimates of uncertainty. His diagnostics do not distinguish between cases where his method works and where it fails. King (1999) raised various objections to our review. After summarizing the issues, we will respond to his main points and a few of the minor ones. The objections have little substance.

6.2 Model comparisons

Our review compared King's method to ecological regression and the neighborhood model. In our test data, the neighborhood model was the most accurate, while King's method was no better than ecological regression. To implement King's method, we used his software package EZIDOS, which we downloaded from his web site. For a brief description of the EI and EZIDOS software packages, see (King 1997, p. xix).

King (1999) contends that we (i) used a biased sample of data sets and (ii) suppressed "estimates for non-Hispanic behavior, about which there is typically more information of the type EI [King's method] would have extracted." Grofman (1991) and Lichtman (1991) are cited for support. Our answer to claim (i) is simple: We used the data that we had. Of course, Grofman and Lichtman made other arguments too; our response is in Freedman et al. (1991).

We turn to claim (ii). It is by no means clear what sort of additional information would be available to King for non-Hispanics. Moreover, the neighborhood model and King's method get totals right for each geographical unit: Thus, any error on the Hispanic side must be balanced by an error of the same size but the opposite sign on the non-Hispanic side. In short, despite King's theorizing, his method is unlikely to beat the neighborhood model on the non-Hispanics.

Empirical proof will be found in Tables 6.1 and 6.2, which show results for the non-Hispanics in the real data sets we considered. (The artificial data will be discussed later.) These tables, and similar ones in our review, show King's method to be inferior to the neighborhood model, for non-Hispanics as well as Hispanics. In the Los Angeles data, his method is also inferior to ecological regression.

King (1997) tried his model on five data sets. These are not readily available, but we were able to get one of them—poverty status by sex in South Carolina block groups—directly from the Census Bureau. We ran the three ecological-inference procedures on this data set (Tables 6.1 and 6.2). King's method succeeds only in the sense that the estimate is

Table 6.1 The non-Hispanics. Comparison of three methods for making ecological inferences, in situations where the truth is known. Results for non-Hispanics in Stockton and Los Angeles, and for men and women in South Carolina.

	Nbd	ER	King	Truth	Z
Stockton					
Exit poll	39.8	25.8	36.5 ± 3.6	42.0	−1.5
Los Angeles					
Education	76.4	81.6	82.9 ± 0.2	78.1	24.0
High Hispanic	60.1	71.9	73.1 ± 1.0	66.3	6.7
Income	53.5	55.4	56.4 ± 0.2	53.2	14.2
Ownership	56.1	57.4	57.5 ± 0.3	56.4	3.9
Party affiliation	58.6	57.2	54.6 ± 0.1	57.3	−33.0
High Hispanic	68.1	54.5	53.5 ± 0.4	61.5	−18.2
South Carolina					
Men in poverty	15.0	−13.3	5.8 ± 6.6	12.9	−1.1
Women in poverty	15.7	43.7	24.2 ± 6.1	17.7	1.1

Note: "Nbd" is the neighborhood model; "ER" is ecological regression. Values in percentages. King's method gives an estimate and a standard error, reported in the format "estimate ± SE"; $Z =$ (estimate − truth)/SE, computed before rounding. In South Carolina, block groups with fewer than twenty-five inhabitants are excluded from the data.

within 1.1 standard errors of truth; the neighborhood model comes much closer to the mark, both for men and women. Where comparisons are feasible, the neighborhood model has been more accurate than King's method on the real data sets, even in his own South Carolina example.

King says that the neighborhood model is not a reliable method of inferring the behavior of subgroups from aggregate data; it is unreasonable, politically naive, and paints "a picture of America that no one would recognize." Perhaps so. However, the neighborhood model demonstrates that ecological inferences are driven largely by assumptions, not by data—a point that King almost concedes. Moreover, when confronted with data, the neighborhood model outperforms the competition, including King's method (Tables 6.1 and 6.2). What are the implications of his remarks for his own model?

Table 6.2 Which estimation procedure comes closer to truth?

	Group	
	Hispanics	Non-Hispanics
Stockton		
Exit poll	Nbd	Nbd
Los Angeles		
Education	Nbd	Nbd
High Hispanic	Nbd	ER
Income	Nbd	Nbd
Ownership	Nbd	Nbd
Party affiliation	Nbd	ER
High Hispanic	Nbd	Nbd
	Males	Females
South Carolina		
Poverty	Nbd	Nbd

Note: "Nbd" is the neighborhood model and "ER" is ecological regression. King's method does not appear in the table because in each case it does less well than the neighborhood model; furthermore, in each of the Los Angeles data sets, it does less well than ecological regression.

6.3 Diagnostics

King contends that we (i) "misinterpret warning messages ... generated by choosing incorrect specifications," and (ii) "use irrelevant tests like whether the regression of T_i on X_i is significant ...." (In the South Carolina example, T_i would be the fraction of persons in block group i who are below the poverty line, and X_i would be the fraction of persons in that block group who are male.)

With respect to (i), we interpreted the warning messages as evidence of error in specifications that analysts, including King himself, often use: see below. With respect to (ii), consider for instance figure 2 in our review [Figure 5.2]. The vertical axis shows $\hat{p}_i$ not T_i—an estimated propensity for a group rather than an observed fraction. This figure is one of King's "bias plots" (King 1997, p. 183). It is one of his standard diagnostics.

The issue that concerned us was the regression of $\hat{p}_i$ on X_i, not the regression of T_i on X_i. On both points, King simply misread what we wrote.

The bottom line: King's diagnostics, like the warning messages printed out by his software, raise warning flags even when the standard errors are reasonable, as in Stockton. Conversely, there are many examples in the Los Angeles data where the method fails—but diagnostics are passed and warning messages disappear.

King's South Carolina data illustrates other possibilities. Figure 6.1 plots for each block group the estimated fraction of men in poverty against the fraction of men in the population. Figure 6.2 repeats the analysis for women. (Every tenth block group is shown; estimates are computed using King's software package EZIDOS.) The regression line for men has a shallow but statistically significant slope; the line for women falls quite steeply. Thus, King's assumption of IID propensities is strongly rejected by the data. Likewise, the warning messages point to specification error.

Warning: Some bounds are very far from distribution mean.
Forcing 2163 simulations to their closest bound.

King (1997, p. 225) insists that "even in [the South Carolina] data set, chosen for its difficulty in making ecological inferences, the inferences are accurate." But warning messages and signals from the diagnostics have been ignored. Perhaps his idea is that when the method succeeds, it succeeds *despite* the difficulties; when it fails, it fails *because of* the difficulties.

King imputes to us the "claim that EI cannot recover the right parameter values from data simulated from EI's model." That is also a misreading. Of course King's method should work if its assumptions are satisfied—as we said on p. 1518 of our review [Chapter 5, p. 85], and demonstrated with two artificial data sets (pp. 1519–20) [Chapter 5,

Figure 6.1 Bias plot for men.   Figure 6.2 Bias plot for women.

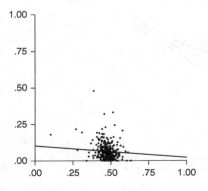

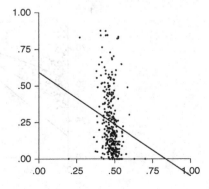

pp. 86–87]. We still think there is a bug in King's software, because the diagnostics sometimes indicate problems where none can exist (p. 1520) [Chapter 5, pp. 87–88].

Here is another example. Applied to the Los Angeles data on party affiliation, King's method estimates the five parameters of the untruncated normal distribution (two means, two standard deviations, and r) as 1.0456, 0.2853, 0.1606, 0.3028, −0.9640. We generated pairs of propensities from this bivariate distribution, kept only pairs that fell into the unit square, computed corresponding tract-level observations, and fed the resulting data back into EZIDOS. The parameter estimates were fine— 1.0672, 0.2559, 0.1607, 0.3024, −0.9640.

The trouble comes in the diagnostics. Figure 6.3 shows our simulated data for every fifth tract. The figure also shows the 80% confidence bands for the tract-level "observations" (the simulated fraction who register democratic); the middle line is the conditional mean. We used EZIDOS to estimate the conditional mean and the confidence bounds from the artificial data generated by the model.

Clearly, something is wrong. The midline should more or less cut through the middle of the scatter diagram, and the band should cover about 80% of the dots. However, most of the dots are above the midline: Indeed, about half of them spill over the top of the band. Similar errors are discussed by McCue (1998).

King presents artificial data for which his model does not hold and the diagnostics pick up the failure in assumptions. This is an existence proof: There are some data sets for which the diagnostics work. In the examples we considered, both real and artificial, the diagnostics were not reliable guides to the performance of King's method. Figures 6.1–6.3

Figure 6.3 $E\{t|x\}$ plot: Artificial Los Angeles data.

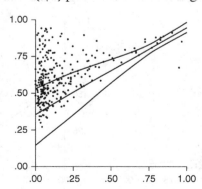

above reinforce this point, for one of his own data sets (South Carolina), and for artificial data generated from his model.

6.4 Other issues

King emphasizes throughout his reply that qualitative information needs to be used, the "50+ options" in his code being tuned accordingly. (Some options in EZIDOS allow for Bayesian inference rather than likelihood methods; others change the numerical algorithms that will be used; still others control print formats.) However, it is hard to see how qualitative information plays any role in the real examples presented by King (1997); and we saw nothing there about the 50+ options. On the contrary, the discussion of the real examples suggests straight-ahead use of maximum-likelihood estimation.

King contends that our description of the constancy assumption is a "caricature." However, equation (2) in our review is exactly the one that is estimated by proponents of ecological regression, like Grofman and Lichtman. Moreover, King appears to misread Goodman (1953), who delineates the narrow circumstances under which ecological inference may be expected to succeed. We can agree that, coldly stated, the assumptions underlying ecological regression are unbelievable.

King denies any "fiducial twist" to his argument. However, there he is, computing a posterior without putting a prior on the parameters of the normal distribution. Apparently, he converts sampling distributions for estimators into posterior distributions for parameters. Isn't that fiducial inference?

According to King, our review of the "extended model" demonstrates error in Freedman et al. (1991). He does not explain the logic. Obviously, different neighborhoods in Los Angeles show different social characteristics—for both Hispanic and non-Hispanic inhabitants. That was true in 1991, and it is true today. What our review adds is this: If you know the answer, one of King's extended models may find it. But if you don't know the answer, the models are just shots in the dark.

6.5 Making the data available

King takes us to task for not providing data underlying our review. Although his other claims are all mistaken, we did decline his request for data. His reaction seems disingenuous. After all, we had previously asked him for his data: He refused, sending us to the web. To read the files he pointed to, you need an HP workstation running UNIX and GAUSS. Even then, all you get is a long string of unidentified numbers. Apparently, what the claim for replication on p. xix of King (1997) means is that if you run his software on his files, on a platform of his choice, you will get his output.

It would be useful to have all the underlying data available in standard format (flat ASCII files, intelligibly documented). If King agrees to our plan and posts his data that way, we will post ours, along with the little simulation program used in Figure 6.3, and the version of EZIDOS that we used. That way, replication and independent analysis will be possible.

6.6 Summary and conclusions

King (1997) has a handful of data sets where his method succeeds. We have another handful where the method fails. Still other examples are contributed by Cho (1998) and Stoto (1998), with mixed results. Thus, King's method works in some data sets but not others. His diagnostics do not discriminate between probable successes and probable failures. That is the extent of the published empirical information regarding the validity of King's method. As a theoretical matter, inferring the behavior of subgroups from aggregate data is generally impossible: The relevant parameters are not identifiable. On this there seems to be some agreement (Freedman, Klein, Ostland, and Roberts, 1998 [Chapter 5]; King, 1999). Thus, caution would seem to be in order—a characteristic not prominent in King (1997) or King (1999).

7

Black Ravens, White Shoes, and Case Selection: Inference with Categorical Variables

ABSTRACT. *Ideas from statistical theory can illuminate issues in qualitative analysis, such as case selection. Mahoney and Goertz (2004) offer some principles for selecting cases, illustrated by Hempel's Paradox of the Ravens. The paradox disappears if one distinguishes between inference about samples and inference about populations from samples. The Mahoney-Goertz rules have limited utility; it is inadvisable to disregard any cell in a 2 × 2 table.*

7.1 Introduction

How should qualitative researchers select cases? This is an important question, which has been widely canvassed. Mahoney and Goertz (2004) offer some principles to govern case selection, illustrating the argument by Hempel's raven paradox. In this chapter, I suggest the paradox can be resolved by distinguishing between samples and populations. I also sug-suggest that the Mahoney-Goertz rules have limited scope.

Previously unpublished.

7.2 The paradox

The raven paradox is due to Carl Hempel (1945). To explain it, suppose that objects can be classified unambiguously as

(i) raven or not, and
(ii) black or not.

The data can then be presented in a 2 × 2 table, with columns corresponding to the first classification and rows to the second. For reference, the cells are labeled A, B, C, D. All four cells are observed.

	Raven	
Black	Yes	No
Yes	A	B
No	C	D

Now consider the time-honored proposition that all ravens are black. According to Jean Nicod (1930) and many scholars who followed him, data in cell A support the proposition. In other words, a black raven is evidence that all ravens are black. As Hempel notes, however, "all ravens are black" is logically equivalent to "all nonblack objects are nonravens."[1] Thus, by Nicod's rule, data in cell D—nonblack objects that are nonravens—also support the blackness of ravens.

In particular, white shoes provide evidence that ravens are black. Many of us find this paradoxical, although Hempel seems eventually to have accepted the idea. There is an extended philosophical literature on white shoes and ravens, including an exchange between I. J. Good (1967, 1968) and Hempel (1967):

"The White Shoe Is a Red Herring,"
"The White Shoe: No Red Herring,"
"The White Shoe *Qua* Herring Is Pink."

The debate has spilled over into the political science journals (see, for instance, *Political Analysis* (2002) 10: 178–207). The paradox is also discussed by Taleb (2007) in a searching critique of current statistical methodology.[2]

I believe the paradox should be resolved by making the following distinction. The proposition "all ravens are black" can be advanced with respect to

(i) the data at hand; or
(ii) some larger population of objects, the data at hand being viewed as a sample from the larger population.

In the first case, what matters is the raven-nonblack cell—C in the table. If this cell is empty, the proposition is correct; if this cell is nonempty, the proposition is incorrect. Other cells in the table are simply irrelevant.[3] Nicod's rule does not apply, and white shoes are beside the point.

On the other hand, if the assertion is about some larger population, and statistical inferences are to be made from the data to the population, then the nature of the sample and the population must be specified (the "sampling model"). In this scenario, "all" is defined relative to the larger population; so is the set of objects that are not ravens as well as the set of objects that are not black.

Nicod's rule applies in some sampling models but not others. White shoes may be powerful evidence for the blackness of ravens, or against— or shoes may be entirely irrelevant. Good (1967) has a cunning example where seeing a black raven increases the likelihood that white ravens will turn up later: see the Appendix below. Hempel (1967) and the rejoinder by Good (1968) gum up the works with herrings of various colors.

To summarize, the illusion of paradox is created by blurring the distinction between the sample and the population. The illusion is dispelled by deciding whether we are discussing the data at hand or extrapolating from the data to a larger population—although, in the second case, a sampling model is needed.

7.3 Case selection

Enough about ravens, shoes, and herrings; what about principles for case selection? Mahoney and Goertz (2004, p. 653) claim their

> Possibility Principle ... provides explicit, rigorous, and the-
> oretically informed guidelines for choosing a set of negative
> cases.... The Possibility Principle holds that only cases where
> the outcome of interest is *possible* should be included in the set
> of negative cases; cases where the outcome is *impossible* should
> be relegated to a set of uninformative and hence irrelevant ob-
> servations.

The possibility principle is elaborated into a rule of exclusion and a rule of inclusion, the former being primary (Mahoney and Goertz 2004, pp. 657–58). These rules will be explained below. They sometimes provide useful heuristics for case selection. However, if the principles are supposed to have general application, they leave something to be desired. In particular, claims of explicitness and rigor are not justified.

The setting has a binary response variable Y, where $Y = 1$ indicates the presence of an outcome of interest; $Y = 0$ indicates its absence. There are binary independent variables, which may be causes of Y. Thus, $X = 1$ indicates the presence of a causal factor, whereas $X = 0$ indicates absence. Mahoney and Goertz are using language in a specialized way, because "impossible" things occur with some frequency. Impossibility, in their terminology, only means that the likelihood is below a selected cutpoint. Consequently, scholars who want to use the Mahoney-Goertz rules must assign likelihoods, choose cutpoints and then dichotomize. For example, "impossibility" might just mean that the likelihood is below the cutpoint of 0.5 (Mahoney and Goertz, pp. 659, 663).[4]

Claims for explicitness and rigor are therefore questionable. Quantifying likelihoods, even in large-N research, is fraught with difficulty. Logit models can of course be fitted to data, but rigorous justification for such models is rarely to be found.[5] Selecting cutpoints is another famous problem.[6] Smaller N does not make life easier.

With respect to defining likelihoods and cutpoints, Mahoney and Goertz (2004, p. 665) say only, "These tradeoffs underscore the importance of making substantively and theoretically informed choices about where to draw the line . . . ." This sound advice will not help when making hard choices. In short, quantifying likelihoods and choosing cutpoints is not an objective process; the claim to have formulated explicit and rigorous guidelines is not justified. Moreover, contrary to suggestions by Mahoney and Goertz, it would appear that the theory informing their guidelines must be supplied by the scholars who use those guidelines.

Another problem should be mentioned. Presence or absence of an outcome of interest seems clear enough in many circumstances. In other circumstances, however, difficulties abound. For example, consider a study showing that left-wing political power promotes economic growth. Scholars with another orientation will use the same data to prove that left-wing power promotes stagnation. Is the outcome of interest growth—or stagnation?

The answer determines which cases are positive and which are negative. The empirical relationship being tested is substantively the same, but different cases will be deemed relevant and irrelevant by the Mahoney-Goertz rules, according to the way the research hypothesis is framed (see the Appendix below for details). In short, if we follow the rules, the relevance of a case is likely to depend on arbitrary choices.

Suppose, however, that such ambiguities have been resolved. There is a binary response variable Y. The outcome of interest is coded as $Y = 1$; negative cases have $Y = 0$. There is one causal variable X, with $X = 0$ or 1. The data can be presented in the following 2×2 table.

	X	
Y	1	0
1	A	B
0	C	D

Labels for the cells are shown in the body of the table. Our working hypothesis is that X and Y are positively related: Setting X to 1 increases the likelihood that $Y = 1$.

Cases in cell D are irrelevant by the Mahoney-Goertz rule of exclusion (p. 658):

> Cases are irrelevant if their value on any eliminatory independent variable predicts the nonoccurrence of the outcome of interest.

Indeed, cases in cell D (with $X = 0$ and $Y = 0$) are negative. Furthermore, an eliminatory independent variable predicts the nonoccurrence of the outcome of interest ($X = 0$ predicts $Y = 0$). Cell D is therefore irrelevant.

Moreover, cell D is also irrelevant by the rule of inclusion (p. 657):

> Cases are relevant if their value on at least one independent variable is positively related to the outcome of interest.

Indeed, $X = 0$ in cell D. Next, the value 0 for the independent variable X is not positively related to the outcome of interest ($Y = 1$). Finally, in our setup, there are no other variables to consider. Therefore, the Mahoney-Goertz rule of inclusion, like their rule of exclusion, says that cell D is irrelevant.[7]

Cell D may indeed be irrelevant under some circumstances. But a blanket assertion of irrelevance seems hasty. For example, most statisticians and epidemiologists would want to know about all four cells—if only to confirm that the association is positive and to determine its magnitude.

We can make this more interesting (and more complicated). Suppose an observer claims there are two types of cases in cell D. For the first type of case, $X = 0$ causes $Y = 0$. For the second type, $Y = 0$ by necessity: In other words, Y would still have been 0 even if we had set X to 1. This is causal heterogeneity. The best way to test such a claim, absent other information, would seem to be scrutiny of cases with $X = 0$ and $Y = 0$. In this kind of scenario, far from being irrelevant, cell D can be critical.

An example with only one important causal variable may seem unusual, but the reasoning about the rule of exclusion continues to apply if there are several variables. For the rule of inclusion, condition on all the covariates but one; then use the argument given above to conclude that some of the cells in the multi-dimensional cross-tab are irrelevant. This is not a sensible conclusion. (The reasoning stays the same, no matter how many variables are in play.) Therefore, the rules of exclusion and inclusion are not good general rules.

Mahoney and Goertz may be thinking of necessary and sufficient causation, although this is not made clear. Let us assume, which would be highly favorable to the enterprise, that there is only one causal variable and no cases in cell B or cell C. If cell D is empty, there is no variance on X or on Y, which will affect the interpretation of the data for some observers. If cells A and D are both nonempty, qualitative researchers will want to examine some cases in each cell in order to check that the association is causal, and to discern the mechanisms by which $X = 1$ causes $Y = 1$, whereas $X = 0$ causes $Y = 0$. So, the cell with $X = 0$ and $Y = 0$ is worth considering even for necessary and sufficient causation.

A real example might be useful. In their multi-methods research on the probabilistic causes of civil war, Fearon and Laitin (2008) found it illuminating to examine cases in the analog of cell D (low probability of civil war according to the model, and no civil war in historical fact). Fearon and Laitin contradict the Mahoney-Goertz rules. In summary, general advice to disregard any particular cell in the 2×2 table is bad advice.

7.A Appendix

7.A.1 Good's example

We begin by sketching Good's construction. With probability 1/2, the population comprises 100 black ravens and 1,000,000 birds that are not ravens; with probability 1/2, the population comprises 1000 black ravens, 1 white raven, and 1,000,000 birds that are not ravens. The population is chosen at random, then a bird is selected at random from the chosen population. If the bird is a black raven, it is likely to have come from the second population. In short, a black raven is evidence that there is a white raven to be seen (eventually).

7.A.2 Simple random samples

We turn to more familiar sampling models. Suppose that a sample is chosen at random without replacement from a much larger population, each object in the population being classified as U or not-U. For instance, the U's might be the sought-after white ravens, so the not-U's comprise red ravens, green ravens, blue ravens, . . . , and black ravens, together with nonravens.

From a Bayesian perspective, it is easy to test the hypothesis that there are no U's in the population. However, much depends on the prior that is used, and justifying the choice can be difficult (Freedman 1995 [Chapter 1]; Freedman and Stark 2003 [Chapter 8]).

Now take the frequentist perspective. If the fraction of U's in the sample is small, that proves U is rare in the population (modulo the usual qualifications). However, unless we make further assumptions, it is impossible to demonstrate by sampling theory that there are no U's in the population. For instance, if the sample size is 1000 and the fraction of U's in the population is 1/1000, there is a substantial chance that no U's will turn up in the sample: The chance is $\left(1 - \frac{1}{1000}\right)^{1000} \doteq 0.37$. So, if there are no U's in the sample, we are entitled to conclude that U is rare—but we cannot conclude that there no U's in the population.

7.A.3 Other possibilities

The two examples below indicate other logical possibilities. For the sake of variety, white shoes are replaced by red herrings. In the first example, *pace* Hempel, a red herring is decisive evidence that not all ravens are black. In the second, by contrast, a red herring is decisive evidence that all ravens are indeed black.

A "population" consists of objects classified as white ravens, black ravens, red herrings, and other things (neither raven nor herring). Different populations have different compositions; however, there are black ravens and things that are neither raven nor herring in every population.

Each example consists of two populations, labeled Population I and Population II. A sample is drawn at random from one of the two populations. It is unknown which population is being sampled. It is required to decide whether, in the population being sampled, all ravens are black.

Example 1. In Population I, there are both white ravens and red herrings. In Population II, there are neither white ravens nor red herrings. If a red herring turns up in the sample, you must be sampling from Population I containing white ravens. This is a useful clue if there are a lot of red herrings and few white ravens.

Example 2. In Population I, there are white ravens but no red herrings. In Population II, there are no white ravens but there are red herrings. If a red herring turns up in the sample, you must be sampling from Population II, where all ravens are black.

So far, we have considered simple random samples. Different kinds of samples are often used, including convenience samples. Procedures that favor some cells at the expense of others can easily skew the data. Sample design is a crucial piece of the puzzle. If you do not look, you will not find evidence against your hypothesis.

7.A.4 Samples and inductive inference

I have focused on inductive inference by sampling, without meaning to imply that statistical theory is the only basis for induction. On the contrary, I believe that in most cases, statistical theory—whether frequentist or Bayesian—permits inductive inference only by imposing artificial assumptions. The frequentist incantation is "independent and identically distributed." The Bayesian denounces frequentists for incoherence, requiring instead that observations are exchangeable—a distinction of Talmudic subtlety (Freedman 1995 [Chapter 1]; Freedman and Stark 2003 [Chapter 8]). How then are scientists to make inductive inferences? That is a topic for another lifetime, but maybe we could start by thinking about what they actually do.

7.A.5 The ravens and causal inference

As I see it, the paradox of the ravens has to do with description and inductive reasoning. Others may see the paradox as being about logic and semantics. What should be blatantly obvious is that the paradox has nothing to do with causal inference per se—which is not to deny that causal reasoning depends on description, classification, induction, logic, and ordinary language.

7.A.6 Ambiguity in the rules

Finally, let us consider the example of left-wing political power and economic growth. Cases can be arrayed in the familiar 2×2 table:

	Growth	Stagnation
Left-wing power	A	B
Right-wing power	C	D

One perspective is that left-wing power causes growth. Then growth is the outcome of interest. As argued above, the Mahoney-Goertz rules imply that cell D is irrelevant. Another perspective is that left-wing power causes stagnation. Now stagnation is the outcome of interest, and it is cell C (negative on outcome, negative on left-wing power) that is irrelevant. This is untidy at best.

Mahoney and Goertz might agree that positive cases are generally relevant. Now there is something of a contradiction. If the research hypothesis is formulated to please the left wing, cell C is relevant, because it is positive. If the hypothesis is formulated to humor the right, cell C is irrelevant, as shown in the previous paragraph.

7.A.7 The odds ratio

Epidemiologists would use the "odds ratio" to summarize the data in a 2×2 table of the kind we have been considering. Let a denote the number of elements in cell A, and so forth. If there are cases in all four cells, the odds ratio is $(a/c)/(b/d) = (a/b)/(c/d) = (ad)/(bc)$. You need all four numbers to compute the odds ratio. The association is positive when the odds ratio is above 1.0; the association is negative when the odds ratio is below 1.0. For additional information, see Gordis (2008).

If ρ denotes the odds ratio, the causal interpretation is this: Setting X to 1 rather than 0 multiplies the odds that $Y = 1$ by the factor ρ. Equivalently, if $Y = 1$ rather than 0, the odds that $X = 1$ are multiplied by the factor ρ. In the present context, given a, b, and c, it is cell D that determines whether X causes Y or X prevents Y—a substantial difference. Cell D is not to be ignored.

Notes

1. Suppose A and B are sets. Write A^c for the complement of A, i.e., the set of things that are not in A. The logical principle is this:

A is a subset of B

if and only if

B^c is a subset of A^c.

2. Taleb argues that rare events ("Black Swans") have major consequences, and conventional statistical models are ill-suited for analyzing such matters. Efforts by statisticians to refute him have so far been unconvincing (*The American Statistician* (2007) 61: 189–200).

3. We can either assume there is at least one black raven or rely on an irritating logical technicality—an empty set is a subset of all sets. In particular, if there are no ravens, they must all be black (as well as any other color of interest).

4. As Mahoney and Goertz (2004, p. 662) explain, "the impossible . . . is very likely to happen in large-N research," that is, with enough cases. To rephrase the rules in terms of the possible rather than the impossible, you have to quantify the probability that $Y = 1$, then choose a cutpoint, and then declare that $Y = 1$ is "possible" if the probability falls above that cutpoint. Compare Mahoney and Goertz (2004, pp. 659–60, 663–65). "[T]he analyst must decide and justify the exact threshold at which the outcome is considered possible" (p. 659). There are similar considerations for the explanatory variables.

5. See Berk (2004), Brady and Collier (2004), Duncan (1984), Freedman (2009), Lieberson and Lynn (2002), Mahoney and Rueschemeyer (2003), Sobel (1998).

6. Cournot (1843) discusses the impact of choosing categories. See Stigler (1986, p. 199) for a summary, or Shaffer (1995).

7. Mahoney and Goertz (2004, p. 658) might suggest that X is not an eliminatory variable in their sense. This is far from clear, especially in view of the claim that "observations with a zero for all the independent variables will always satisfy causal sufficiency and thus artificially inflate the number of cases where the theory works ..." (p. 664). In any event, this suggestion would not explain the paradoxical implications of the rule of inclusion. Goertz (2008, p. 10) confirms my reading of the Mahoney-Goertz thesis: "Typically, we will focus our attention on the [cell A] cases," whereas cases in cell D "are problematic for qualitative researchers."

Acknowledgments

I would like to thank David Collier, Thad Dunning, Paul Humphreys, Janet Macher, Jay Seawright, Jas Sekhon, and Philip B. Stark for useful comments.

8

What is the Chance
of an Earthquake?

With Philip B. Stark

ABSTRACT. *Making sense of earthquake forecasts is surprisingly difficult. In part, this is because the forecasts are based on a complicated mixture of geological maps, rules of thumb, expert opinion, physical models, stochastic models, and numerical simulations, as well as geodetic, seismic, and paleoseismic data. Even the concept of probability is hard to define in this context. For instance, the U.S. Geological Survey developed a probability model according to which the chance of an earthquake of magnitude 6.7 or greater before the year 2030 in the San Francisco Bay Area is 0.7 ± 0.1. How is that to be understood? Standard interpretations of probability cannot be applied. Despite their careful work, the USGS probability estimate is shaky, as is the uncertainty estimate.*

8.1 Introduction

What is the chance that an earthquake of magnitude 6.7 or greater will occur before the year 2030 in the San Francisco Bay Area? The U.S.

Earthquake Science and Seismic Risk Reduction (2003) NATO Science Series IV. Earth and Environmental Sciences. 21: 201–16. With kind permission of Springer Science and Business Media.

Geological Survey estimated the chance to be 0.7 ± 0.1 (USGS, 1999). In this chapter, we try to interpret such probabilities.

Making sense of earthquake forecasts is surprisingly difficult. In part, this is because the forecasts are based on a complicated mixture of geological maps, rules of thumb, expert opinion, physical models, stochastic models, numerical simulations, as well as geodetic, seismic, and paleoseismic data. Even the concept of probability is hard to define in this context. We examine the problems in applying standard definitions of probability to earthquakes, taking the USGS forecast—the product of a particularly careful and ambitious study—as our lead example. The issues are general and concern the interpretation more than the numerical values. Despite the work involved in the USGS forecast, their probability estimate is shaky, as is the uncertainty estimate.

This chapter is organized as follows. Section 8.2 discusses various interpretations of probability, including relative frequency and degree of belief. Section 8.3 discusses the USGS forecast. Section 8.4 quotes a well-known critique of the relative frequency interpretation. Section 8.5 gives conclusions.

8.2 Interpreting probability

Probability has two aspects. There is a formal mathematical theory, axiomatized by Kolmogorov (1956). And there is an informal theory that connects the mathematics to the world, i.e., defines what "probability" means when applied to real events. It helps to start by thinking about simple cases. For example, consider tossing a coin. What does it mean to say that the chance of heads is $1/2$? In this section, we sketch some of the interpretations—symmetry, relative frequency, and strength of belief.[1] We examine whether the interpretation of weather forecasts can be adapted for earthquakes. Finally, we present Kolmogorov's axioms and discuss a model-based interpretation of probability, which seems the most promising.

8.2.1 Symmetry and equally likely outcomes

Perhaps the earliest interpretation of probability is in terms of "equally likely outcomes," an approach that comes from the study of gambling. If the n possible outcomes of a chance experiment are judged equally likely—for instance, on the basis of symmetry—each must have probability $1/n$. For example, if a coin is tossed, $n = 2$; the chance of heads is $1/2$, as is the chance of tails. Similarly, when a fair die is thrown, the six possible outcomes are equally likely. However, if the die is loaded, this argument does not apply. There are also more subtle difficulties. For example, if two dice are thrown, the total number of spots can be anything from two through twelve—but these eleven outcomes are far from equally likely. In earthquake forecasting, there is no obvious symmetry to exploit. We therefore need a different theory of probability to make sense of earthquake forecasts.

8.2.2 The frequentist approach

The probability of an event is often defined as the limit of the relative frequency with which the event occurs in repeated trials under the same conditions. According to frequentists, if we toss a coin repeatedly under the same conditions,[2] the fraction of tosses that result in heads will converge to 1/2: That is why the chance of heads is 1/2. The frequentist approach is inadequate for interpreting earthquake forecasts. Indeed, to interpret the USGS forecast for the Bay Area using the frequency theory, we would need to imagine repeating the years 2000–2030 over and over again—a tall order, even for the most gifted imagination.

8.2.3 The Bayesian approach

According to Bayesians, probability means degree of belief. This is measured on a scale running from 0 to 1. An impossible event has probability 0; the probability of an event that is sure to happen equals 1. Different observers need not have the same beliefs, and differences among observers do not imply that anyone is wrong.

The Bayesian approach, despite its virtues, changes the topic. For Bayesians, probability is a summary of an opinion, not something inherent in the system being studied.[3] If the USGS says "there is chance 0.7 of at least one earthquake with magnitude 6.7 or greater in the Bay Area between 2000 and 2030," the USGS is merely reporting its corporate state of mind, and may not be saying anything about tectonics and seismicity. More generally, it is not clear why one observer should care about the opinion of another. The Bayesian approach therefore seems to be inadequate for interpreting earthquake forecasts. For a more general discussion of the Bayesian and frequentist approaches, see Freedman (1995) [Chapter 1].

8.2.4 The principle of insufficient reason

Bayesians—and frequentists who should know better—often make probability assignments using Laplace's principle of insufficient reason (Hartigan, 1983, p. 2): If there is no reason to believe that outcomes are not equally likely, take them to be equally likely. However, not believed to be unequal is one thing; known to be equal is another. Moreover, all outcomes cannot be equally likely, so Laplace's prescription is ambiguous.

An example from thermodynamics illustrates the problem (Feller, 1968; Reif, 1965). Consider a gas that consists of n particles, each of which can be in any of r quantum states.[4] The state of the gas is defined by a "state vector." We describe three conventional models for such a gas, which differ only in the way the state vector is defined. Each model takes all possible values of the state vector—as defined in that model—to be equally likely.

1. Maxwell-Boltzman. The state vector specifies the quantum state of each particle; there are

$$r^n$$

possible values of the state vector.

2. Bose-Einstein. The state vector specifies the number of particles in each quantum state. There are

$$\binom{n + r - 1}{n}$$

possible values of the state vector.[5]

3. Fermi-Dirac. As with Bose-Einstein statistics, the state vector specifies the number of particles in each quantum state, but no two particles can be in the same state. There are

$$\binom{r}{n}$$

possible values of the state vector.[6]

Maxwell-Boltzman statistics are widely applicable in probability theory,[7] but describe no known gas. Bose-Einstein statistics describe the thermodynamic behavior of bosons—particles whose spin angular momentum is an integer multiple of $\hbar$, Planck's constant h divided by 2π. Photons and He^4 atoms are bosons. Fermi-Dirac statistics describe the behavior of fermions, particles whose spin angular momentum is a half-integer multiple of $\hbar$. Electrons and He^3 atoms are fermions.[8]

Bose-Einstein condensates—very low temperature gases in which all the atoms are in the same quantum state—were first observed experimentally by Anderson et al. (1995). Such condensates occur for bosons, not fermions—compelling evidence for the difference in thermodynamic statistics. The principle of insufficient reason is not a sufficient basis for physics: It does not tell us when to use one model rather than another. Generally, the outcomes of an experiment can be defined in quite different ways, and it will seldom be clear a priori which set of outcomes—if any—obeys Laplace's dictum of equal likelihood.

8.2.5 Earthquake forecasts and weather forecasts

Earthquake forecasts look similar in many ways to weather forecasts, so we might look to meteorology for guidance. How do meteorologists interpret statements like "the chance of rain tomorrow is 0.7"? The standard interpretation applies frequentist ideas to forecasts. In this view, the chance of rain tomorrow is 0.7 means that 70% of such forecasts are followed by rain the next day.

Whatever the merits of this view, meteorology differs from earthquake prediction in a critical respect. Large regional earthquakes are rare; they have recurrence times on the order of hundreds of years.[9] Weather forecasters have a much shorter time horizon. Therefore, weather prediction does not seem like a good analogue for earthquake prediction.

8.2.6 Mathematical probability: Kolmogorov's axioms

For most statisticians, Kolmogorov's axioms are the basis for probability theory—no matter how the probabilities are to be interpreted. Let Σ be a σ-algebra[10] of subsets of a set S. Let P be a real-valued function on Σ. Then P is a probability if it satisfies the following axioms:

- $P(A) \geq 0$ for every $A \in \Sigma$;
- $P(S) = 1$;
- if $A_j \in \Sigma$ for $j = 1, 2, \ldots$, and $A_j \cap A_k = \emptyset$ whenever $j \neq k$, then

$$P\left(\bigcup_{j=1}^{\infty} A_j\right) = \sum_{j=1}^{\infty} P(A_j). \tag{1}$$

The first axiom says that probability is nonnegative. The second defines the scale: Probability 1 means certainty. The third says that if $A_1, A_2, \ldots$ are pairwise disjoint, the probability that at least one A_j occurs is the sum of their probabilities.

8.2.7 Probability models

Another interpretation of probability seems more useful for making sense of earthquake predictions: Probability is just a property of a mathematical model intended to describe some features of the natural world. For the model to be useful, it must be shown to be in good correspondence with the system it describes. That is where the science comes in.

Here is a description of coin tossing that illustrates the model-based approach. A coin will be tossed n times. There are 2^n possible sequences of heads and tails. In the mathematical model, those sequences are taken to be equally likely: Each has probability $1/2^n$, corresponding to probability $1/2$ of heads on each toss and independence among the tosses.

This model has observational consequences that can be used to test its validity. For example, the probability distribution of the total number X of heads in n tosses is binomial:

$$P(X = k) = \binom{n}{k} \frac{1}{2^n}.$$

If the model is correct, when n is at all large we should see around $n/2$ heads, with an error on the order of $\sqrt{n}$. Similarly, the model gives probability distributions for the number of runs, their lengths, and so forth, which can be checked against data. The model is very good, but imperfect: With many thousands of tosses, the difference between a real coin and the model coin is likely to be detectable. The probability of heads will not be exactly $1/2$ and there may be some correlation between successive tosses.

This interpretation—that probability is a property of a mathematical model and has meaning for the world only by analogy—seems the most appropriate for earthquake prediction. To apply the interpretation, one posits a stochastic model for earthquakes in a given region and interprets a number calculated from the model to be the probability of an earthquake in some time interval. The problem in earthquake forecasts is that the models—unlike the models for coin tossing—have not been tested against relevant data. Indeed, the models cannot be tested on a human time scale, so there is little reason to believe the probability estimates. As we shall see in the next section, although some parts of the earthquake models are constrained by the laws of physics, many steps involve extrapolating rules of thumb far beyond the data they summarize; other steps rely on expert judgment separate from any data; still other steps rely on ad hoc decisions made as much for convenience as for scientific relevance.

8.3 The USGS earthquake forecast

We turn to the USGS forecast for the San Francisco Bay Area (USGS, 1999). The forecast was constructed in two stages. The first stage built a collection of 2000 models for linked fault segments, consistent with regional tectonic slip constraints, in order to estimate seismicity rates. The models were drawn by Monte Carlo from a probability distribution defined using data and expert opinion.[11] We had trouble understanding the details, but believe that the models differed in the geometry and dimensions of fault segments, the fraction of slip released aseismically on each fault segment, the relative frequencies with which different combinations of fault segments rupture together, the relationship between fault area and earthquake size, and so forth.

Each model generated by the Monte Carlo was used to predict the regional rate of tectonic deformation; if the predicted deformation was not close enough to the measured rate of deformation, the model was discarded.[12] This was repeated until 2000 models met the constraints. That set of models was used to estimate the long-term recurrence rate of earthquakes of different sizes and to estimate the uncertainties of those rate estimates for use in the second stage.

The second stage of the procedure created three generic stochastic models for fault segment ruptures, estimating parameters in those models from the long-term recurrence rates developed in the first stage. The stochastic models were then used to estimate the probability that there will be at least one magnitude 6.7 or greater earthquake by 2030.

We shall try to enumerate the major steps in the first stage—the construction of the 2000 models—to indicate the complexity.

1. Determine regional constraints on aggregate fault motions from geodetic measurements.

2. Map faults and fault segments; identify fault segments with slip rates of at least 1 mm/y. Estimate the slip on each fault segment principally from paleoseismic data, occasionally augmented by geodetic and other data. Determine (by expert opinion) for each segment a "slip factor," the extent to which long-term slip on the segment is accommodated aseismically. Represent uncertainty in fault segment lengths, widths, and slip factors as independent Gaussian random variables with mean 0.[13] Draw a set of fault segment dimensions and slip factors at random from that probability distribution.

3. Identify (by expert opinion) ways in which segments of each fault can rupture separately and together.[14] Each such combination of segments is a "seismic source."

4. Determine (by expert opinion) the extent to which long-term fault slip is accommodated by rupture of each combination of segments for each fault.

5. Choose at random (with probabilities of 0.2, 0.2, and 0.6 respectively) one of three generic relationships between fault area and moment release to characterize magnitudes of events that each combination of fault segments supports. Represent the uncertainty in the generic relationship as Gaussian with zero mean and standard deviation 0.12, independent of fault area.[15]

6. Using the chosen relationship and the assumed probability distribution for its parameters, determine a mean event magnitude for each seismic source by Monte Carlo simulation.

7. Combine seismic sources along each fault "in such a way as to honor their relative likelihood as specified by the expert groups" (USGS, 1999, p. 10); adjust the relative frequencies of events on each source so that every fault segment matches its geologic slip rate—as estimated previously from paleoseismic and geodetic data. Discard the combination of sources if it violates a regional slip constraint.

8. Repeat the previous steps until 2000 regional models meet the slip constraint. Treat the 2000 models as equally likely for the purpose of estimating magnitudes, rates, and uncertainties.

9. Steps 1–8 model events on seven identified fault systems, but there are background events not associated with those faults. Estimate the background rate of seismicity as follows. Use an (unspecified) Bayesian procedure to categorize historical events from three catalogs either as associated or not associated with the seven fault systems. Fit a generic Gutenberg-Richter magnitude-frequency relation $N(M) = 10^{a-bM}$ to the events deemed not to be associated with the seven fault systems. Model this background seismicity as a marked Poisson process. Extrapolate the Poisson model to $M \geq 6.7$, which gives a probability of 0.09 of at least one event.[16]

This first stage in the USGS procedure generates 2000 models and estimates long-term seismicity rates as a function of magnitude for each seismic source. We now describe the second stage—the earthquake forecast itself. Our description is sketchy because we had trouble understanding the details from the USGS report. The second stage fits three types of stochastic models for earthquake recurrence—Poisson, Brownian passage time (Ellsworth et al., 1998), and "time-predictable"—to the long-term seismicity rates estimated in the first stage.[17] Ultimately, those stochastic models are combined to estimate the probability of a large earth-quake.

The Poisson and Brownian passage time models were used to es-timate the probability that an earthquake will rupture each fault seg-ment. Some parameters of the Brownian passage time model were fitted to the data, and some were set more arbitrarily; for example, aperiodic-ity (standard deviation of recurrence time, divided by expected recurrence time) was set to three different values, 0.3, 0.5, and 0.7. The Poisson model does not require an estimate of the date of last rupture of each segment, but the Brownian passage time model does; those dates were estimated from the historical record. Redistribution of stress by large earthquakes was modeled; predictions were made with and without adjustments for stress redistribution. Predictions for each segment were combined into predic-tions for each fault using expert opinion about the relative likelihoods of different rupture sources.

A "time-predictable model" (stress from tectonic loading needs to reach the level at which the segment ruptured in the previous event for the segment to initiate a new event) was used to estimate the probability that an earthquake will originate on each fault segment. Estimating the state of stress before the last event requires knowing the date of the last event and the slip during the last event. Those data are available only for the 1906 earthquake on the San Andreas Fault and the 1868 earthquake on the southern segment of the Hayward Fault (USGS, 1999, p. 17), so the time-predictable model could not be used for many Bay Area fault segments.

The calculations also require estimating the loading of the fault over time, which in turn relies on viscoelastic models of regional geological structure. Stress drops and loading rates were modeled probabilistically (USGS, 1999, p. 17); the form of the probability models is not given. The loading of the San Andreas Fault by the 1989 Loma Prieta earth-quake and the loading of the Hayward Fault by the 1906 earthquake were modeled. The probabilities estimated using the time-predictable model were converted into forecasts using expert opinion about the relative like-lihoods that an event initiating on one segment will stop or will propagate to other segments. The outputs of the three types of stochastic models for each fault segment were weighted according to the opinions of a panel of fifteen experts. When results from the time-predictable model were not available, the weights on its output were in effect set to zero.

There is no straightforward interpretation of the USGS probability forecast. Many steps involve models that are largely untestable; modeling choices often seem arbitrary. Frequencies are equated with probabilities, fiducial distributions are used, outcomes are assumed to be equally likely, and subjective probabilities are used in ways that violate Bayes rule.[18]

8.3.1 What does the uncertainty estimate mean?

The USGS forecast is 0.7 ± 0.1, where 0.1 is an uncertainty estimate (USGS, 1999). The 2000 regional models produced in stage 1 give an estimate of the long-term seismicity rate for each source (linked fault segments), and an estimate of the uncertainty in each rate. By a process we do not understand, those uncertainties were propagated through stage 2 to estimate the uncertainty of the estimated probability of a large earthquake. If this view is correct, 0.1 is a gross underestimate of the uncertainty. Many sources of error have been overlooked, some of which are listed below.

1. Errors in the fault maps and the identification of fault segments.[19]

2. Errors in geodetic measurements, in paleoseismic data, and in the viscoelastic models used to estimate fault loading and subsurface slip from surface data.

3. Errors in the estimated fraction of stress relieved aseismically through creep in each fault segment and errors in the relative amount of slip assumed to be accommodated by each seismic source.

4. Errors in the estimated magnitudes, moments, and locations of historical earthquakes.

5. Errors in the relationships between fault area and seismic moment.

6. Errors in the models for fault loading.

7. Errors in the models for fault interactions.

8. Errors in the generic Gutenberg-Richter relationships, not only in the parameter values but also in the functional form.

9. Errors in the estimated probability of an earthquake not associated with any of the faults included in the model.

10. Errors in the form of the probability models for earthquake recurrence and in the estimated parameters of those models.

8.4 A view from the past

Littlewood (1953) wrote:

Mathematics (by which I shall mean pure mathematics) has no grip on the real world; if probability is to deal with the real world it must contain elements outside mathematics; the *meaning* of "probability" must relate to the real world, and there must be one or more "primitive" propositions about the real world, from which we can then proceed deductively (i.e. mathematically). We will suppose (as we may by lumping several primitive propositions together) that there is just one primitive proposition, the "probability axiom", and we will call it A for short. Although it has got to be *true*, A is by the nature of the case incapable of deductive proof, for the sufficient reason that it is about the real world . . . .

There are 2 schools. One, which I will call mathematical, stays inside mathematics, with results that I shall consider later. We will begin with the other school, which I will call philosophical. This attacks directly the "real" probability problem; what are the axiom A and the meaning of "probability" to be, and how can we justify A? It will be instructive to consider the attempt called the "frequency theory". It is natural to believe that if (with the natural reservations) an act like throwing a die is repeated n times the proportion of 6's will, *with certainty*, tend to a limit, p say, as $n \to \infty$. (Attempts are made to sublimate the limit into some Pickwickian sense—"limit" in inverted commas. But either you *mean* the ordinary limit, or else you have the problem of explaining how "limit" behaves, and you are no further. You do not make an illegitimate conception legitimate by putting it into inverted commas.) If we take this proposition as "A" we can at least settle off-hand the other problem of the *meaning* of probability; we define its measure for the event in question to be the number p. But for the rest this A takes us nowhere. Suppose we throw 1000 times and wish to know what to expect. Is 1000 large enough for the convergence to have got under way, and how far? A does not say. We have, then, to add to it something about the rate of convergence. Now an A cannot assert a *certainty* about a particular number n of throws, such as "the proportion of 6's will *certainly* be within $p \pm \epsilon$ for large enough n (the largeness depending on ϵ)". It can only say "the proportion will lie between $p \pm \epsilon$ *with at least such and such probability (depending on ϵ and n_0) whenever $n \gg n_0$*".

The vicious circle is apparent. We have not merely failed to *justify* a workable A; we have failed even to *state* one which would work if its truth were granted. It is generally agreed that the frequency theory won't work. But whatever the theory it is clear that the vicious circle is very deep-seated: certainty being impossible, whatever A is made to state can be stated only in terms of "probability".

8.5 Conclusions

Making sense of earthquake forecasts is difficult, in part because standard interpretations of probability are inadequate. A model-based interpretation is better, but lacks empirical justification. Furthermore, probability models are only part of the forecasting machinery. For example, the USGS San Francisco Bay Area forecast for 2000–2030 involves geological mapping, geodetic mapping, viscoelastic loading calculations, paleoseismic observations, extrapolating rules of thumb across geography and magnitude, simulation, and many appeals to expert opinion. Philosophical difficulties aside, the numerical probability values seem rather arbitrary.

Another large earthquake in the San Francisco Bay Area is inevitable, and imminent in geologic time. Probabilities are a distraction. Instead of making forecasts, the USGS could help to improve building codes and to plan the government's response to the next large earthquake. Bay Area residents should take reasonable precautions, including bracing and bolting their homes as well as securing water heaters, bookcases, and other heavy objects. They should keep first aid supplies, water, and food on hand. They should largely ignore the USGS probability forecast.

Notes

1. See Stigler (1986) for history prior to 1900. Currently, the two main schools are the frequentists and the Bayesians. Frequentists, also called objectivists, define probability in terms of relative frequency. Bayesians, also called subjectivists, define probability as degree of belief. We do not discuss other theories, such as those associated with Fisher, Jeffreys, and Keynes, although we touch on Fisher's 'fiducial probabilities" in note 11.

2. It is hard to specify precisely which conditions must be the same across trials, and, indeed, what "the same" means. Within classical physics, for instance, if all the conditions were exactly the same, the outcome would be the same every time—which is not what we mean by randomness.

3. A Bayesian will have a prior belief about nature. This prior is updated as the data come in, using Bayes rule: In essence, the prior is reweighted according to the likelihood of the data (Hartigan, 1983, pp. 29ff). A

Bayesian who does not have a proper prior—that is, whose prior is not a probability distribution—or who does not use Bayes rule to update, is behaving irrationally according to the tenets of his own doctrine (Freedman, 1995 [Chapter 1]). For example, the Jeffreys prior is generally improper, because it has infinite mass; a Bayesian using this prior is exposed to a money-pump (Eaton and Sudderth, 1999, p. 849; Eaton and Freedman, 2004). It is often said that the data swamp the prior: The effect of the prior is not important if there are enough observations (Hartigan, 1983, pp. 34ff). This may be true when there are many observations and few parameters. In earthquake prediction, by contrast, there are few observations and many parameters.

4. The number of states depends on the temperature of the gas, among other things. In the models we describe, the particles are "non-interacting." For example, they do not bond with each other chemically.

5. To define the binomial coefficients, consider m things. How many ways are there to choose k out of the m? The answer is given by the binomial coefficient

$$\binom{m}{k} = \binom{m}{m-k} = \frac{m!}{k!(m-k)!}$$

for $k = 0, 1, \ldots, m$. Let n and r be positive integers. How many sequences $(j_1, j_2, \ldots, j_r)$ of nonnegative integers are there with $j_1 + j_2 + \cdots + j_r = n$? The answer is

$$\binom{n+r-1}{n}.$$

For the argument, see Feller (1968). To make the connection with Bose-Einstein statistics, think of $\{j_1, j_2, \ldots, j_r\}$ as a possible value of the state vector, with j_i equal to the number of particles in quantum state i.

6. That is the number of ways of selecting n of the r states to be occupied by one particle each.

7. In probability theory, we might think of a Maxwell-Boltzman "gas" that consists of $n = 2$ coins. Each coin can be in either of $r = 2$ quantum states—heads or tails. In Maxwell-Boltzman statistics, the state vector has two components, one for each coin. The components tell whether the corresponding coin is heads or tails. There are

$$r^n = 2^2 = 4$$

possible values of the state vector: HH, HT, TH, and TT. These are equally likely.

To generalize this example, consider a box of r tickets, labeled $1, 2, \ldots, r$. We draw n tickets at random with replacement from the box. We can think of the n draws as the quantum states of n particles, each of which has r possible states. This is "ticket-gas." There are r^n possible outcomes, all equally likely, corresponding to Maxwell-Boltzman statistics. The case $r = 2$ corresponds to coin-gas; the case $r = 6$ is "dice-gas," the standard model for rolling n dice.

Let $X = \{X_1, \ldots, X_r\}$ be the occupancy numbers for ticket-gas: In other words, X_i is the number of particles in state i. There are

$$\binom{n + r - 1}{n}$$

possible values of X. If ticket-gas were Bose-Einstein, those values would be equally likely. With Maxwell-Boltzman statistics, they are not: Instead, X has a multinomial distribution. Let $j_1, j_2, \ldots, j_r$ be nonnegative integers that sum to n. Then

$$P(X_1 = j_1, X_2 = j_2, \ldots, X_r = j_r) = \frac{n!}{j_1! j_2! \cdots j_r!} \times \frac{1}{r^n}.$$

The principle of insufficient reason is not sufficient for probability theory, because there is no canonical way to define the set of outcomes which are to be taken as equally likely.

8. The most common isotope of Helium is He^4; each atom consists of two protons, two neutrons, and two electrons. He^3 lacks one of the neutrons, which radically changes the thermodynamics.

9. There is only about one earthquake of magnitude 8+ per year globally. In the San Francisco Bay Area, unless the rate of seismicity changes, it will take on the order of a century for a large earthquake to occur, which is not a relevant time scale for evaluating predictions.

10. The collection Σ must contain S and must be closed under complementation and countable unions. That is, Σ must satisfy the following conditions: $S \in \Sigma$; if $A \in \Sigma$, then $A^c \in \Sigma$; and if $A_1, A_2, \ldots, \in \Sigma$, then $\cup_{j=1}^\infty A_j \in \Sigma$.

11. Some parameters were estimated from data. The Monte Carlo procedure treats such parameters as random variables whose expected values are the estimated values, and whose variability follows a given parametric form (Gaussian). This is "fiducial inference" (Lehmann, 1986, pp. 229–30), which is neither frequentist nor Bayesian. There are also several

competing theories for some aspects of the models, such as the relation-ship between fault area and earthquake magnitude. In such cases, the Monte Carlo procedure selects one of the competing theories at random, according to a probability distribution that reflects "expert opinion as it evolved in the study." Because the opinions were modified after analyzing the data, these were not prior probability distributions; nor were opinions updated using Bayes rule. See note 3.

12. About 40% of the randomly generated models were discarded for violating a constraint that the regional tectonic slip be between 36 mm/y and 43 mm/y.

13. The standard deviations are zero—no uncertainty—in several cases where the slip is thought to be accommodated purely seismically; see Table 2 of (USGS, 1999). Even the non-zero standard deviations seem to be arbitrary.

14. It seems that the study intended to treat as equally likely all $2^n - 1$ ways in which at least one of n fault segments can rupture; however, the example on p. 9 of USGS (1999) refers to six possible ways a three-segment fault can rupture, rather than $2^3 - 1 = 7$, but then adds the possibility of a "floating earthquake," which returns the total number of possible combinations to seven. Exactly what the authors had in mind is not clear. Perhaps there is an implicit constraint: Segments that rupture must be contiguous. If so, then for a three-segment fault where the seg-ments are numbered in order from one end of the fault (segment 1) to the other (segment 3), the following six rupture scenarios would be possi-ble: $\{1\}, \{2\}, \{3\}, \{1, 2\}, \{2, 3\},$ and $\{1, 2, 3\}$; to those, the study adds the seventh "floating" earthquake.

15. The relationships are all of the functional form $M = k + \log A$, where M is the moment magnitude and A is the area of the fault. There are few relevant measurements in California to constrain the relationships (only seven "well-documented" strike-slip earthquakes with $M \geq 7$, dating back as far as 1857), and there is evidence that California seismicity does not follow the generic model (USGS, 1999).

16. This probability is added at the end of the analysis, and no uncertainty is associated with this number.

17. Stage 1 produced estimates of rates for each source; apparently, these are disaggregated in stage 2 into information about fault segments by using expert opinion about the relative likelihoods of segments rupturing separately and together.

18. See notes 3 and 11.

19. For example, the Mount Diablo Thrust Fault, which slips at 3 mm/y, was not recognized in 1990 but is included in the 1999 model (USGS, 1999, p. 8). Moreover, seismic sources might not be represented well as linked fault segments.

9

Salt and Blood Pressure:
Conventional Wisdom Reconsidered

With Diana B. Petitti

ABSTRACT. *The "salt hypothesis" is that higher levels of salt in the diet lead to higher levels of blood pressure, increasing the risk of cardiovascular disease. Intersalt, a cross-sectional study of salt levels and blood pressures in fifty-two populations, is often cited to support the salt hypothesis, but the data are somewhat contradictory. Four of the populations (Kenya, Papua, and two Indian tribes in Brazil) do have low levels of salt and blood pressure. Across the other forty-eight populations, however, blood pressures go down as salt levels go up—contradicting the hypothesis. Experimental evidence suggests that the effect of a large reduction in salt intake on blood pressure is modest and that health consequences remain to be determined. Funding agencies and medical journals have taken a stronger position favoring the salt hypothesis than is warranted, raising questions about the interaction between the policy process and science.*

It is widely believed that dietary salt leads to increased blood pressure and higher risks of heart attack or stroke. This is the "salt hypothesis." The corollary is that salt intake should be drastically reduced. There are three main kinds of evidence: (i) animal experiments, (ii) observational

Evaluation Review (2001) 25: 267–87.

studies on humans, and (iii) human experiments. Animal experiments are beyond the scope of the present chapter, although we give a telegraphic summary of results. A major observational study cited by those who favor salt reduction is Intersalt (1986, 1988). Intersalt is the main topic of the present chapter, and we find that the data do not support the salt hypothesis. The other major observational study is Smith et al. (1988), and this contradicts the salt hypothesis.

There have been many intervention studies on humans, and several meta-analyses. Although publication bias is a concern, the experiments do suggest some reduction in blood pressure for hypertensive subjects from aggressive reduction in salt intake; the effect for normotensives is smaller. Recently, the DASH studies manipulated diet and salt intake. Both have an effect, and there is an interaction. Intervention studies on humans are a second topic of our chapter. To document the effect of salt reduction on morbidity or mortality, much larger intervention studies would be needed, with much longer followup. This point is discussed too. Finally, implications for policy analysis are noted.

9.1 Animal studies

Rodents, the best-studied species, show strain-specific effects of salt intake on blood pressure. In some strains, a diet high in salt leads to a marked increase in pressure; but in other strains, there is no effect. Studies of non-human primates, which are more limited, suggest that some animals are salt-sensitive and some are not. In other words, for some animals, blood pressure increases when salt is added to the diet; for other animals, there is no response.

9.2 The Intersalt study

Intersalt was an observational study conducted at fifty-two centers in thirty-two countries; about 200 subjects age twenty to fifty-nine were recruited in each center. The two Brazilian centers were Indian tribes, the Yanomamo and Xingu. There was a center in Kenya and one in Papua New Guinea. In Canada, there were centers in Labrador and in St. John's (Newfoundland). In the United States, there was a center in Hawaii, a center in Chicago, and four centers in Mississippi.

Blood pressure (systolic and diastolic) was measured for each subject, along with urinary sodium and potassium (mmols/24 hours), and various confounders such as body mass index (weight/height^2). Other confounders (like alcohol consumption) were obtained from questionnaires. Replicate urine measurements were obtained for a sub-sample of the subjects. Table 9.1 indicates some of the data available for the various centers; units are explained below.

Within each center, the subjects' blood pressures were regressed on their ages: The slope of the resulting line indicates how rapidly blood pressure increases with age. (Complications will be discussed later.) Slopes were then correlated with salt levels across centers. The correlation was significant, and seems to be the major finding of Intersalt as well as the basis for much advice to restrict salt intake.

In each center, the subjects' blood pressures were also regressed on their urinary salt levels. The within-center regression coefficients were variable, some being positive, some negative, and some insignificant. Within-center regression coefficients were "pooled"—averaged—across centers, with weights inversely proportional to estimated variances. Generally, the within-center coefficients were adjusted for age and sex; sometimes, for age, sex, body mass index, alcohol, and potassium intake; the likely size of measurement error in urinary salt was estimated from the replicate measurements, and statistical procedures were sometimes used to adjust results of cross-center regressions for measurement error.

Pooled results were highly significant, especially after correction for measurement error. The estimated effect of salt on blood pressure depends on the statistical adjustments: Reduction of salt amounting to 100 mmol per day is estimated to lead to a reduction in systolic pressure in the range

Table 9.1 Intersalt data. Systolic blood pressure. Selected centers. Median urinary salt (mmol Na/24 hours); median blood pressure (mm Hg); slope of blood pressure on age (mm Hg/ year); slope of blood pressure on urinary salt (mm Hg/mmol Na/ 24 hours).

	Na	BP	BP on age	BP on Na
Yanomamo, Brazil	0.2	95	.079	−.173
Xingu, Brazil	6	99	.052	−.037
Papua New Guinea	27	108	.149	+.037
Kenya	51	110	.206	+.033
⋮	⋮	⋮	⋮	⋮
Hawaii	130	124	.638	+.044
Chicago	134	115	.287	+.001
Labrador	149	119	.500	+.043
⋮	⋮	⋮	⋮	⋮
Tianjin, PRC	242	118	.640	+.035

from 1 to 6 mm Hg; for diastolic pressure, the estimated reduction ranges from 0.03 to 2.5 mm Hg. See Intersalt (1988, Table 9.1) and Elliott et al. (1996, Table 9.1). By way of comparison, the urinary salt level in the Chicago center was 134 mmol, not far from the current U.S. average; a reduction of 100 mmol gets down to the level in Kenya or Papua New Guinea (Table 9.1).

9.3 Units for salt and blood pressure

The units in Table 9.1 may be unfamiliar and irritating, but they are standard in the field. Relatively little salt is retained or excreted other than in the urine and dietary measurements are quite troublesome, so intake is measured by urinary excretion. Table salt is sodium chloride (NaCl), and urinary salt levels are measured in terms of sodium content, by weight. The unit of weight is the millimole (mmol), that is, 1/1000 of the gram molecular weight. Sodium (Na) has atomic weight nearly 23; so a mole of Na weighs 23 grams, and 1 gram of Na is $1/23 = 0.0435$ moles $= 43.5$ mmols. A dietary intake of 2.5 grams per day of table salt corresponds to 1 gram per day of sodium and 43.5 mmols per day of urinary sodium excretion; the other 1.5 grams is the chlorine. By way of calibration, a typical American dietary intake is 8.5 grams per day of salt, which corresponds to $8.5/2.5 = 3.4$ grams per day of sodium, and $3.4 \times 43.5 \doteq 150$ mmols per day of urinary sodium.

BP is blood pressure, measured in two phases—systolic and diastolic. The systolic phase corresponds to blood being pumped out of the heart, and the pressure is higher; the diastolic phase corresponds to blood flowing back into the heart, and pressure is lower. Pressure is measured relative to the height of a column of mercury; units are millimeters of mercury (mm Hg). Average U.S. systolic pressure for persons over the age of eighteen is about 125 mm Hg; average diastolic pressure is about 75 mm Hg: Standard deviations are about twenty and twelve, respectively.

Averages and standard deviations for BP are computed from the third replication of the National Health and Nutrition Examination Survey (NHANES III). Each replication is based on a large probability sample of the U.S. population; subjects fill out questionnaires describing diet, socioeconomic status, and so forth; they also undergo a thorough medical examination. The NHANES data will come up again later.

A blood pressure of 140/75 means 140 systolic and 75 diastolic. "Normotensive" persons have normal blood pressures, and "hypertensives" have high blood pressures. Precise definitions vary from one study to another, but 160/95 would generally be considered diagnostic of hypertension. In some studies, even 140/90 would be classified as hypertension.

9.4 Patterns in the Intersalt data

The correlational pattern across the Intersalt centers between salt level and blood pressure is complex and has not received the attention it deserves. Figure 9.1a plots the median systolic blood pressure against the median level of urinary salt. The data are clearly nonlinear, because there are four outliers—centers with extremely low levels of salt and blood pressure. These are the two Brazilian tribes, Papua New Guinea, and Kenya; see Table 9.1. The four outliers show the expected upward trend. In the other forty-eight centers, the trend is downward, although not significantly. (The adjustments contemplated by Intersalt create a positive slope, but significance is not achieved; with forty-eight points, the adjusted slope is .0251 and $P = .33$; if all 52 points are used, the adjusted slope is .0446 and $P < .01$; Intersalt 1988, Figure 9.3).

Figure 9.1b plots the rate of change of systolic blood pressure with age at each center against the median level of urinary salt. There is a

Figure 9.1 Panel (a) Systolic blood pressure vs urinary salt. Median levels. Excluding the two Brazilian tribes, Papua New Guinea, and Kenya, the trend is downward but not significant ($n = 48, r = -.14, P = .34$, two-sided).
Panel (b) Rate of increase of systolic blood pressure with age, plotted against the median level of salt in the urine for subjects at that center. Even in the forty-eight centers, there is a significant upward trend ($n = 48, r = .27, P = .05$, two-sided).

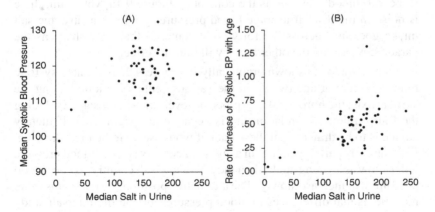

Note: The units for the horizontal axis in Figures 9.1–9.6 are mmols per day of urinary sodium—not sodium chloride. The data are from summary statistics reported by Intersalt (1988, Appendix I).

significant, positive relationship: At centers with higher levels of salt, systolic blood pressure generally increases more rapidly with age. In combination, however, Figures 9.1a and 9.1b lead to a paradox. For each of the forty-eight study centers, the regression line of blood pressure on age must pass through the middle of the scatter diagram, so that blood pressure at middle age should equal the average blood pressure. In middle age, there is at best no cross-center relationship between salt and blood pressures (Figure 9.1a). Since blood pressures increase more rapidly in the centers with higher salt levels (Figure 9.1b), it follows that young people in the high-salt centers must have lower blood pressures than their counterparts in the centers with lower salt intake.

In more detail, suppose (i) there is a linear relationship between age (x) and blood pressure (y) for subjects within each of the forty-eight centers; (ii) across the centers, as average salt intake goes up, the slope of the line goes up; (iii) subjects in all forty-eight centers have the same average age ($\bar{x}$) and average blood pressure ($\bar{y}$). As always, the regression line for each center has to go through the point of averages ($\bar{x}, \bar{y}$) for that center. The point of averages is the same for all the centers—assumption (iii). Therefore, the lines for the high-salt centers have to start lower than the lines for the low-salt centers, in order not to pass over them at $\bar{x}$.

Assumption (i), with random error around the line, seems to be a driving force behind the analyses presented by Intersalt. Assumption (ii), again with some noise, is just Figure 9.1b. Assumption (iii), at least with respect to blood pressure, is the content of Figure 9.1a; yet again, there is noise in the data. If average blood pressures go down as average salt intake goes up—across the forty-eight centers—that only sharpens the paradox. Noise, on the other hand, will blur the effect.

The paradox is shown graphically in Figure 9.2. Estimated systolic blood pressure at age twenty in the various centers is plotted along the vertical axis; the horizontal axis plots the levels of urinary salt. Excluding the four outliers, the relationship is negative and significant. If dietary advice is to be drawn from these data, it would seem to be the following. Live the early part of your life in a high-salt country so your blood pressure will be low as a young adult; then move to a low-salt country so your blood pressure will increase slowly. The alternative position, which seems more realistic, is that differences in blood pressures among the Intersalt study populations are mainly due to uncontrolled confounding—not variations in salt intake.

The underlying Intersalt data do not seem to be available, as discussed below, so Figure 9.2 takes the average age at each center as the midpoint of the age range, namely, forty. Blood pressure at age twenty in each center can then be estimated (by regression) as the overall median at that center, less twenty times the slope of blood pressure on age. There is an annoying numerical coincidence here: Age twenty is twenty below the midrange of forty. If A denotes age, the difference $40 - A$ should be multiplied by the slope of the regression line, to get the estimated amount by which blood pressure at age A is below blood pressure at age forty. Theoretically, of course, such regression adjustments should be based on arithmetic averages: If y is regressed on x, the regression line goes through the point of averages $(\bar{x}, \bar{y})$, not the point of medians. Medians are used as in Intersalt (1988), but there would be little difference in results if means were used.

Figure 9.3 repeats the analysis for diastolic pressure, with similar results. In Figure 9.3a, the downwards slope among the forty-eight centers is significant; after adjustments recommended by Intersalt (1988, Figure 9.4), the slope is still downwards, although it is no longer significant. In Figure 9.3b, the slopes of diastolic blood pressure on age are strongly related to salt levels. In Figure 9.3c, the downwards slope among the forty-eight centers is highly significant: For young people in those centers, estimated diastolic blood pressure is negatively related to salt intake, contradicting the salt hypothesis.

Figure 9.2   Estimated systolic blood pressure at age twenty plotted against median urinary salt levels. In the forty-eight centers—excluding the two Brazilian tribes, Papua New Guinea, and Kenya—there is a downward trend, which is significant ($n = 48$, $r = -.31$, $P = .02$, two-sided).

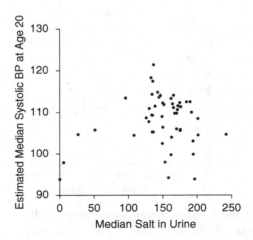

Figure 9.3 Panel (a) Diastolic blood pressure vs urinary salt. In forty-eight centers—excluding the two Brazilian tribes, Papua New Guinea, and Kenya—the downward trend is significant ($n = 48, r = -.31, P = .02$, two-sided).

Panel (b) Rate of increase of diastolic blood pressure with age, plotted against the median level of salt in the urine for subjects at that center. Even in the forty-eight centers, there is a highly significant positive trend ($n = 48, r = .40, P < .01$, two-sided).

Panel (c) Estimated diastolic blood pressure at age twenty plotted against median urinary salt levels. In the forty-eight centers, there is a downward trend which is highly significant ($n = 48, r = -.42, P < .01$, two-sided).

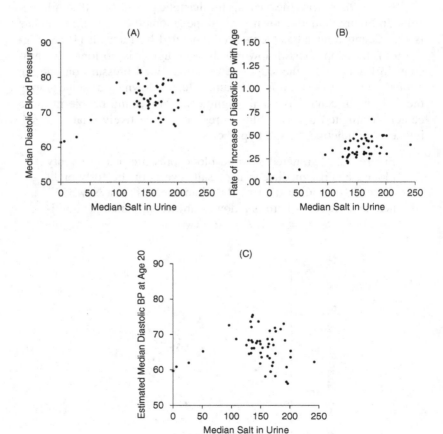

Generally, the Intersalt investigators favor results obtained by combining data from all fifty-two centers. Any such analysis, however, only serves to underline what is already obvious: Subjects in the four outlying centers have much lower blood pressures than subjects in the other forty-eight centers, somewhat less rapid increase of blood pressure with age, and dramatically lower salt intake.

9.5 P-values

The Intersalt investigators use P-values to assess their results. We follow suit, although the interpretation of P may be somewhat problematic in these contexts. (i) The forty-eight study centers might be viewed as a random sample from some imaginary collection of potential study centers. Additional statistical assumptions (such as linearity and homoscedasticity) may need to be imposed on this hypothetical super-population, depending on the analysis that is to be rationalized. (ii) It might be assumed that the data were generated in accordance with some linear regression model, with a null hypothesis specifying that a certain coefficient vanishes. Although options (i) and (ii) have their aficionados, we find them unattractive (Abbott 1997; Berk and Freedman 2003 [Chapter 2]; Goldthorpe 1999; Freedman, 1995 [Chapter 1], 1999, with further citations).

There is at least one other possibility: For scatter diagrams like those presented here, with the four outliers set aside, P approximates the probability of obtaining larger correlations than the observed ones—if the x- and y-coordinates are randomly paired (Freedman and Lane 1983). In any event, our test statistic was $t = \sqrt{48 - 2}\, r/\sqrt{1 - r^2}$, referred to a normal distribution; equivalently, a straight line is fitted to the forty-eight points, and the slope is examined to see if it is significantly different from 0.

9.6 The protocol

The Intersalt investigators offered a large number of analyses of the data and have returned to the topic more than once. See Intersalt (1988), Elliott et al. (1996), and Stamler (1997). For additional detail, see the *Journal of Human Hypertension* (1989) 3(5). The results are not entirely consistent, and the protocol (Intersalt 1986) must now be considered.

(1) "The primary hypothesis to be tested in INTERSALT is that average blood pressure and prevalence of hypertension are linearly related across populations to the average levels of sodium intake, potassium intake (inversely) and the sodium-potassium intake ratio" (p. 781).

(2) "The variation in electrolyte intake across the study population is judged to be large enough to permit, as a second

hypothesis, examining also these same relationships at the level of individuals, despite well known within-individual variability in such intake" (p. 782).

(3) "It is not expected that useful estimates will be possible ... at the level of particular study populations; but it will be possible to look at the relations in individuals across the study as a whole.... The individual and group relationships will be jointly explored by multi-level analytic techniques" (p. 785).

(4) Adjustment for (random) measurement error is suggested within center but not across center (p. 783).

(5) Possible confounders include height, weight, physical activity, type of work, socioeconomic status, alcohol, family history, and medication (pp. 783–84).

The primary Intersalt hypothesis—point (1) above—is rejected by the data. As Figures 9.1 and 9.4 demonstrate, average blood pressure levels are not linearly related to salt intake across the study populations: (i) the four outliers are different from the other forty-eight centers; and (ii) the relationship between blood pressure and salt is different in the two groups of data—positive in the first, negative in the second. In short, the relationship does not even seem to be monotone. The Intersalt investigators have paid comparatively little attention to prevalence of hypertension, also mentioned as a primary variable in point (1), but the relationship between prevalence and salt is much like that shown in Figures 9.1–9.3 for blood pressure and salt.

With respect to potassium intake, Intersalt (1988, p. 324) acknowledges that "potassium was inconsistently related to blood pressure in these cross-center analyses." What they mean is that blood pressure is positively related rather than negatively related to potassium levels; the correlation is either highly significant or not significant, depending on the details. In the forty-eight centers, $r = .40$, $P < .01$ for the systolic phase, and $r = .19$, $P = .19$ for diastolic. For all fifty-two centers, the correlations are .15 and .03. (Dropping the four outliers makes a difference, because the Xingu and Yanomamo have very high potassium levels and very low blood pressures.) In any event, the primary study hypothesis is rejected by the data, for potassium as well as sodium.

Adjusting cross-center regressions for measurement error appears to be a post hoc exercise—point (4). Pooling the within-center coefficients is also post hoc, and seems to replace more obvious multi-level regression analyses suggested by (2) and (3). The protocol (Intersalt 1986) does not mention the idea of pooling within-center regression coefficients. Furthermore, these post-hoc analyses are of doubtful validity, even on their own terms: The weights used to compute the overall average effect depend critically on unverified assumptions about the error structure in the regressions, and there are equally unverified assumptions about the nature of the measurement error in the urine variables. (Taking an average may be harmless, but the force of the assumptions will be felt when deriving standard errors and P-values.)

No adjustment is made for measurement error in confounders such as alcohol consumption. Moreover, numerous confounders remain completely uncontrolled. Diet—apart from its sodium or potassium content— would seem to be one major unmeasured confounding variable, as discussed below. Other potential confounders are listed in the protocol— point (5) —but not controlled in the data analysis: for example, physical activity, type of work, and socioeconomic status. More generally, Intersalt's chief analytic idea is that people in Chicago can be converted to Yanomamo Indians by running a regression with a few control variables, a vision that will commend itself to some observers but not others.

The rate of increase of blood pressure with age versus the salt level is also a post hoc analysis. This has been acknowledged, if indirectly, by the principal figure in the Intersalt group—Stamler (1997, p. 634S). At scientific meetings where these issues are raised, Intersalt investigators respond that age by blood pressure was to have been the primary analysis, according to minutes of the working group. The response is peculiar— what else is in those minutes? Moreover, Intersalt (1988, p. 320) clearly states that results "were assessed both within and across centres in accordance with prior plans," citing the published protocol (Intersalt 1986). Finally, the investigators cannot so easily brush aside the paradoxical implications of their models: For young people, blood pressure is negatively related to salt intake.

9.7 Human experiments

This section turns to human experiments, where salt intake is manipulated and the effect on blood pressure is determined. There have been many such experiments, and three recent meta-analyses—by Midgley et al. (1996), Cutler et al. (1997), and Graudal et al. (1998). Midgley et al. and Cutler et al. both regress blood pressure reduction on salt reduction and look for a significant slope; reductions are measured by comparing data in the treatment and control conditions. Cutler et al. find significance, Midgley et al. do not. By contrast with Midgley et al., Cutler et al. force their line to go through the origin. Apparently, the decision to force the line through the origin is what leads to significance (Graudal et al. 1998, p. 1389).

Presumably, the idea behind the constraint is that zero reduction in blood pressure corresponds to zero reduction in salt intake. Notably, however, the control groups in the experiments generally achieve some reduction in blood pressure. Thus, zero reduction in salt intake may well have an effect, depending on attendant circumstances. Generally, confounding due to flaws in experimental design—for instance, lack of blinding—can push the line away from the origin (Cutler et al. 1997, p. 644S; Midgley et al. 1996, pp. 1592–94; Graudal et al. 1998, p. 1389; Swales 2000, p. 4).

Table 9.2 shows the estimated reduction in systolic and diastolic blood pressure (mm Hg) for normotensive and hypertensive subjects, corresponding to a 100 mmol per day reduction in urinary sodium. There is a larger effect on systolic than diastolic pressure, and hypertensives are more affected than normotensives. However, agreement among the three studies is not good. Indeed, Midgley et al. and Graudal et al. report only a minimal effect for normotensives, while Cutler et al. find a bigger effect. As noted before, a typical American dietary intake is 8.5 grams per day of salt (NaCl), which corresponds to 3.4 grams per day of sodium (Na), and 150 mmols per day of urinary sodium excretion. On this scale, a 100 mmol reduction in sodium is striking.

Given the lack of concordance in Table 9.2, it will not come as a surprise that the three meta-analyses differ at the bottom line. Cutler et al. are strongly anti-sodium, while the other two papers are relatively neutral. Thus, Cutler et al. (1997, p. 648S) find "conclusive evidence that moderate sodium reduction lowers systolic and diastolic blood pressure . . . ." However, according to Midgley et al. (1996, p. 1590), "dietary sodium

Table 9.2 Estimated reduction in blood pressure (mm Hg) due to reduction in urinary sodium by 100 mmols per day; three meta-analyses.

| | Normotensive | | Hypertensive | |
	Systolic	Diastolic	Systolic	Diastolic
Cutler et al. (1997)	2.3	1.4	5.8	2.5
Midgley et al. (1996)	1.1	0.1	3.7	0.9
Graudal et al. (1998)	0.8	0.2	3.6	1.6

Note: "Normotensives" have normal blood pressure, "hypertensives" have high blood pressure.

restriction might be considered for older hypertensive individuals, but . . . the evidence in the normotensive population does not support current recommendations for universal dietary sodium restriction." Similarly, Graudal et al. (1998, p. 1383) conclude that the data "do not support a general recommendation to reduce sodium intake."

9.8 Publication bias

Cutler et al. (1997, p. 648S) say there was "no indication for diastolic blood pressure from graphic and regression analysis that small negative studies were underrepresented"; for systolic blood pressure, "the graphic plot was more suggestive," although significance is not reached. Midgley et al. conclude that publication bias is evident, using a funnel plot to make the assessment.

Figure 9.4 is a funnel plot showing changes in systolic blood pressure plotted against sample size. (Occasionally, treatment and control groups were of slightly different sizes; then the average of the two was used.) Studies on hypertensives and normotensives are represented by different symbols; data are from Cutler et al. Most of the studies find a reduction in blood pressure, plotted as a negative value. In a few studies, salt reduction leads to increased blood pressure, plotted as a positive value. The smaller studies generally find more dramatic decreases in blood pressure.

Figure 9.4 Funnel plot. Studies with hypertensive subjects are marked by dots; normotensives, by crosses. Change in systolic blood pressure plotted against square root of sample size. In some studies, treatment increases blood pressure, plotted as positive values on the y-axis. Smaller studies show bigger effects, suggesting publication bias.

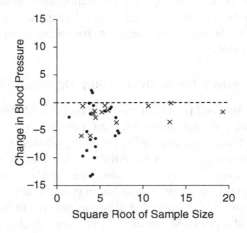

The difference between estimated effect sizes in the large studies and the small ones is what indicates publication bias: Unpublished small studies cannot make it into the picture.

It may be helpful to describe the funnel plot more abstractly. The effect measure is plotted on the vertical axis and a measure of sample size on the horizontal axis. In the absence of publication bias, the graph should—because of sampling variability—have the shape of a funnel with the wide opening to the left. The tip should point to the right and center on the true effect size. The funnel should be horizontal. The large studies and the small ones should be symmetrically distributed around the true effect size. If there is bias against the publication of small studies with null results or results that are unexpected, the wide part of the funnel will be distorted. For more discussion, see Petitti (1999) or Swales (2000).

Some analysts assess publication bias by estimating the number of imaginary zero-effect trials that would be needed to change the results from significant to nonsignificant. If the number is large, that is evidence against publication bias. However, this "file-drawer" approach assumes that the missing estimates are centered on zero and ignores the possibility that smaller studies with contrarian findings—significant or insignificant—are the ones that have been withheld from publication. See Rosenthal (1979), Oakes (1990), Iyengar and Greenhouse (1988), or Petitti (1999). The funnel plot seems preferable.

After a systematic review of nonpharmacologic interventions to lower blood pressure, including salt reduction, Ebrahim and Davey-Smith (1998, pp. 441, 444) find the evidence to be "surprisingly inadequate," in part because "the majority of RCTs were of low methodological quality and bias often tended to increase the changes observed." Swales (2000) makes a similar point with respect to non-randomized studies which suggest large effects and are frequently cited. For additional discussion of meta-analysis in the medical context see, for instance, Shapiro (1994) or Bailar (1997, 1999).

9.9 DASH—Dietary Approaches to Stop Hypertension

DASH-1 assessed the effect on blood pressure of three diets: a control diet, a fruit-and-vegetables diet, and a combination diet. The latter was rich in fruit and vegetables, dairy products, whole grains, with limited amounts of fish, poultry, and meat. All three diets had the same moderate salt levels, 3 grams per day of sodium. The DASH-1 combination diet achieved quite striking reductions in blood pressure among hypertensive subjects (11.4 mm Hg systolic, 5.5 diastolic, relative to the control diet). See Harsha et al. (1999), Moore et al. (1999), or Appel et al. (1997).

The DASH-2 trial has a factorial design with two diets and three levels of daily sodium: 3.3 grams, 2.4 grams, and 1.5 grams. The control diet is meant to resemble what typical Americans eat; the other diet is like the DASH-1 combination diet: compare Svetkey et al. (1999). Before publication of study results, the investigators issued a press release on May 17, 2000 (http://www.nhlbi.nih.gov/new/press/may17-00.htm). The impact of salt reduction was emphasized—

NHLBI Study Shows Large Blood Pressure Benefit From Reduced Dietary Sodium

The lower the amount of sodium in the diet, the lower the blood pressure, for both those with and without hypertension, according to a National Heart, Lung, and Blood Institute (NHLBI)-supported clinical study.

But diet has a considerable impact too, and there are interactions (Sacks et al. 2001, Figure 9.1). For normotensives on the DASH diet, according to charts presented at scientific meetings, cutting salt in half reduces blood pressure only by 1 or 2 mm—an effect which does not reach statistical significance, and is minor at best. The charts do not appear in the published article (compare Sacks et al. 2001, Figure 9.2; also see Taubes 2000). The published article contends that the "results should be applicable to most people in the United States," although the study population was chosen from groups that are relatively sensitive to changes in salt intake: high blood pressure at baseline, 134/86 compared to an age-adjusted U.S. average of 122/76; overweight, 85 kg compared to 77 kg; 56% African-American, compared to 12% (Sacks et al. 2001, p. 8, Tables 9.1 and 9.2; NHANES III). Such complications have so far been ignored. Further comment must await publication of more details on the experiment and the statistical analysis.

9.10 Health effects of salt

In essence, the Intersalt investigators argue that substantially reducing salt intake will make a small reduction in blood pressure. Other epidemiologic evidence suggests that lowering blood pressure by small amounts in normotensive populations reduces the risk of heart attack and stroke. However, even if both propositions are accepted, the link between salt and risk remains to be established. See, for instance, the exchange between Psaty et al. (1999) and Temple (1999) on the general usefulness of surrogate endpoints.

There is a huge literature on the health effects of salt; some of the more recent and salient papers will now be mentioned. Smith, Crombie,

Tavendale et al. (1988) ran a large observational study in Scotland (7354 men and women age forty to fifty-nine), and found no effect of salt on blood pressure after adjusting for potassium intake. He et al. (1999) find adverse health effects from high salt intake for overweight persons. However, for persons of normal weight, there is no association between health risks and salt intake. Data are from long-term followup of subjects in NHANES I, and salt intake was measured by dietary questionnaire. Of course, with better measures of salt intake, the study might have turned out differently. In other observational studies, Alderman et al. (1991, 1995) find risks in salt reduction; Kumanyika and Cutler (1997) disagree. Also see Graudal et al. (1998) on health risks from salt reduction. Resnick (1999) stresses the role of calcium; also see McCarron and Reusser (1999).

Port et al. (2000) discuss nonlinearities in risk due to blood pressure. Their reanalysis of the Framingham data suggests that risk rises more slowly with increasing blood pressure among normotensives and more rapidly among hypertensives. The U.S. Preventive Services Task Force (1996, p. 625) finds "There is insufficient evidence to recommend for or against counseling the general population to reduce dietary sodium intake ... to improve health outcomes, but recommendations to reduce sodium intake may be made on other grounds." Taubes (1998) has a scathing review of the salt epidemiology.

To determine the effect of salt reduction or dietary interventions on mortality or morbidity, large-scale long-term intervention studies would be needed, and diet seems more promising. The DASH trials had a two- or three-month study period, with several hundred subjects, which is adequate only for assessing effects on surrogate endpoints like blood pressure or chemistry. Also see Graudal et al. (1998, p. 1389), Ebrahim and Davey-Smith (1998, p. 4).

9.11 Back to Intersalt

Hanneman (1996) notes the paradox in the Intersalt data, by estimating the blood pressure of infants. Law (1996) and Stamler et al. (1996) find this argument "bizarre" and think "it is incorrect" to extrapolate beyond the ages in the study (the present analysis uses age twenty). The latter authors call attention to the large range in average blood pressures across centers for subjects age fifty to fifty-nine. The range may be large, but its relevance is obscure. More to the point, predicted blood pressures at age sixty show no relationship to salt levels, when the four outliers in the data are excluded ($n = 48, r = .04$ systolic, $r = -.10$ diastolic). If high salt intake leads to high blood pressure at old age, the correlations should be strongly positive. On the other hand, if the data are nonlinear and predictions from regression models are not trustworthy, the investigators should not be using regressions to generate summary statistics, or drawing biological conclusions from model parameters.

The difficulties in correcting for measurement error are discussed by Smith and Phillips (1996), with a response by Dyer et al. (1996). MacGregor and Sever (1996) defend Intersalt by reference to other data, but this begs a salient question: Do the Intersalt data speak for or against the salt hypothesis? The Intersalt investigators have declined to make the underlying data public, "because of the need to preserve the independence of scientific investigation, the integrity of the data, and the confidentiality of information ..." (Elliott et al. 1996, p. 1249). We cannot see how releasing data threatens integrity or compromises scientific independence; reversing these propositions makes them more plausible. Moreover, data can be released without identifying subjects, so confidentiality need not be an issue.

Our review of the literature is no doubt incomplete in various respects, but it is sufficient to provide context for questions about the Intersalt data.

9.12 The salt epidemiologists respond

The National Heart Lung and Blood Institute convened a workshop to address criticisms of the salt hypothesis, as in Taubes (1998). However, these criticisms are barely acknowledged in the official report on the workshop (Chobanian and Hill 2000), according to which

> [S]tudies show unequivocally that lowering high blood pressure can reduce the likelihood of developing or dying from CVD [cardiovascular disease]. Second, dietary factors in individuals and in the population at large have important effects on blood pressure levels, which are generally assumed to translate to CVD risk .... An abundance of scientific evidence indicates that higher sodium consumption is associated with higher levels of blood pressure. This evidence is found in animal studies, observational epidemiologic studies, and clinical studies and trials.

> The INTERSALT findings support similar studies that show a relationship between sodium intake and blood pressure. The discussion relative to INTERSALT emphasized that its strengths are its large sample size and sophisticated statistical analyses ... it was noted that difficult statistical issues are involved in the interpretation of the INTERSALT data.

If this is the concession, it is too subtle. And the language is hauntingly similar to Stamler's (1997, p. 626S) defense of his study:

> The INTERSALT results, which agree with findings from other diverse studies, including data from clinical observations, therapeutic interventions, randomized controlled trials, animal experiments, physiologic investigations, evolutionary biology research, anthropologic research, and epidemiologic studies, support the judgment that habitual high salt intake is one of the quantitatively important preventable mass exposures causing the unfavorable population-wide blood pressure pattern that is a major risk factor for epidemic cardiovascular disease.

Next, we quote from the editors of the *British Medical Journal*. The sentiments seem eminently reasonable to many proponents of the salt hypothesis. Persons not in the fold may react differently.

> Like any group with vested interests, the food industry resists regulation. Faced with a growing scientific consensus that salt increases blood pressure and the fact that most dietary salt (65–85%) comes from processed foods, some of the world's major food manufacturers have adopted desperate measures to try to stop governments from recommending salt reduction. Rather than reformulate their products, manufacturers have lobbied governments, refused to cooperate with expert working parties, encouraged misinformation campaigns, and tried to discredit the evidence. (Godlee 1996, p. 1239)

Drafts of our critique have been circulated in the community of salt epidemiologists. Reactions can be paraphrased as follows.

- The regression of blood pressure on age within center doesn't indicate how rapidly blood pressure increases with age because the data aren't longitudinal. [Fair enough, but then what were the Intersalt people doing?]

- Epidemiologists can never wait for final proof. Instead, recommendations must be made in the interest of promoting good health for the public.

- The effect of salt reduction may be detectable only in hypertensives, but today's normotensives are tomorrow's hypertensives.

- Public health guidelines to reduce sodium consumption from three grams to one gram will hurt no one and may benefit thousands.

- Access to data can distort, confuse, intimidate, and muddy the waters of medical care and public health.

In summary, the public must be protected from salt, from the machinations of industry, and above all from the data.

9.13 Policy implications

One segment of the public health community—funded by the National Heart Lung and Blood Institute and endorsed by many journals in the field—has decided that salt is a public health menace. Therefore, salt consumption must be drastically curtailed. The force with which this conclusion is presented to the public is not in any reasonable balance with the strength of the evidence. Programs, once in place, develop a life of their own; the possibility of health benefits becomes probability, and probability becomes certainty. After all, the public is easily confused by complications, only professionals can weigh the evidence, and where is the harm in salt reduction?

The harm is to public discourse. The appearance of scientific unanimity is a powerful political tool, especially when the evidence is weak. Dissent becomes a threat, which must be marginalized. If funding agencies and journals are unwilling to brook opposition, rational discussion is curtailed. There soon comes about the pretense of national policy based on scientific inquiry—without the substance. In our view, salt is only one example of this phenomenon.

Acknowledgments

We thank Jamie Robins for help that borders on collaboration.

10

The Swine Flu Vaccine and Guillain-Barré Syndrome: A Case Study in Relative Risk and Specific Causation

With Philip B. Stark

ABSTRACT. *Epidemiologic methods were developed to prove general causation: identifying exposures that increase the risk of particular diseases. Courts often are more interested in specific causation: On balance of probabilities, was the plaintiff's disease caused by exposure to the agent in question? Some authorities have suggested that a relative risk greater than 2.0 meets the standard of proof for specific causation. Such a definite criterion is appealing, but there are difficulties. Bias and confounding are familiar problems; and individual differences must also be considered. The issues are explored in the context of the swine flu vaccine and Guillain-Barré syndrome. The conclusion: There is a considerable gap between relative risks and proof of specific causation.*

10.1 Introduction

This article discusses the role of epidemiologic evidence in toxic tort cases, especially, relative risk: Does a relative risk above 2.0 show specific causation? Relative risk compares groups in an epidemiologic study: One group is exposed to some hazard—like a toxic substance; the other "control" group is not exposed. For present purposes, relative risk is the ratio

$$RR = Observed/Expected.$$

The numerator in this fraction is the number of injuries observed in the exposed group. The expected number in the denominator is computed on the theory that exposure has no effect, so that injury rates in the exposed group should be the same as injury rates in the control group. Adjustments are often made to account for known differences between the two groups, for instance, in the distribution of ages.

The basic intuition connecting relative risk and probability of causation can be explained as follows. Suppose that the exposed and unexposed groups in an epidemiologic study are similar except for the exposure of interest so that confounding is not an issue. For simplicity, suppose also that the two groups are the same size. To have specific numbers, there are 400 injuries among the exposed and only 100 among the unexposed. In other words, the observed number of injuries is 400, compared to an expected 100—the two groups being comparable by assumption. The relative risk is 400/100 = 4.

The implication: But for exposure, there would be only 100 injuries among the exposed instead of 400, so 300 of the 400 injuries are attributable to the exposure and 100 to other factors. Apparently, then, each injury among the exposed has chance 3/4 of being attributable to exposure. (That is the point to watch.) Likewise, a relative risk of three corresponds to a chance of 2/3, while a relative risk of two corresponds to a chance of 1/2, which is the breakpoint.[1]

The object here is to explore the scientific logic behind these intuitions. Of course, any epidemiologic study is likely to have problems of bias: Uncontrolled confounding appears to be the rule rather than the exception.[2] When effects are large, such problems may not be material; when relative risk is near the critical value of 2.0, potential biases need to be assessed more carefully.

Individual differences also play an important role: Plaintiff may not resemble typical members of the study population, and the effects of such differences need to be considered. This is a salient difficulty in connecting relative risk to specific causation. With a randomized controlled experiment, for instance, treatment and control groups are balanced in the aggregate but not at the level of individuals. Thus, even with the best research designs—where general causation is easily demonstrated—specific causation remains troublesome.

We wanted to consider such issues in the context of a real example, in part to see how well the courtroom evidence stands up when examined retrospectively. Mike Green kindly provided a list of legal opinions where relative risk and specific causation come together.[3] Generally, the evidence of harm was shaky. In one case—*Manko v. United States*[4]—there turned out to be a substantial body of epidemiologic evidence, showing that the swine flu vaccine caused Guillain-Barré syndrome. And the vaccine campaign of 1976 is itself a fascinating case study.

Guillain-Barré syndrome (GBS) is a rare neurological disorder. GBS is sometimes triggered by vaccination or by infection. Paralysis is a sequel, although most patients make a complete recovery in a few weeks or months. The epidemiology of swine flu vaccine and GBS will be summarized below. Then *Manko* will be discussed as well as the use of relative risk to demonstrate specific causation. Although the plaintiff prevailed, his proof of specific causation seems questionable, due in part to differences between him and typical members of the study population.

There is a simple probability model where intuitions about relative risk and causation can be analyzed. The model sets aside all problems of confounding and bias, and considers only difficulties created by individual differences. For any particular plaintiff, the probability of causation is not identifiable from the data. Even the average probability of causation can be much lower than intuition suggests. For instance, if 4% of the exposed group suffers injury compared to 1% among the unexposed, the relative risk is four and the probability of causation would seem to be 3/4; but the average probability of causation in the model can be as low as 3%, the difference in injury rates.[5]

10.2 The swine flu vaccine and GBS

This section reviews the swine flu vaccination campaign of 1976 and the epidemiology of Guillain-Barré syndrome as background for the discussion of *Manko*. The story begins in 1918, with an influenza pandemic that killed some twenty million people worldwide. In February of 1976, a soldier in training at Fort Dix, New Jersey, died of influenza; the virus turned out to be similar in antigenic type to the 1918 virus. With public health professionals at the CDC (Centers for Disease Control and Prevention) taking the lead, the Federal Government organized a massive immunization campaign. Vaccination began on October 1, 1976. The vaccine was targeted at the 151 million people age eighteen and over; some forty-three million were eventually vaccinated. However, beyond the initial cluster at Fort Dix, only a handful of additional cases materialized, and several public health figures wanted the campaign stopped.

A moratorium was declared on December 16, 1976—in part because an epidemic seemed increasingly unlikely, and in part because there were sporadic reports of GBS cases following vaccination.[6]

The CDC set up a nationwide surveillance system to collect case reports on GBS from state health authorities, who in turn worked with local authorities, hospitals, and doctors. Using these data, Langmuir et al.[7] analyzed the incidence rate of GBS among the vaccinated by weeks since vaccination; this rate is shown as the highly peaked solid line in panel (a) of Figure 10.1.[8] Rates are "per million person-weeks" of observation; these are incidence rates, not relative risks. (Ten persons followed for one week count as ten person-weeks of observation; so does one person followed for ten weeks; the incidence rate is the number of new cases during a week, divided by the number of persons observed that week.)

Shown for comparison is the "background rate": The incidence rate of GBS among the unvaccinated by calendar week from October 1st (lower broken line, also computed from Langmuir et al.'s data). Notice that two time scales are involved: weeks from vaccination to onset for the vaccinated group, and weeks from start of program to onset for the unvaccinated. The sizes of the vaccinated and unvaccinated populations are changing rapidly over time due to the vaccination campaign; size is taken into account in computing the rates through adjustments to the denominator—the number of person-weeks of observation.[9]

Panel (a) in Figure 10.1 shows that for some weeks after vaccination, the incidence rate of GBS rises sharply, becoming much larger than the background rate; later, there is a reversion to background levels. In other words, there is a clear association between vaccination and GBS, provided the onset of GBS is within a few weeks of vaccination.

Is the association causal? That is still controversial. No excess risk for GBS was observed in the military or with previous vaccines much like the swine flu vaccine. Further arguments and counter-arguments will not be discussed here.[10] After reviewing the data and the literature, we think that a finding of general causation is reasonable: On balance of evidence, the swine flu vaccine could well have increased the risk for GBS for a period of several weeks after vaccination.

The background rate in Figure 10.1a is shown on a magnified scale in panel (b). After the moratorium, there is a precipitous drop in the "raw" (i.e., reported) background rate. This drop is best explained as an artifact of data collection. After the moratorium, it seems probable that GBS was less in the news, neurologists were less likely to make the diagnosis among unvaccinated persons, and state health departments were less diligent in collecting the data and reporting to CDC.[11]

Figure 10.1 Panel (a) shows the incidence rate among the vaccinated by week since vaccination (highly peaked solid line). This rate is compared to the background rate (lower broken line) among the unvaccinated by week since the start of vaccination campaign. Two time scales are involved. The moratorium occurred in the eleventh week after the start of the campaign, indicated by a vertical line.

Panel (b) shows the background rate in more detail, both truncated and raw. (The "truncated" background rate is prevented from falling below a lower bound of 0.24 cases per million person-weeks.)

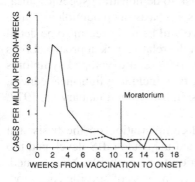

(A) INCIDENCE OF GBS AMONG
VACCINATED (HIGHLY PEAKED SOLID LINE)
& UNVACCINATED (LOWER BROKEN LINE).

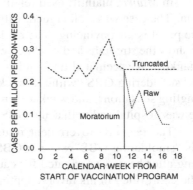

(B) BACKGROUND:
INCIDENCE OF GBS AMONG UNVACCINATED.
VERTICAL SCALE IS MAGNIFIED TENFOLD.

The background rate of GBS (among unvaccinated persons) is a critical baseline statistic: The incidence rate of GBS among the vaccinated persons is compared to this baseline in Figure 10.1 and in computations of relative risk. GBS is not a reportable disease nor is the diagnosis easy. Thus, considerable uncertainty attaches to the background rate. Langmuir et al. did not believe the background could be below 0.24 per million person-weeks.[12]

Following their lead, Figure 10.1a takes the background rate as 0.24 after the moratorium ("truncation"): The lower broken line is horizontal after week eleven. Current literature suggests a background rate of 0.2 to 0.4 per million person-weeks, with only minor seasonal variation—confirming the estimates of Langmuir et al.[13]

Another feature of the data analysis in Langmuir et al.[14] will be relevant. They distinguished between cases with extensive and limited paralysis. The association was strong for the extensive cases, but there was little evidence of association for the limited cases.[15] A change in the legal situation should also be noted. Before the 1976 swine flu campaign got under way, the insurance companies refused to issue coverage for adverse events resulting from vaccination, and the drug companies refused to produce the vaccine without coverage.

To resolve this impasse, the Federal Government accepted liability.[16] Thus, GBS victims applied for compensation not to the vaccine providers but to the Federal Government.[17] There were roughly 500 GBS victims among the vaccinated and a similar number among the unvaccinated. About 4000 claims were filed against the Federal Government as a result of the swine flu campaign, alleging $4 billion in damages.[18] One of the claims—*Manko*—is the topic of the next section.

10.3 The Manko case

In *Manko*, plaintiff used relative risk to demonstrate specific causation. The case was well argued, with a solid basis in epidemiology. Still, the proof is unconvincing. The evidence will be reviewed in some detail to show the strengths and weaknesses of the relative-risk approach. Louis Manko was vaccinated on October 20, 1976, and developed symptoms of "smoldering GBS" within a week or two, including light-headedness, tingling sensations, and weakness in his limbs. Around January 15, 1977, he was hospitalized with acute GBS.

The Federal Government refused compensation on the basis that his "smoldering GBS" was not GBS, and his acute GBS developed too long after he was vaccinated for causation to be probable. Manko sued. The court ruled in his favor, adopting two theories of specific causation. (i) If "smoldering GBS" is indeed GBS, then causation follows from the epidemiologic evidence. (ii) If on the other hand plaintiff contracted GBS in mid-January of 1977, some thirteen weeks after vaccination, specific causation still follows because the relative risk for such late-onset cases is well above the threshold value of 2.0.

The arguments on causation for late-onset cases[19] are the most interesting. Plaintiff introduced expert testimony from Nathan Mantel and Martin Goldfield. Mantel was a well-known biostatistician at the National Institutes of Health. Goldfield was the county medical officer who worked on the Fort Dix outbreak. He was one of the first to identify the disease as influenza and one of the first to advise against mass vaccination. Defendants' epidemiology experts were Leonard Kurland of the Mayo Clinic and Neal Nathanson of the Pennsylvania Medical School. They were co-authors of the "Langmuir report."[20]

Panel (a) in Figure 10.1—essentially the case for the defense on late-onset GBS cases—shows only a small excess risk after the eighth week. However, Goldfield and Mantel argued that in order to compare like with like, it was necessary to "stratify" on time of vaccination and time since vaccination when computing relative risks. (Stratification will be explained below.) The rationale was ingenious. They hypothesized a decrease in reporting of vaccinated GBS cases parallel to the decline in reporting of the unvaccinated cases.

As discussed in Section 10.1, relative risk compares the observed number of GBS cases with the number expected on the theory that vaccination does not cause GBS:

$$RR = Observed/Expected.$$

Goldfield and Mantel computed the expected numbers for each week from vaccination to onset, separately for each vaccination cohort—those vaccinated in week one, those vaccinated in week two, and so forth. Finally, they summed the contributions from the various cohorts to get the expected number of cases in each week after vaccination.[21] In effect, this synchronizes the two time scales in Figure 10.1.

Goldfield and Mantel used the raw (untruncated) background rates to compute the relative risk, as in Figure 10.2a. Late-onset cases are now being compared to the very small number of background cases reported after the moratorium, and the relative risk is large.[22] For comparison,

Figure 10.2 Relative risk for GBS among the vaccinated, plotted by time since vaccination. Panel (a) shows the Goldfield-Mantel analysis with stratification by time of vaccination as well as time since vaccination; raw background rates are used. Panel (b) stratifies the same way, but background rates below 0.24 per million person-weeks are replaced by 0.24 (truncation). The short horizontal line pools the data in weeks eleven to sixteen to stabilize the estimates.

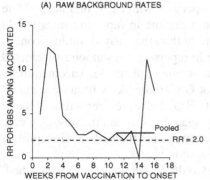
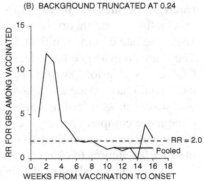

(A) RAW BACKGROUND RATES

(B) BACKGROUND TRUNCATED AT 0.24

panel (b) in Figure 10.2 shows relative risks computed by the Goldfield-Mantel procedure, stratifying both on time of vaccination and time since vaccination, but with background rates truncated at 0.24 per million person-weeks of exposure.

The threshold relative risk of 2.0 is marked by dashed horizontal lines. There were no cases in the fourteenth week after vaccination, only four in the fifteenth week, and one in the sixteenth week. The tail of the curve is quite shaky, so plaintiff's experts pooled the data for weeks eleven to sixteen as indicated by the solid horizontal lines in both panels.

Both panels in Figure 10.2 use the same observed numbers and compute expected numbers the same way—except for truncation. The issue is not stratification but truncation. The crucial question: Was there a drop in reporting of vaccinated GBS cases after the moratorium, parallel to the drop in background rates? If so, Figure 10.2a is persuasive and the relative risk for late-onset cases is well above 2.0. If not, panel (b) is the one to use and excess risk is minimal.

10.3.1 Completeness of reporting

Both sides in *Manko* agreed that the drop in background rates was artifactual.[23] The issue was the plaintiff's hypothesis of a parallel drop in reporting of vaccinated cases. To validate that hypothesis, Goldfield and Mantel[24] compared the incidence rate of GBS among the vaccinated before and after the decline in background rates. However, the numbers are small. Furthermore, a real decline in the incidence rate is only to be expected, because the attack rate decreases with time since vaccination (Figure 10.1), and most vaccinations occurred fairly early in the sequence of events. Thus, it is not easy to demonstrate a decline in reported incidence rates over and above the expected real decline, although there may be something to the idea.[25]

To address the completeness of reporting, Langmuir et al.[26] compared attack rates for three cohorts—persons with early, middle, and late vaccinations—the theory being that a decline in reporting rates would affect the late cohort significantly more than the early or middle cohort. They saw no evidence for a decline in reporting rates among vaccinated GBS cases. A priori, such a decline seems implausible. Vaccination by itself could have made a diagnosis of GBS more likely because vaccination was seen as a leading cause of GBS. Moreover, reporting is likely to be more complete among the vaccinated cases than the unvaccinated: Vaccinated cases generally had to be reported to the Federal Government in order for victims to claim compensation.[27]

For an empirical test, following Goldfield and Mantel, we "smoothed" the relative risks in Figure 10.2b to make the curve decline more slowly and regularly after the first four weeks. The objective was to reduce the impact of chance fluctuations and potential misreporting. The smoothed curve was then used to estimate the likely number of post-moratorium GBS cases among the vaccinated.

Although specific results depend on the smoothing, the reporting of vaccinated GBS cases seems to have dropped after the moratorium by no more than 20%; there also seems to have been overreporting for a couple of weeks just prior to the moratorium—which suggests that onset dates were advanced by a week or two in the CDC's database around that time.[28] If the relative risk for late-onset GBS in Figure 10.2b is biased downward, the effect is small. Current medical literature does not support the hypothesis of swine flu vaccination as a cause of late-onset GBS.[29]

10.3.2 Discovery issues

In pre-trial discovery proceedings, the Federal Government declined to produce the CDC's detailed medical records on GBS victims.[30] For some of these cases, critical information on the date of vaccination or the date of onset of GBS was missing in the summary sheets that were made public and used both by plaintiffs and defense. To resolve this discovery issue, the court imposed an information sanction. Langmuir et al.[31] had excluded from their analyses some twenty-eight cases with missing dates. Plaintiff's experts were allowed to count eight of these cases as having late onset.[32]

Table 10.1 shows the relative risk for GBS with onset eleven to sixteen weeks after vaccination, computed on various sets of assumptions; the Goldfield-Mantel stratification procedure is used to compute all the expected values and relative risks in the table. When background rates are truncated, stratification and discovery sanctions only bring the relative risk up to 1.66. As the table confirms, stratification is a sideshow; the critical issue is the truncation used to correct for incomplete reporting.[33]

The table also shows that relative risk depends on severity of illness: With late-onset GBS, paradoxically, relative risk is lower for severe cases. In general, "the" relative risk in an epidemiologic study is the average of relative risks for various subgroups. Differences are only to be expected, and that is the topic of the next section.

10.3.3 Individual differences

Individual differences are the next topic. Prior infection is a risk factor for GBS: About 62% of the unvaccinated GBS cases had some illness in the month before onset. For the vaccinated cases, only 33% had prior illness.[34] A somewhat informal calculation[35] suggests that prior ill-

Table 10.1 Relative risks for GBS cases, with onsets in weeks eleven to sixteen after vaccination. RR = Observed/Expected. The first column computes the "Expected" using the raw background rates; the second column truncates the background rate at 0.24 per million person-weeks. Row 1 shows data for cases with extensive paralysis; row 2, for all cases; row 3 adds eight cases to the numerator, as a consequence of sanctions imposed by the court on defendants. The Goldfield-Mantel stratification procedure is used throughout.

	Raw	Truncated
Extensive cases	$9/4.41 = 2.04$	$9/10.2 = 0.88$
All cases	$21/7.40 = 2.84$	$21/17.5 = 1.20$
Sanctions	$29/7.40 = 3.92$	$29/17.5 = 1.66$

ness multiplies the relative risk by about $33\%/62\% \doteq 0.53$. Manko had an infection with respiratory and gastrointestinal symptoms a week or two before his hospitalization for acute GBS,[36] and multiplying the relative risk of 3.92 by 0.53 brings it very close to the critical value of 2.0.

Goldfield and Mantel argued, however, that the 0.53 includes a selection effect because people are advised against vaccination immediately following illness. To avoid the selection effect, Goldfield and Mantel based the numerator of their correction only on the late-onset GBS cases among vaccinated persons, where 53% were preceded by illness;[37] the relative risk should now be multiplied by $53\%/62\% \doteq 0.85$.[38]

The number of late-onset cases is rather small (Table 10.1), and the experience of this group should probably not be compared to all unvaccinated cases but only to cases with onsets in a similar time period, namely late December and early January: The pattern of respiratory infections, for example, is seasonal (by contrast with the pattern of background GBS). Plaintiff's argument is therefore not wholly convincing. Current literature confirms that about 2/3 of GBS cases are triggered by previous illness.[39] With respect to one pathogen—*Campylobacter jejuni*, which causes gastrointestinal symptoms—the molecular basis for subsequent GBS is now reasonably well understood.[40]

Age is another factor to consider. Manko was sixty-four years old at vaccination.[41] That would reduce the relative risk by perhaps 25%, if it is fair to average across onset times.[42] Finally, the clinical course of the disease should be mentioned. About 95% of patients reach their nadir within a month of onset, and roughly 70% recover completely within a year.[43] In this respect too, Manko was quite unlike the bulk of the GBS victims,[44] so data about them may not help very much in deciding the cause of his injury. (These arguments apply as well to smoldering GBS, although the issues in court turned more on the medical definitions.)

   Manko was a well-argued case with a solid empirical base, thoroughly reported in the epidemiologic literature. Even so, the proof of specific causation—starting from a relative risk of four—seems unconvincing.[45] That gives us pause, and the issue goes well beyond *Manko*.

10.4 Summary and conclusions

   The scientific connection between specific causation and a relative risk of 2.0 is doubtful. If the relative risk is near 2.0, problems of bias and confounding in the underlying epidemiologic studies may be serious, perhaps intractable. Problems created by individual differences may be equally difficult. Bias and confounding affect the estimation of relative risk from the underlying data. By contrast, individual differences affect the interpretation of relative risk—namely, the application to any specific individual.

   With *Manko*, at least in retrospect, it is difficult to establish an elevated relative risk for late-onset cases. Moreover, the plaintiff is in crucial detail remarkably unlike the other GBS victims. So the connection between him and the data stays rather loose. Mathematical models show how the effect of individual differences can be represented in a more general—but more abstract—setting.[46] The results confirm one of the central points about *Manko*: Epidemiologic data usually cannot determine the probability of causation in any meaningful way, because of individual differences.

Notes

1. For previous discussions from various perspectives, see the American Medical Association (1987); Black and Lilienfeld (1984); Green, Freedman, and Gordis (2000); Hart and Honoré (1985), especially p. 104 on the idea of "but-for" causation; Kaye and Freedman (2000), note 38; Petitti (1996); and Robins and Greenland (1989).

   Also see *Cimino v. Raymark*, 151 F.3d 297, 301–02 (5th Cir. 1998); In re *Fibreboard*, 893 F.2d 706 (5th Cir. 1990). "It is evident that these statistical estimates deal only with general causation, for population-based probability estimates do not speak to a probability of causation in any one case; the estimate of relative risk is a property of the studied population, not of an individual's case. This type of procedure does not allow proof that a particular defendant's asbestos really caused a particular plaintiff's disease; the only fact that can be proved is that in most cases the defendant's asbestos would have been the cause." *Id.* at p. 712; footnotes, citations, italics, and internal quote marks omitted.

2. Confounding means that the exposed and unexposed groups in a study differ systematically on factors related to the probability of injury. Confounding leads to bias in estimated relative risk, when the calculation of the expected number of injuries among the exposed fails to reckon with systematic differences: For instance, measurements may not be available on some important confounder, or the impact on risks may be underestimated. For discussion and citations to the literature, see Freedman (1999) and Kaye and Freedman (2000), section IIA.

3. Personal communication. Also see Green et al. (2000), at note 140, and Kaye and Freedman (2000), at note 38.

4. 636 F.Supp. 1419 (W.D. Mo. 1986), aff'd in part, 830 F.2d 831 (8th Cir. 1987). In other cases with fact patterns similar to *Manko*, the defendant prevailed: see, e.g., *In re Swine Flu Immunization Products Liability Litigation, Alvarez v. United States*, 495 F.Supp. 1188 (D. Co. 1980) and *Lima v. United States*, 508 F.Supp. 897 (D. Co. 1981). There is a useful summary of the medical and legal background in *Alvarez*, at pp. 1190–91, 1194–96.

5. The intuitive arguments for probability of causation in effect assume uniformity of risk across people or random selection of persons to consider. By contrast, our probability model views some people as more susceptible to injury, others less; each individual has his or her own specific probability of causation. The average of these individualized probabilities is small when most of the injuries due to exposure are likely to occur in a relatively small subgroup of the exposed population. Furthermore, the probability of bringing suit may vary with susceptibility to injury, a relationship which is also considered in the model. The present article is adapted from Freedman and Stark (1999); the probability model is developed in an appendix to that paper.

6. There are two different accounts of the vaccine campaign, Neustadt and Fineberg (1981) and Silverstein (1981). The latter was written to correct the former; but there is broad agreement on the central points. Also see Kolata (1999).

7. Langmuir, Bregman, Kurland, Nathanson, and Victor (1984) (the "Langmuir report"). Also see Langmuir (1979) and Schonberger, Bregman, Sullivan-Bolyai, Keenlyside, et al. (1979). Langmuir was the founder of the Epidemic Intelligence Service at the CDC.

8. Rates were computed by us from their data; Freedman and Stark (1999), appendix B.

9. Langmuir et al. (1984) had data covering October 2, 1976 to January 31, 1977, sub-divided into seventeen "periods": A period is generally a week, but period one is nine days long and period seventeen is eight days. The length of the period is taken into account when computing person-weeks of observation. The horizontal scale for the background rate in Figure 10.1b should really be labeled "calendar period" rather than "calendar week." The number of GBS cases, vaccinated and unvaccinated, was determined through the CDC's surveillance program.

10. See, for instance, Hahn (1998). Also see Beghi, Kurland, Mulder, and Wiederholt (1985); Hughes (1990); Hughes and Rees (1997); Kurland, Wiederholt, Kirkpatrick, Potter, and Armstrong (1985); Ropper, Wijdicks, and Truax (1991); and Safranek, Lawrence, Kurland, Culver, et al. (1991). Hughes and Rees find the evidence less ambiguous than do other authors. In subsequent mass vaccinations excess risk is minimal although statistical significance is achieved if data for 1992–93 and 1993–94 are pooled. See Lasky, Terracciano, Magder, Koski, et al. (1998). Also see Hurwitz, Schonberger, Nelson, and Holman (1981) and Kaplan, Katona, Hurwitz, and Schonberger (1982).

If the hypothesis of causation is rejected, the patterns in Figures 10.1 and 10.2 are explicable as statistical artifacts. GBS is not easily distinguished from certain other neurological conditions: The publicity about swine flu and GBS could increase the reporting rate; there would be more of a tendency for ambiguous cases following vaccination to be classified as GBS, by comparison with similar cases among the unvaccinated.

11. Larry Schonberger, who was doing surveillance at the CDC, reports that a number of states put significantly less effort into data collection after the moratorium (personal communication); also see Schonberger et al. (1979), at p. 197. Some of the drop may also be due to increasing delays in reporting cases to the CDC.

12. Langmuir et al. (1984), at pp. 856–59. In the classification used by Langmuir et al. (1984), cases of type A and B have "extensive" paralysis, type C and D are "limited," while type E means "insufficient information." Langmuir et al. give lower bounds of 0.14 and 0.07 for cases with extensive and limited paralysis; we have added 0.03 for cases with insufficient information, computed from data in their table 8. See Freedman and Stark (1999), appendix B, for more detail on background rates and the calculations in Figures 10.1 and 10.2.

13. See Hahn (1998), at p. 635; Hughes (1990), at p. 101; and Ropper et al. (1991), at p. 19. But also see Lasky et al. (1998), who found a rate of about 0.15 per million person-weeks. A rate of 0.24 per million person-weeks translates to one case per 100,000 persons per year, approximately; both scales are used in the literature. Certain forms of GBS, rare in North America but prevalent elsewhere, do show seasonal variation.

14. Langmuir et al. (1984).

15. Here we have to differ with Langmuir et al. (1984). On our reckoning, the relative risk for cases of type C-D-E is about seven in the first week after vaccination, with a fairly smooth decline to background levels by the eleventh week. See *supra* note 12 for the classification.

16. The National Influenza Immunization Program of 1976 (P.L. 94-380) and the Swine Flu Act (42 U.S.C. §247b) provide that claims are brought under the Federal Tort Claims Act (28 U.S.C. §§1346(b), 2671, *et seq.*).

17. Current legal procedures for handling vaccine-related injuries are discussed in Johnson, Drew, and Miletich (1998).

18. Langmuir et al. (1984), at p. 842; Silverstein (1981), at p. 127; and Nathanson and Alexander (1996). The total number of GBS cases was computed by us from data in Langmuir et al. (1984).

19. *Manko*, 636 F.Supp. 1433ff.

20. Langmuir et al. (1984).

21. For details, see appendix B to Freedman and Stark (1999). Separating the contributions from the various cohorts is an instance of what epidemiologists call "stratification." The observed number of cases is not affected by stratification, but the expected number is—because the background rates used in the calculation depend on time.

22. Since the raw background rate is low after the moratorium, the expected number of cases will be low in that time period, and the ratio of observed to expected (i.e., the relative risk) will be correspondingly high.

23. Goldfield, Tr. 6.44, for the plaintiff; Langmuir et al. (1984), at p. 856, for the defense. ("Tr. 6.44" is p. 44 of vol. 6 of the *Manko* trial transcript.) Nathanson states the issue quite clearly at Tr. 18.113–15. However, the court found "no significant decline in reporting cases of GBS" after the moratorium. *Manko*, 636 F.Supp. 1435.

24. Tr. 6.61–67, especially Tr. 6.66.

25. See figure 1 in Schonberger et al. (1979); also see Langmuir et al. (1984), table 5. See generally Retailliau, Curtis, Storr, Caesar, et al. (1980).

26. Langmuir et al. (1984), at pp. 860ff.

27. Also see Marks and Halpin (1980), at pp. 2490, 2493.

28. For discussion, see Freedman and Stark (1999), appendix B. Since most vaccinations occur fairly early and most GBS cases among the vaccinated occur soon after vaccination, as noted *supra*, the details of the smoothing do not have a major impact on results.

29. Hahn (1998), at p. 636; Hughes (1990), at p. 102; and Ropper et al. (1991), at pp. 28–29, 57.

30. The government took the position that there were binding nationwide discovery rules, which did not require production of the disputed records. The trial court disagreed, and the appeals court declined to review that issue. 830 F.2d 831, 834–35.

31. Langmuir et al. (1984), table 1. Langmuir et al. began with about 1300 case reports; 100 were excluded because onsets fell outside the study period (*Id.* at p. 843); another 100 were excluded for reasons that are not made clear (*Id.*). Reasons for other exclusions are detailed in the table, e.g., vaccination was recommended only for persons over the age eighteen, and 121 cases were below that age.

32. According to the sanction, plaintiff's experts were allowed to fill in the missing dates any way that did not contradict other information on the summary sheet. Certain other ambiguities could also be resolved in favor of plaintiff's statistical theories. See *Manko*, 636 F.Supp. 1438, 1453 on sanctions, and 1436–37 on the calculation of relative risk. We infer the figure of eight additional cases to reconcile the numbers in notes 10 and 11 of the opinion (*Id.*) with the data in Langmuir et al. (1984).

33. The numbers in the table are computed by us from data in Langmuir et al. (1984); for details, see appendix B in Freedman and Stark (1999). For the classification of cases by extent of paralysis, see *supra* note 12.

34. See Schonberger et al. (1979), at p. 116, and Langmuir (1979), at p. 663.

35. This calculation is like the one used by plaintiff's experts Goldfield and Mantel, and starts from Bayes' rule; it is reconstructed in Freedman and Stark (1999), appendix B.

36. Plaintiff's exhibit 401.

37. Tr. 7.39.

38. In *Manko*, 636 F.Supp. 1419, at note 12, the multiplier is given as 0.87. Different experts—even on the same side—seem to have been using slightly different versions of the CDC database. And there is an annoying numerical coincidence, as 0.53 crops up twice with two different meanings.

39. See, for instance, Hahn (1998), at p. 636; Hughes (1990), at p. 106; or Ropper et al. (1991), at p. 57.

40. See Nachamkin, Allos, and Ho (1998). Also see Asbury (2000) and Hughes, Hadden, Gregson, and Smith (1999). Among other things, these papers indicate that GBS comprises several different diseases, each with a characteristic etiology.

41. Plaintiff's exhibit 401, Tr. 16.193.

42. See Schonberger et al. (1979), at p. 114, and Lasky et al. (1998), table 1.

43. See Hahn (1998), at p. 639, and Hughes (1990), at pp. 122–23; compare *Manko* 636 F.Supp. 1427.

44. As noted in Section 10.3, Manko contracted a mild form of the illness within a week or two of vaccination; his condition gradually deteriorated, and acute illness struck three months later. Even at the time of trial—seven years after vaccination—he was severely incapacitated. *Manko* 636 F.Supp. 1429, 1441.

45. See *supra* Table 10.1, showing a relative risk of 3.92. The opinion quotes the relative risk as 3.89. *Manko* 636 F.Supp. 1437.

46. The impact of individual differences on the probability of specific causation is discussed analytically in appendix A, Freedman and Stark (1999).

Acknowledgments

We would like to thank the following persons for useful discussions: Michael Berger, Richard Berk, Joe Cecil, John Conley, Mike Finkelstein, Mike Green, Angelika Hahn, Paul Humphreys, Jamie Robins, and Larry Schonberger. Many of the participants in the case shared their knowledge with us, including some of the epidemiology experts (Leonard Kurland,

Nathan Mantel, and Neal Nathanson) and the lawyers who presented the epidemiologic evidence (Leslie Ohta and Charles Thomas). The Department of Justice provided surviving portions of the trial transcript. Part of this work was completed while Philip B. Stark was on appointment as a Miller Research Professor in the Miller Institute for Basic Research in Science.

11

Survival Analysis: An Epidemiological Hazard?

ABSTRACT. *Proportional-hazards models are frequently used to analyze data from randomized controlled trials. This is a mistake. Randomization does not justify the models, which are rarely informative. Simpler methods work better. This discussion is salient because the misuse of survival analysis has introduced a new hazard in epidemiology: It can lead to serious mistakes in medical treatment. Life tables, Kaplan-Meier curves, and proportional-hazards models, aka "Cox models," all require strong assumptions, such as stationarity of mortality and independence of competing risks. Where the assumptions fail, the methods also tend to fail. Justifying those assumptions is fraught with difficulty. This is illustrated with examples: the impact of religious feelings on survival and the efficacy of hormone replacement therapy. What are the implications for statistical practice? With observational studies, the models could help disentangle causal relations if the assumptions behind the models can be justified.*

In this chapter, I will discuss life tables and Kaplan-Meier estimators, which are similar to life tables. Then I turn to proportional-hazards models, aka "Cox models." Along the way, I will look at the efficacy of screening for lung cancer, the impact of negative religious feelings on survival, and the efficacy of hormone replacement therapy.

The American Statistician (2008) 62: 110–19. Copyright © 2008 by the American Statistical Association. Reprinted with permission. All rights reserved.

What are the conclusions about statistical practice? Proportional-hazards models are frequently used to analyze data from randomized controlled trials. This is a mistake. Randomization does not justify the models, which are rarely informative. Simpler analytic methods should be used first.

With observational studies, the models would help us disentangle causal relations *if* the assumptions behind the models could be justified. Justifying those assumptions, however, is fraught with difficulty.

11.1 Cross-sectional life tables

Cross-sectional life tables date back to John Graunt and Edmond Halley in the 17th century. There were further developments by Daniel Bernoulli in 1760, when he computed what life expectancy would be—if smallpox were eliminated. His calculations make a key assumption to be discussed later: the independence of competing risks.

Here is a simple discrete case to illustrate the idea behind cross-sectional life tables. (These tables are called "cross-sectional" because they can be computed from vital statistics available at one point in time, covering people of all ages.) There are N_t people alive at the beginning of age t, but n_t of them die before reaching age $t + 1$. The death probability in year t of life is n_t/N_t, the survival probability is $1 - n_t/N_t$. The probability at birth ("age 0") of surviving T years or more is estimated as

$$(1) \qquad \prod_{t=0}^{T-1} \left(1 - \frac{n_t}{N_t}\right).$$

There are corrections to make if you want to get from discrete time to continuous time; this used to be a major topic in applied mathematics. However, the big assumption in constructing the life table is that death rates do not change over time. If there is a trend, the life table will be biased. From Bernoulli's day onwards, death rates have been going down in the Western world; this was the beginning of the demographic transition (Kirk 1996). Therefore, cross-sectional life tables understate life expectancy.

11.2 Hazard rates

Let τ be a positive random variable—the waiting time for failure. Suppose τ has a continuous positive density f. The distribution function is $F(t) = \int_0^t f(u)\,du$, with $F' = f$. The survival function is $S = 1 - F$. The hazard rate is

$$(2) \qquad h(t) = \frac{f(t)}{1 - F(t)}.$$

The intuition behind the formula is that $h(t)\,dt$ represents the conditional probability of failing in the interval $(t, t + dt)$, given survival until time t.

We can recover f, S, and F from the hazard rate:

$$(3) \qquad S(t) = 1 - F(t) = \exp\left(-\int_0^t h(u)\,du\right),$$

$$(4) \qquad f(t) = h(t)S(t).$$

A consequence of (2) or (3) that $\int_0^\infty h(u)\,du = \infty$. In many studies, the failure rate is low. Then $F(t) \approx 0$, $S(t) \approx 1$, and $f(t) \approx h(t)$ over the observable range of t's.

Technical notes. (i) To derive $\int_0^\infty h(u)\,du = \infty$ from (2): if $0 \leq t_n < t_{n+1}$, then

$$\int_{t_n}^{t_{n+1}} h(u)\,du > [S(t_n) - S(t_{n+1})]/S(t_n).$$

Choose the t_n inductively, with $t_0 = 0$ and t_{n+1} so large that $S(t_{n+1}) < S(t_n)/2$. Then sum over n. Also see Rudin (1976, p. 79). The derivation from (3) is clear, again because $S(\infty) = 0$.

(ii) Equation (2) says that $S'/S = -h$. Solving for S with the constraint $S(0) = 1$ gives $S(t) = \exp\left(-\int_0^t h(u)\,du\right)$.

Here are four types of failure, the first two drawn from consulting projects, the others to be discussed later on. (i) A light bulb burns out. (This may seem too trite to be true, but the client was buying a lot of bulbs: Which brand to buy, and when to relamp?) (ii) A financial institution goes out of business. (iii) A subject in a clinical trial dies. (iv) A subject in a clinical trial dies of a pre-specified cause, for instance, lung cancer.

Some examples may help to clarify the mathematics.

Example 1. If τ is standard exponential, $P(\tau > t) = \exp(-t)$ is the survival function, and the hazard rate is $h \equiv 1$.

Example 2. If τ is Weibull, the survival function is by definition

$$(5) \qquad P(\tau > t) = \exp(-at^b).$$

The density is

$$(6) \qquad f(t) = abt^{b-1}\exp(-at^b),$$

and the hazard rate is

$$(7) \qquad h(t) = abt^{b-1}.$$

Here, $a > 0$ and $b > 0$ are parameters. The parameter b controls the shape of the distribution, and a controls the scale. If $b > 1$, the hazard rate keeps going up: The longer you live, the shorter your future life will be. If $b < 1$, the hazard rate goes down: The longer you live, the longer your future life will be. The case $b = 1$ is the exponential: If you made it to time t, you still have the same exponential amount of lifetime left ahead of you.

Example 3. If c and d are positive constants and U is uniform on the unit interval, then $c(-\log U)^d$ is Weibull: $a = (1/c)^{1/d}$ and $b = 1/d$.

Example 4. If τ_i are independent with hazard rates h_i, the minimum of the τ's has hazard rate $\sum_i h_i$.

Turn now to the independence of competing risks. We may have two kinds of failure, like death from heart disease or death from cancer. Independence of competing risks means that the time to death from heart disease is independent of the time to death from cancer.

There may be a censoring time c as well as the failure time τ. Independence of competing risks means that c and τ are independent. The chance that $\tau > t + s$ given $\tau > t$ and $c = t$ equals the chance that $\tau > t + s$ given $\tau > t$, without the c. If they lose track of you, that doesn't change the probability distribution of your time to failure. (Independence of c and τ is often presented as a separate condition, rather than being folded into the independence of competing risks.)

11.3 The Kaplan-Meier estimator

In a clinical trial, t is usually time on test, that is, time from randomization. Time on test is to be distinguished from age and calendar time ("period"). The analysis here assumes stationarity: Failure times are determined by time on test and are not influenced by age or period.

We also have to consider censoring, which occurs for a variety of reasons. For instance, one subject may withdraw from the study. Another subject may get killed by an irrelevant cause. If failure is defined as death from heart disease, and the subject gets run over by a bus, this is not failure, this is censoring. (At least, that's the party line.) A third subject may be censored because he survived until the end of the study.

Subjects may be censored at late times if they were early entrants to the trial. Conversely, early censoring is probably common among late entrants. We're going to lump all forms of censoring together, and we're going to assume independence of competing risks.

Suppose there are no ties (no two subjects fail at the same time). At any particular time t with a failure, let N_t be the number of subjects on test "at time $t-$," that is, just before time t. The probability of surviving from $t-$ to $t+$ is $1 - 1/N_t$. You just multiply these survival probabilities to get a monotone decreasing function, which is flat between failures but goes down a little bit at each failure:

$$(8) \qquad T \to \prod_{t \leq T} \left(1 - \frac{1}{N_t}\right).$$

This is the Kaplan-Meier (1958) survival curve. Notice that N_t may go down between failures, at times when subjects are censored. However, the Kaplan-Meier curve does not change at censoring times. Of course, censored subjects are excluded from future N_t's, and do not count as failures either. The modification for handling ties is pretty obvious.

In a clinical trial, we would draw one curve for the treatment group and one for the control group. If treatment postpones time to failure, the survival curve for the treatment group will fall off more slowly. If treatment has no effect, the two curves will be statistically indistinguishable.

What is the curve estimating? If subjects in treatment are independent with a common survival function, that is what we will be getting, and likewise for the controls. What if subjects aren't independent and identically distributed? Under suitable regularity conditions, with independent subjects, independence of competing risks, and stationarity, the Kaplan-Meier curve for the treatment group estimates the average curve we would see if all subjects were assigned to treatment. Similarly for the controls.

Kaplan-Meier estimators are subject to bias in finite samples. Technical details behind consistency results are not simple; references will be discussed below. Among other things, the times t at which failures occur are random. The issue is often finessed (in this paper, too).

The Kaplan-Meier curve is like a cross-sectional life table, but there is some difference in perspective. The context for the life table is grouped cross-sectional data. The context for the Kaplan-Meier curve is longitudinal data on individual subjects.

How would we estimate the effect on life expectancy of eliminating smallpox? In Bernoulli's place, we might compute the Kaplan-Meier curve, censoring the deaths from smallpox. What he did was to set up differential equations describing the hazard rate ("force of mortality") due to various causes. Independence of competing risks is assumed. If the people who died of smallpox were likely to die shortly thereafter of something else anyway ("frailty"), we would all be over-estimating the impact of eliminating smallpox.

Using data from Halley (1693), Bernoulli estimated that life expectancy at birth was around twenty-seven years; eliminating smallpox would add three years to this figure. In 2007, life expectancy at birth was eighty years or thereabouts, in the United States, the United Kingdom, France, Germany, the Netherlands, and many other European countries—compared to thirty-five years or so in Swaziland and some other very poor countries.

11.4 An application of the Kaplan-Meier estimator

If cancer can be detected early enough, before it has metastasized, there may be improved prospects for effective therapy. That is the situation for breast cancer and cervical cancer, among other examples. Claudia Henschke et al. (2006) tried to make the case for lung cancer. This was an intriguing but unsuccessful application of survival analysis.

Henschke and her colleagues screened 31,567 asymptomatic persons at risk for lung cancer using low-dose CT (computerized tomography), resulting in a diagnosis of lung cancer in 484 participants. These 484 subjects had an estimated ten-year survival rate of 80%. Of the 484 subjects, 302 had stage I cancer and were resected within one month of diagnosis. The resected group had an estimated ten-year survival rate of 92%. The difference between 92% and 80% was reported as highly significant.

Medical terminology. Cancer has *metastasized* when it has spread to other organs. *Stage* describes the extent to which a cancer has progressed. Stage I cancer is early-stage cancer, which usually means small size, limited invasiveness, and a good prognosis. In a *resection*, the surgeon opens the chest cavity, and removes the diseased portion of the lung. *Adenocarcinomas* (referred to below) are cancers that appear to have originated in glandular tissue.

Survival curves (figure 2 in the paper) were computed by the Kaplan-Meier method. Tick marks are used to show censoring. Deaths from causes other than lung cancer were censored, but a lot of the censoring is probably because the subjects survived until the end of the study. In this respect among others, crucial details are omitted. The authors conclude:

> [that] CT screening . . . can detect clinical stage I lung cancer in a high proportion of persons when it is curable by surgery. In a population at risk for lung cancer, such screening could prevent some 80% of deaths from lung cancer. (p. 1769)

The evidence is weak. For one thing, conventional asymptotic confidence intervals on the Kaplan-Meier curve are shaky, given the limited number of data after month sixty. (Remember, late entrants to the trial will only be at risk for short periods of time.) For another thing, why are the authors looking only at deaths from lung cancer rather than total mortality? Next, stage I cancers—the kind detected by the CT scan—are small. This augurs well for long-term survival, treatment or no treatment. Even more to the point, the cancers found by screening are likely to be slow-growing. That is "length bias."

Table 3 in Henschke et al. shows that most of the cancers were adenocarcinomas; these generally have a favorable prognosis. Moreover, the cancer patients who underwent resection were probably healthier to start with than the ones who didn't. In short, the comparison between the resection group and all lung cancers is uninformative. One of the things lacking in this study is a reasonable control group.

If screening speeds up detection, that will increase the time from detection to death—even if treatment is ineffective. The increase is called "lead time" or "lead-time bias." (To measure the effectiveness of screening, you might want to know the time from detection to death, net of lead time.) Lead time and length bias are discussed in the context of breast cancer screening by Shapiro et al. (1988).

When comparing their results to population data, Henschke et al. measure benefits as the increase in time from diagnosis to death. This is misleading, as we have just noted. CT scans speed up detection, but we do not know whether that helps the patients live longer because we do not know whether early treatment is effective. Henschke et al. are assuming what needs to be proved. For additional discussion, see Patz et al. (2000) and Welch et al. (2007).

Lead time bias and length bias are problems for observational studies of screening programs. Well-run clinical trials avoid such biases, if benefits are measured by comparing death rates among those assigned to screening and those assigned to the control group. This is an example of the intention-to-treat principle (Hill 1961, p. 259).

A hypothetical will clarify the idea of lead time. "Crypto-megalo-grandioma" (CMG) is a dreadful disease, which is rapidly fatal after diagnosis. Existing therapies are excruciating and ineffective. No improvements are on the horizon. However, there is a screening technique that can reliably detect the disease ten years before it becomes clinically manifest. Will screening increase survival time from diagnosis to death? Do you want to be screened for CMG?

11.5 The proportional-hazards model in brief

Assume independence of competing risks; subjects are independent of one another; there is a baseline hazard rate $h > 0$, which is the same for all subjects. There is a vector of subject-specific characteristics X_{it}, which is allowed to vary with time. The subscript i indexes subjects and

t indexes time. There is a parameter vector β, which is assumed to be the same for all subjects and constant over time. Time can be defined in several ways. Here, it means time on test; but see Thiébaut and Bénichou (2004). The hazard rate for subject i is assumed to be

$$(9) \qquad h(t) \exp(X_{it}\beta).$$

No intercept is allowed: The intercept would get absorbed into h. The most interesting entry in X_{it} is usually a dummy for treatment status. This is 1 for subjects in the treatment group, and 0 for subjects in the control group. We pass over all technical regularity conditions in respectful silence.

The likelihood function is not a thing of beauty. To make this clear, we can write down the log-likelihood function $L(h, \beta)$, which is a function of the baseline hazard rate h and the parameter vector β. For the moment, we will assume there is no censoring and the X_{it} are constant (not random). Let τ_i be the failure time for subject i. By (3)-(4),

$$(10a) \qquad L(h, \beta) = \sum_{i=1}^{n} \log f_i(\tau_i | h, \beta),$$

where

$$(10b) \qquad f_i(t|h, \beta) = h_i(t|\beta) \exp\left(-\int_0^t h_i(u|\beta)\, du\right),$$

and

$$(10c) \qquad h_i(t|\beta) = h(t) \exp(X_{it}\beta).$$

This is a mess, and maximizing over the infinite-dimensional parameter h is a daunting prospect.

Cox (1972) suggested proceeding another way. Suppose there is a failure at time t. Remember, t is time on test, not age or period. Consider the set R_t of subjects who were on test just before time t. These subjects have not failed yet, or been censored. So they are eligible to fail at time t. Suppose it was subject j who failed. Heuristically, the chance of it being subject j rather than anybody else in the risk set is

$$(11) \qquad \frac{h(t) \exp(X_{jt}\beta)\, dt}{\sum_{i \in R_t} h(t) \exp(X_{it}\beta)\, dt} = \frac{\exp(X_{jt}\beta)}{\sum_{i \in R_t} \exp(X_{it}\beta)}.$$

Subject j is in numerator and denominator both, and by assumption there are no ties: Ties are a technical nuisance. The baseline hazard rate $h(t)$ and the dt cancel! Now we can do business.

Multiply the right side of (11) over all failure times to get a "partial-likelihood function." This is a function of β. Take logs and maximize to get $\hat{\beta}$. Compute the Hessian—the second derivative matrix of the log-partial-likelihood—at $\hat{\beta}$. The negative of the Hessian is the "observed partial information." Invert this matrix to get the estimated variance-covariance matrix for the $\hat{\beta}$'s. Take the square root of the diagonal elements to get asymptotic standard errors.

Partial likelihood functions are not real likelihood functions. The harder you think about (11) and the multiplication, the less sense it makes. The chance of what event, exactly? Conditional on what information? Failure times are random, not deterministic. This is ignored by (11). The multiplication is bogus. For example, there is no independence: If Harriet is at risk at time T, she cannot have failed at an earlier time t. Still, there is mathematical theory to show that $\hat{\beta}$ performs like a real MLE, under the regularity conditions that we have passed over; also see Example 5 below.

Proportional-hazards models are often used in observational studies and in clinical trials. The latter fact is a real curiosity. There is no need to adjust for confounding if the trial is randomized. Moreover, in a clinical trial, the proportional-hazards model makes its calculations conditional on assignment. The random elements are the failure times for the subjects. As far as the model is concerned, the randomization is irrelevant. Equally, randomization does not justify the model.

11.5.1 A mathematical diversion

Example 5. Suppose the covariates $X_{it} \equiv X_i$ do not depend on t and are non-stochastic; for instance, covariates are measured at recruitment into the trial and are conditioned out. Suppose there is no censoring. Then the partial likelihood function is the ordinary likelihood function for the ranks of the failure times. Kalbfleisch and Prentice (1973) discuss more general results.

Sketch proof. The argument is not completely straightforward, and all the assumptions will be used. As a matter of notation, subject i has failure time τ_i. The hazard rate of τ_i is $h(t) \exp(X_i \beta)$, the density is $f_i(t)$, and the survival function is $S_i(t)$. Let $c_i = \exp(X_i \beta)$. We start with the case $n = 2$. Let $C = c_1 + c_2$. Use (3)-(4) to see that

$$(12) \qquad P(\tau_1 < \tau_2) = \int_0^\infty S_2(t) f_1(t)\, dt$$

$$= c_1 \int_0^\infty h(t) S_1(t) S_2(t) \, dt$$

$$= c_1 \int_0^\infty h(t) \exp\left(-C \int_0^t h(u) du\right) dt.$$

Last but not least,

(13) $$C \int_0^\infty h(t) \exp\left(-C \int_0^t h(u) du\right) dt = 1$$

by (4). So

(14) $$P(\tau_1 < \tau_2) = \frac{c_1}{c_1 + c_2}.$$

That finishes the proof for $n = 2$.

Now suppose $n > 2$. The chance that τ_1 is the smallest of the τ's is

$$\frac{c_1}{c_1 + \cdots + c_n},$$

as before: Just replace τ_2 by $\min\{\tau_2, \ldots, \tau_n\}$. Given that $\tau_1 = t$ and τ_1 is the smallest of the τ's, the remaining τ's are independent and concentrated on (t, ∞). If we look at the random variables $\tau_i - t$, their conditional distributions will have hazard rates $c_i h(t + \cdot)$, so we can proceed inductively. A rigorous treatment might involve regular conditional distributions (Freedman 1971, pp. 347ff). This completes the sketch proof.

Another argument, suggested by Russ Lyons, is to change the time scale so the hazard rate is identically 1. Under the conditions of Example 5, the transformation $t \rightarrow \int_0^t h(u) \, du$ reduces the general case to the exponential case. Indeed, if H is a continuous, strictly increasing function that maps $[0, \infty)$ onto itself, then $H(\tau_i)$ has survival function $S_i \circ H^{-1} = S_i(H^{-1})$.

The mathematics does say something about statistical practice. At least in the setting of Example 5, and contrary to general opinion, the model does not use time-to-event data. It uses only the ranks: Which subject failed first, which failed second, and so forth. That, indeed, is what enables the fitting procedure to get around problems created by the intractable likelihood function.

11.6 An application of the proportional-hazards model

Pargament et al. (2001) report on religious struggle as a predictor of mortality among very sick patients. Subjects were 596 mainly Baptist and Methodist patients age 55+, hospitalized for serious illness at the Duke Medical Center and the Durham Veterans' Affairs Medical Center. There was a two-year followup, with 176 deaths and 152 subjects lost to followup. Key variables of interest were positive and negative religious feelings. There was adjustment by proportional hazards for age, race, gender, severity of illness, . . . , and for missing data.

The main finding reported by Pargament et al. is that negative religious feelings increase the death rate. The authors say:

> Physicians are now being asked to take a spiritual history . . . . Our findings suggest that patients who indicate religious struggle during a spiritual history may be at particularly high risk . . . . Referral of these patients to clergy to help them work through these issues may ultimately improve clinical outcomes; further research is needed . . . . (p. 1885)

The main evidence is a proportional-hazards model. Variables include age (in years), education (highest grade completed), race, gender, and . . .

Religious feelings

> Positive and negative religious feelings were measured on a seven-item questionnaire, the subject scoring 0–3 points on each item. The following are two representative items (quoted from the paper).
>
> + "collaboration with God in problem solving"
> − "decided the devil made this happen"

Physical health

> Number of current medical problems, 1–18.
> ADL—Activities of Daily Life.
> Higher scores mean less ability to function independently.
> Patient self-rating, poor to excellent.
> Anesthesiologist rating of patient, 0–5 points.
> 0 is healthy, 5 is very sick.

Mental health

> MMSE—Mini-Mental State Examination.
> Higher scores indicate better cognitive functioning.
> Depression, measured on a questionnaire with eleven items.
> Quality of life is observer-rated on five items.

To review briefly, the baseline hazard rate in the model is a function of time t on test; this baseline hazard rate gets multiplied by $e^{X\beta}$, where X can vary with subject and t. Estimation is by partial likelihood.

Table 11.1 Hazard ratios. Pargament et al. (2001).

Religious feelings −	1.06	**
Religious feelings +	0.98	
Age (years)	1.39	**
Black	1.21	
Female	0.71	*
Hospital	1.14	
Education	0.98	
Physical health		
Diagnoses	1.04	
ADL	0.98	
Patient	0.71	* * *
Anesthesiologist	1.54	* * *
Mental health		
MMSE	0.96	
Depression	0.95	
Quality of life	1.03	

$* \ P < .10 \quad ** \ P < .05 \quad *** \ P < .01$

Table 11.1 shows estimated hazard ratios, that is, ratios of hazard rates. Age is treated as a continuous variable. The hazard ratio of 1.39 reported in the table is $\exp(\hat{\beta}_A)$, where $\hat{\beta}_A$ is the estimated coefficient for age in the model. The interpretation would be that each additional year of age multiplies the hazard rate by 1.39. This is a huge effect.

Similarly, the 1.06 is $\exp(\hat{\beta}_N)$, where $\hat{\beta}_N$ is the estimated coefficient of the "negative religious feelings" score. The interpretation would be that each additional point on the score multiplies the hazard rate by 1.06.

The proportional-hazards model is linear on the log scale. Effects are taken to be constant across people, and multiplicative rather than additive or synergistic. Thus, in combination, an extra year of age and an extra point on the negative religious feelings scale are estimated to multiply the hazard rate by 1.39×1.06.

11.6.1 The crucial questions

The effect is so small—the hazard ratio of interest is only 1.06— that bias should be a real concern. Was the censoring really independent? Were there omitted variables? Were the measurements too crude? What about reverse causation? For example, there may well be income effects; income is omitted. We might get different answers if age was measured in months rather than years; health at baseline seems to be crudely measured as well. Finally, the model may have causation backwards, if severe illness causes negative religious feelings.

This is all taken care of by the model. But what is the justification for the model? Here is the authors' answer:

> This robust semiparametric procedure was chosen for its flexibility in handling censored observations, time-dependent predictors, and late entry into the study. (p. 1883)

The paper has a large sample and a plan for analyzing the data. These positive features are not as common as might be hoped. However, as the quote indicates, there is scant justification for the statistical model. (This is typical; the research hypothesis is atypical.)

11.7 Does HRT prevent heart disease?

There are about 50 observational studies that, on balance, say yes: HRT (hormone replacement therapy) cuts the risk of heart disease. Several experiments say no: There is no protective effect, and there may even be harm. The most influential of the observational studies is the Nurses' Health Study, which claims a reduction in risk by a factor of two or more.

11.7.1 Nurses' Health Study: Observational

Results from the Nurses' Health Study have been reported by the investigators in numerous papers. We consider Grodstein, Stampfer, Manson et al. (1996). In that paper, 6224 postmenopausal women on combined HRT are compared to 27,034 never-users. (Former users are considered separately.) There are 0–16 years of followup, with an average of eleven years. Analysis is by proportional hazards. Failure was defined as either a non-fatal heart attack or death from coronary heart disease.

The treatment variable is HRT. The investigators report seventeen confounders, including age, age at menopause, height, weight, smoking, blood pressure, cholesterol, . . . , exercise. Eleven of the confounders make it into the main model. Details are a little hazy, and there may be some variation from one paper to another. The authors say:

> Proportional-hazards models were used to calculate relative risks and 95 percent confidence intervals, adjusted for confounding variables . . . . We observed a marked decrease in the risk of major coronary heart disease among women who took estrogen with progestin, as compared with the risk among women who did not use hormones (multivariate adjusted relative risk 0.39; 95 percent confidence interval, 0.19 to 0.78) . . . . (p. 453)

The authors do not believe that the protective effect of HRT can be explained by confounding:

> Women who take hormones are a self-selected group and usually have healthier lifestyles with fewer risk factors.... However, ... participants in the Nurses' Health Study are relatively homogeneous.... Unknown confounders may have influenced our results, but to explain the apparent benefit on the basis of confounding variables, one must postulate unknown risk factors that are extremely strong predictors of disease and closely associated with hormone use. (p. 458)

11.7.2 Women's Health Initiative: Experimental

The biggest and most influential experiment is WHI, the Women's Health Initiative. Again, there are numerous papers, but the basic one is Rossouw et al. (2002). In the WHI experiment, 16,608 postmenopausal women were randomized to HRT or control. The study was stopped early, with an average followup period of only five years, because HRT led to excess risk of breast cancer.

The principal result of the study can be summarized as follows. The estimated hazard ratio for CHD (Coronary Heart Disease) is 1.29, with a nominal 95% confidence interval of 1.02 to 1.63: "Nominal" because the confidence level does not take multiple comparisons into account. The trialists also reported a 95% confidence interval from 0.85 to 1.97, based on a Bonferroni correction for multiple looks at the data.

The analysis is by proportional hazards, stratified by clinical center, age, prior disease, and assignment to diet. (The effects of a low-fat diet were studied in another overlapping experiment.) The estimated hazard ratio is $\exp(\hat{\beta}_T)$, where $\hat{\beta}_T$ is the coefficient of the treatment dummy. The confidence intervals are asymmetric because they start on the log scale. The theory produces confidence intervals for β_T, but the parameter of interest is $\exp(\beta_T)$. So you have to exponentiate the endpoints of the intervals.

For a first cut at the data, let us compare the death rates over the followup period (per woman randomized) in the treatment and control groups:

$$231/8506 = 27.2/1000 \text{ vs } 218/8102 = 26.9/1000,$$
$$\text{crude rate ratio} = 27.2/26.9 = 1.01.$$

HRT does not seem to have much of an effect.

The trialists' primary endpoint was CHD. We compute the rates of CHD in the treatment and control groups:

$$164/8506 = 19.3/1000 \quad \text{vs} \quad 122/8102 = 15.1/1000,$$
$$\text{crude rate ratio} = 19.3/15.1 = 1.28.$$

MI (myocardial infarction) means the destruction of heart muscle due to lack of blood—a heart attack. CHD is coronary heart disease, operationalized here as fatal or non-fatal MI. The rate ratios are "crude" because they are not adjusted for any imbalances between treatment and control groups.

If you want standard errors and confidence intervals for rate ratios, use the delta method, as explained in the Appendix (Section 11.A). On the log scale, the delta method gives a standard error of $\sqrt{1/164 + 1/122} = 0.12$. To get the 95% confidence interval for the hazard ratio, multiply and divide the 1.28 by $\exp(2 \times 0.12) = 1.27$. You get 1.01 to 1.63 instead of 1.02 to 1.63 from the proportional-hazards model. What did the model bring to the party?

Our calculation ignores blocking and time-to-event data. The trialists have ignored something too: the absence of any logical foundation for the model. The experiment was very well done. The data summaries are unusually clear and generous. The discussion of the substantive issues is commendable. The modeling, by contrast, seems ill-considered—although it is by no means unusual. (The trialists did examine the crude rate ratios.)

Agreement between crude rate ratios and hazard ratios from multivariate analysis is commonplace. Indeed, if results were substantively different, there would be something of a puzzle. In a large randomized controlled experiment, adjustments should not make much difference because the randomization should balance the treatment and control groups with respect to prognostic factors. Of course, if P is close to 5% or 1%, multivariate analysis can push results across the magic line, which has some impact on perceptions.

11.7.3 Were the observational studies right, or the experiments?

If you are not committed to HRT or to observational epidemiology, this may not seem like a difficult question. However, efforts to show the observational studies got it right are discussed in three journals:

2004 *International Journal of Epidemiology* 33(3),

2005 *Biometrics* 61(4),

2005 *American Journal of Epidemiology* 162(5).

For the Nurses' study, the argument is that HRT should start right after menopause, whereas in the WHI experiment, many women in treatment started HRT later.

The WHI investigators ran an observational study in parallel with the experiment. This observational study showed the usual benefits. The argument here is that HRT creates an initial period of risk, after which the benefits start. Neither of these timing hypotheses is fully consistent with the data, nor are the two hypotheses entirely consistent with each other (Petitti and Freedman 2005). Results from late followup of WHI show an increased risk of cancer in the HRT group, which further complicates the timing hypothesis (Heiss et al. 2008).

For reviews skeptical of HRT, see Petitti (1998, 2002). If the observational studies got it wrong, confounding is the likely explanation. An interesting possibility is "prevention bias" or "complier bias" (Barrett-Connor 1991; Petitti 1994). In brief, subjects who follow doctors' orders tend to do better, even when the orders are to take a placebo. In the Nurses' study, taking HRT seems to be thoroughly confounded with compliance.

In the clofibrate trial (Freedman, Pisani, and Purves 2007, pp. 14, A-4), compliers had half the death rate of non-compliers—in the drug group and the placebo group both. Interestingly, the difference between compliers and non-compliers could not be predicted using baseline risk factors.

Another example is the HIP trial (Freedman 2009, pp. 4–5). If you compare women who accepted screening for breast cancer to women who refused, the first group had a 30% lower risk of death from causes other than breast cancer. Here, the compliance effect can be explained, to some degree, in terms of education and income. Of course, the Nurses' Health Study rarely adjusted for such variables.

Many other examples are discussed in Petitti and Chen (2008). For instance, using sunblock reduces the risk of heart attacks by a factor of two; this estimate is robust when adjustments are made for covariates.

Women who take HRT are women who see a doctor regularly. These women are at substantially lower risk of death from a wide variety of diseases (Grodstein et al. 1997). The list includes diseases where HRT is not considered to be protective. The list also includes diseases like breast cancer, where HRT is known to be harmful. Grodstein et al. might object that, in their multivariate proportional-hazards model, the hazard ratio for breast cancer isn't quite significant—either for current users or former users, taken separately.

11.8 Simulations

If the proportional-hazards model is right or close to right, it works pretty well. Precise measures of the covariates are not essential. If the model is wrong, there is something of a puzzle: What is being estimated by fitting the model to the data? One possible answer is the crude rate ratio in a very large study population. We begin with an example where the model works, then consider an example in the opposite direction.

11.8.1 The model works

Suppose the baseline distribution of time to failure for untreated subjects is standard exponential. There is a subject-specific random variable W_i which multiplies the baseline time and gives the time to failure for subject i if untreated. The hazard rate for subject i is therefore $1/W_i$ times the baseline hazard rate. By construction, the W_i are independent and uniform on $[0, 1]$. Treatment doubles the failure time, that is, cuts the hazard rate in half—for every subject. We censor at time 0.10, which keeps the failure rates moderately realistic.

We enter $\log W_i$ as the covariate. This is exactly the right covariate. The setup should be duck soup for the model. We can look at simulation data on 5000 subjects, randomized to treatment or control by the toss of a coin. The experiment is repeated 100 times.

The crude rate ratio is 0.620 ± 0.037. (In other words, the average across the repetitions is 0.620, and the standard deviation is 0.037.)

The estimated hazard ratio for the model without the covariate is 0.581 ± 0.039.

The estimated hazard ratio for the model with the covariate $\log W_i$ is 0.498 ± 0.032.

The estimated hazard ratio is $\exp(\hat{\beta}_T)$, where $\hat{\beta}_T$ is the coefficient of the treatment dummy in the fitted model. The "real" ratio is 0.50. If that's what you want, the full model looks pretty good. The no-covariate model goes wrong because it fails to adjust for $\log W_i$. This is complicated: $\log W_i$ is nearly balanced between the treatment and control groups, so it is not a confounder. However, without $\log W_i$, the model is no good: Subjects do not have a common baseline hazard rate. The Cox model is not "collapsible."

The crude rate ratio (the failure rate in the treatment arm divided by the failure rate in the control arm) is very close to the true value, which is

$$(15) \qquad \frac{1 - E[\exp(0.05/W_i)]}{1 - E[\exp(0.10/W_i)]}.$$

The failure rates in treatment and control are about 17% and 28%, big enough so that the crude rate ratio is somewhat different from the hazard ratio: $1/W_i$ has a long, long tail. In this example and many others, the crude rate ratio seems to be a useful summary statistic.

The model is somewhat robust against measurement error. For instance, suppose there is a biased measurement of the covariate: We enter $\sqrt{-\log W_i}$ into the model, rather than $\log W_i$. The estimated hazard ratio is 0.516 ± 0.030, so the bias in the hazard ratio—created by the biased measurement of the covariate—is only 0.016. Of course, if we degrade the measurement further, the model will perform worse. If the covariate is $\sqrt{-\log W_i} + \log U_i$ where U_i is an independent uniform variable, the estimate is noticeably biased: 0.574 ± 0.032.

11.8.2 The model does not work

We modify the previous construction a little. To begin with, we drop W_i. The time to failure if untreated, τ_i, is still standard exponential; and we still censor at time 0.10. As before, the effect of treatment is to double τ_i, which cuts the hazard rate in half. So far, so good: We are still on home ground for the model.

The problem is that we have a new covariate,

$$(16) \qquad\qquad Z_i = \exp(-\tau_i) + cU_i,$$

where U_i is an independent uniform variable and c is a constant. Notice that $\exp(-\tau_i)$ is itself uniform. The hapless statistician in this fable will have the data on Z_i, but will not know how the data were generated.

The simple proportional-hazards model, without covariates, matches the crude rate ratio. If we enter the covariate into the model, all depends on c. Here are the results for $c = 0$.

The crude rate ratio is 0.510 ± 0.063. (The true value is $1.10/2.10 \approx 0.524$.)

The estimated hazard ratio for the model without the covariate is 0.498 ± 0.064.

The estimated hazard ratio for the model with the covariate defined by (16) is 0.001 ± 0.001.

The crude rate ratio looks good, and so does the no-covariate model. However, the model with the covariate says that treatment divides the hazard rate by 1000. Apparently, this is the wrong kind of covariate to put into the model.

If $c = 1$, so that noise offsets the signal in the covariate, the full model estimates a hazard ratio of about 0.45—somewhat too low. If

$c = 2$, noise swamps the (bad) signal, and the full model works fine. There is actually a little bit of variance reduction.

Some observers may object that Z in (16) is not a confounder, because (on average) there will be balance between treatment and control. To meet that objection, change the definition to

$$(17) \qquad Z_i = \exp(-\tau_i) + \zeta_i \exp(-\tau_i/2) + cU_i,$$

where ζ_i is the treatment dummy. The Z defined by (17) is unbalanced between treatment and control groups. It is related to outcomes, so it contains valuable information. In short, it is a classic example of a confounder. But, for the proportional-hazards model, it's the wrong kind of confounder—poison, unless c is quite large.

For the proof, here are the results for $c = 2$. Half the variance is accounted for by noise, so there is a lot of dilution. Even so—

The crude rate ratio is 0.522 ± 0.056.

The estimated hazard ratio for the model without the covariate is 0.510 ± 0.056.

The estimated hazard ratio for the model with the covariate defined by (17) is 0.165 ± 0.138.

(We have independent randomization across examples, which is how 0.510 in the previous example changed to 0.522 here.) Putting the covariate defined by (17) into the model biases the hazard ratio downwards by a factor of three.

What is wrong with these covariates? The proportional-hazards model is not only about adjusting for confounders, it is also about *hazards that are proportional to the baseline hazard*. The key assumption in the model is something like this. Given that a subject is alive and uncensored at time t, and given the covariate history up to time t, the probability of failure in $(t, t + dt)$ is $h(t) \exp(X_{it}\beta) \, dt$, where h is the baseline hazard rate. In (16) with $c = 0$, the conditional failure time will be known, because Z_i determines τ_i. So the key assumption in the model breaks down. If c is small, the situation is similar, as it is for the covariate in (17).

Some readers may ask whether problems can be averted by judicious use of model diagnostics. No doubt, if we start with a well-defined type of breakdown in modeling assumptions, there are diagnostics that will detect the problem. Conversely, if we fix a suite of diagnostics, there are problems that will evade detection (Freedman 2008e).

11.9 Causal inference from observational data

Freedman (2009) reviews a logical framework, based on Neyman (1923), in which regression can be used to infer causation. There is a straightforward extension to the Cox model with non-stochastic covariates. Beyond the purely statistical assumptions, the chief additional requirement is "invariance to intervention." In brief, manipulating treatment status should not change the statistical relations.

For example, suppose a subject chose the control condition, but we want to know what would have happened if we had put him into treatment. Mechanically, nothing is easier: Just switch the treatment dummy from 0 to 1, and compute the hazard rate accordingly. Conceptually, however, we are assuming that the intervention would not have changed the baseline hazard rate, or the values of the other covariates, or the coefficients in the model.

Invariance is a heroic assumption. How could you begin to verify it without actually doing the experiment and intervening? That is one of the essential difficulties in using models to make causal inferences from non-experimental data.

11.10 What is the bottom line?

There needs to be some hard thinking about the choice of covariates, the proportional-hazards assumption, the independence of competing risks, and so forth. In the applied literature, these issues are rarely considered in any depth. That is why the modeling efforts, in observational studies as in experiments, are often unconvincing.

Cox (1972) grappled with the question of what the proportional-hazards model was good for. He ends up by saying:

> [i] Of course, the [model] outlined here can be made much more specific by introducing explicit stochastic processes or physical models. The wide variety of possibilities serves to emphasize the difficulty of inferring an underlying mechanism indirectly from failure times alone rather than from direct study of the controlling physical processes. [ii] As a basis for rather empirical data reduction, [the model] seems flexible and satisfactory. (p. 201)

The first point is undoubtedly correct, although it is largely ignored by practitioners. The second point is at best debatable. If the model is wrong, why are the estimates of fictitious parameters a good summary of the data? In any event, questions about summary statistics seem largely irrelevant:

Practitioners fit the model to the data without considering assumptions, and leap to causal conclusions.

11.11 Where do we go from here?

I will focus on clinical trials. Altman et al. (2001) document persistent failures in the reporting of the data, and make detailed proposals for improvement. The following recommendations are complementary; also see Andersen (1991).

(i) As is usual, measures of balance between the group assigned to treatment and the group assigned to control should be reported.

(ii) After that should come a simple intention-to-treat analysis, comparing rates (or averages and standard deviations) among those assigned to the treatment group and those assigned to the control group.

(iii) Crossover and deviations from protocol should be discussed.

(iv) Subgroup analyses should be reported, and corrections for crossover if that is to be attempted. Two sorts of corrections are increasingly common. (a) Per-protocol analysis censors subjects who cross over from one arm of the trial to the other, for instance, subjects who are assigned to control but insist on treatment. (b) Analysis by treatment received compares those who receive treatment with those who do not, regardless of assignment. These analyses require special justification (Freedman 2006b).

(v) Regression estimates (including logistic regression and proportional hazards) should be deferred until rates and averages have been presented. If regression estimates differ from simple intention-to-treat results, and reliance is placed on the models, that needs to be explained. The usual models are not justified by randomization, and simpler estimators may be more robust.

(vi) The main assumptions in the models should be discussed. Which ones have been checked. How? Which of the remaining assumptions are thought to be reasonable? Why?

(vii) Authors should distinguish between analyses specified in the trial protocol and other analyses. There is much to be said for looking at the data. But readers need to know how much looking was involved before that significant difference popped out.

(viii) The exact specification of the models used should be posted on journal websites, including definitions of the variables. The underlying data should be posted too, with adequate documentation. Patient confidentiality would need to be protected, and authors may deserve a grace period after first publication to further explore the data.

Some studies make data available to selected investigators under stringent conditions (Geller et al. 2004), but my recommendation is different. When data-collection efforts are financed by the public, the data should be available for public scrutiny.

11.12 Some pointers to the literature

Early publications on vital statistics and life tables include Graunt (1662), Halley (1693), and Bernoulli (1760). Bernoulli's calculations on smallpox may seem a bit mysterious. For discussion, including historical context, see Gani (1978) or Dietz and Heesterbeek (2002). A useful book on the early history of statistics, including life tables, is Hald (2005).

Freedman (2008a, 2008b [Chapter 12], 2008c [Chapter 13]) discusses the use of models to analyze experimental data. In brief, the advice is to do it late if at all. Fremantle et al. (2003) have a critical discussion on use of "composite endpoints," which combine data on many distinct endpoints. An example, not much exaggerated, would be fatal MI + nonfatal MI + angina + heartburn.

Typical presentations of the proportional-hazards model (this one included) involve a lot of handwaving. It is possible to make math out the handwaving. But this gets very technical very fast, with martingales, compensators, left-continuous filtrations, and the like. One of the first rigorous treatments was Odd Aalen's Ph.D. thesis at Berkeley, written under the supervision of Lucien LeCam. See Aalen (1978) for the published version, which builds on related work by Pierre Bremaud and Jean Jacod.

Survival analysis is sometimes viewed as a special case of "event history analysis." Standard mathematical references include Andersen et al. (1996) and Fleming and Harrington (2005). A popular alternative is Kalbfleisch and Prentice (2002). Some readers like Miller (1998); others prefer Lee and Wang (2003). Jewell (2003) is widely used. Technical details in some of these texts may not be in perfect focus. If you want mathematical clarity, Aalen (1978) is still a paper to be recommended.

For a detailed introduction to the subject, look at Andersen and Keiding (2006). This book is organized as a one-volume encyclopedia. Peter Sasieni's entry on the "Cox Regression Model" is a good starting point; after that, just browse. Lawless (2003) is another helpful reference.

11.A Appendix: The delta method in more detail

The context for this discussion is the Women's Health Initiative, a randomized controlled experiment on the effects of hormone replacement therapy. Let N and N' be the numbers of women randomized to treatment and control. Let ξ and ξ' be the corresponding numbers of failures (that is, for instance, fatal or non-fatal heart attacks).

The crude rate ratio is the failure rate in the treatment arm divided by the rate in the control arm, with no adjustments whatsoever. Algebraically, this is $(\xi/N)/(\xi'/N')$. The log of the crude rate ratio is

$$(18) \qquad \log \xi - \log \xi' - \log N + \log N'.$$

Let $\mu = E(\xi)$. So

$$(19) \qquad \log \xi = \log \left[\mu \left(1 + \frac{\xi - \mu}{\mu} \right) \right]$$

$$= \log \mu + \log \left(1 + \frac{\xi - \mu}{\mu} \right)$$

$$\approx \log \mu + \frac{\xi - \mu}{\mu},$$

because $\log(1 + h) \approx h$ when h is small. The delta-method $\approx$ a one-term Taylor series.

For present purposes, we can take ξ to be approximately Poisson. So $\mathrm{var}(\xi) \approx \mu \approx \xi$ and

$$(20) \qquad \mathrm{var}\left(\frac{\xi - \mu}{\mu} \right) \approx \frac{1}{\mu} \approx \frac{1}{\xi}.$$

A similar calculation can be made for ξ'. Take ξ and ξ' to be approximately independent, so the log of the crude rate ratio has variance approximately equal to $1/\xi + 1/\xi'$.

The modeling is based on the idea that each subject has a small probability of failing during the trial. This probability is modifiable by treatment. Probabilities and effects of treatment may differ from one subject to another. Subjects are assumed to be independent, and calculations are conditional on assignment.

Exact combinatorial calculations can be made. These would be based on the permutations used in the randomization, and would be "unconditional." The random element is the assignment. (The contrast is with model-based calculations, which are conditional on assignment.) To take blocking, censoring, or time-to-failure into account, you would usually need a lot more data than the summaries published in the articles.

For additional information on the delta method, see van der Vaart (1998). Many arguments for asymptotic behavior of the MLE turn out to depend on more rigorous (or less rigorous) versions of the delta method. Similar comments apply to the Kaplan-Meier estimator.

Acknowledgments

Charles Kooperberg, Russ Lyons, Diana Petitti, Peter Sasieni, and Peter Westfall were very helpful. Kenneth Pargament generously answered questions about his study.

Part III

New Developments:
Progress or Regress?

12

On Regression Adjustments in Experiments with Several Treatments

ABSTRACT. *Regression adjustments are often made to experimental data to address confounders that may not be balanced by randomization. Since randomization does not justify the models, bias is likely; nor are the usual variance calculations to be trusted. Here, we evaluate regression adjustments using Neyman's non-parametric model. Previous results are generalized, and more intuitive proofs are given. A bias term is isolated, and conditions are given for unbiased estimation in finite samples.*

12.1 Introduction

Data from randomized controlled experiments (including clinical trials) are often analyzed using regression models and the like. The behavior of the estimates can be calibrated using the non-parametric model in Neyman (1923), where each subject has potential responses to several possible treatments. Only one response can be observed, according to the subject's assignment; the other potential responses must then remain unobserved. Covariates are measured for each subject and may be entered into the

Annals of Applied Statistics (2008) 2: 176–96.

regression, perhaps with the hope of improving precision by adjusting the data to compensate for minor imbalances in the assignment groups.

As discussed in Freedman (2006b [Chapter 17], 2008a), randomization does not justify the regression model, so that bias can be expected, and the usual formulas do not give the right variances. Moreover, regression need not improve precision. Here, we extend some of those results, with proofs that are more intuitive. We study asymptotics, isolate a bias term of order $1/n$, and give some special conditions under which the multiple-regression estimator is unbiased in finite samples.

What is the source of the bias when regression models are applied to experimental data? In brief, the regression model assumes linear additive effects. Given the assignments, the response is taken to be a linear combination of treatment dummies and covariates with an additive random error; coefficients are assumed to be constant across subjects. The Neyman model makes no assumptions about linearity and additivity. If we write the expected response given the assignments as a linear combination of treatment dummies, coefficients will vary across subjects. That is the source of the bias (algebraic details are given below).

To put this more starkly, in the Neyman model, inferences are based on the random assignment to the several treatments. Indeed, the only stochastic element in the model *is* the randomization. With regression, inferences are made conditional on the assignments. The stochastic element is the error term, and the inferences depend on assumptions about that error term. Those assumptions are not justified by randomization. The breakdown in assumptions explains why regression comes up short when calibrated against the Neyman model.

For simplicity, we consider three treatments and one covariate, the main difficulty in handling more variables being the notational overhead. There is a finite population of n subjects, indexed by $i = 1, \ldots, n$. Defined on this population are four variables a, b, c, z. The value of a at i is a_i, and so forth. These are fixed real numbers. We consider three possible treatments, A, B, C. If, for instance, i is assigned to treatment A, we observe the response a_i but do not observe b_i or c_i.

The population averages are the parameters of interest here:

$$(1) \qquad \bar{a} = \frac{1}{n} \sum_{i=1}^{n} a_i, \quad \bar{b} = \frac{1}{n} \sum_{i=1}^{n} b_i, \quad \bar{c} = \frac{1}{n} \sum_{i=1}^{n} c_i.$$

For example, $\bar{a}$ is the average response if all subjects are assigned to A. This could be measured directly, at the expense of losing all information about $\bar{b}$ and $\bar{c}$. To estimate all three parameters, we divide the population at random into three sets A, B, C, of fixed sizes n_A, n_B, n_C. If $i \in A$, then i receives treatment A; likewise for B and C. We now have a simple model for a clinical trial. As a matter of notation, A stands for a random set as well as a treatment.

Let U, V, W be dummy variables for the sets. For instance, $U_i = 1$ if $i \in A$ and $U_i = 0$ otherwise. In particular, $\sum_i U_i = n_A$, and so forth. Let x_A be the average of x over A, namely,

$$(2) \qquad x_A = \frac{1}{n_A} \sum_{i \in A} x_i.$$

Plainly, $a_A = \sum_{i \in A} a_i / n_A$ is an unbiased estimator, called the "ITT estimator," for $\bar{a}$. Likewise for B and C. "ITT" stands for intention-to treat. The idea, of course, is that the sample average is a good estimator for the population average. The intention-to-treat principle goes back to Hill (1961); for additional discussion, see Freedman (2006b). One flaw in the notation (there are doubtless others): x_A is a random variable, being the average of x over the random set A. By contrast, n_A is a fixed quantity, being the number of elements in A.

In the Neyman model, the observed response for subject $i = 1, \dots,$ n is

$$(3) \qquad Y_i = a_i U_i + b_i V_i + c_i W_i,$$

because a, b, c code the responses to the treatments. If, for instance, i is assigned to A, the response is a_i. Furthermore, $U_i = 1$ and $V_i = W_i = 0$, so $Y_i = a_i$. In this circumstance, b_i and c_i would not be observable.

We come now to multiple regression. The variable z is a covariate. It is observed for every subject, and is unaffected by assignment. Applied workers often estimate the parameters in (1) by a multiple regression of Y on U, V, W, z. This is the multiple-regression estimator whose properties are to be studied. The idea seems to be that estimates are improved by adjusting for random imbalance in assignments.

The standard regression model assumes linear additive effects, so that

$$(4) \qquad E(Y_i | U, V, W, z) = \beta_1 U_i + \beta_2 V_i + \beta_3 W_i + \beta_4 z_i,$$

where β is constant across subjects. However, the Neyman model makes no assumptions about linearity or additivity. As a result, $E(Y_i | U, V, W, z)$ is given by the right hand side of (3), with coefficients that vary across subjects. The variation in the coefficients contradicts the basic assumption needed to prove that regression estimates are unbiased (Freedman 2009, p. 43). The variation in the coefficients is the source of the bias.

Analysts who fit (4) to data from a randomized controlled experiment seem to think of $\hat{\beta}_1$ as estimating the effect of treatment A, namely, $\bar{a}$ in (1). Likewise, $\hat{\beta}_3 - \hat{\beta}_1$ is used to estimate $\bar{c} - \bar{a}$, the differential effect of treatment C versus A. Similar considerations apply to other effects. However, these estimators suffer from bias and other problems to be explored below.

We turn for a moment to combinatorics. Proposition 1 is a well known result. (All proofs are deferred to a technical appendix, Section 12.A.)

Proposition 1. *Let $\tilde{p}_S = n_S/n$ for $S = A, B,$ or C.*

(i) $E(x_A) = \bar{x}.$

(ii) $\text{var}(x_A) = \dfrac{1}{n-1} \dfrac{1 - \tilde{p}_A}{\tilde{p}_A} \text{var}(x).$

(iii) $\text{cov}(x_A, y_A) = \dfrac{1}{n-1} \dfrac{1 - \tilde{p}_A}{\tilde{p}_A} \text{cov}(x, y).$

(iv) $\text{cov}(x_A, y_B) = -\dfrac{1}{n-1} \text{cov}(x, y).$

Here, $x, y = a, b, c,$ or z. Likewise, A in (i-ii-iii) may be replaced by B or C. And A, B in (iv) may be replaced by any other distinct pair of sets. By, e.g., $\text{cov}(x, y)$ we mean

$$\frac{1}{n} \sum_{i=1}^{n} (x_i - \bar{x})(y_i - \bar{y}).$$

Curiously, the result in (iv) does not depend on the fractions of subjects allocated to the three sets. We can take $x = z$ and $y = z$. For instance,

$$\text{cov}(z_A, z_B) = -\frac{1}{n-1} \text{var}(z).$$

The finite-sample multivariate CLT in Theorem 1 below is a minor variation on results in Höglund (1978). The theorem will be used to prove the asymptotic normality of the multiple-regression estimator. There are several regularity conditions for the theorem.

Condition #1. There is an a priori bound on fourth moments. For all $n = 1, 2, \ldots$ and $x = a, b, c,$ or z,

$$(5) \qquad \frac{1}{n} \sum_{i=1}^{n} |x_i|^4 < L < \infty.$$

Condition #2. The first- and second-order moments, including mixed moments, converge to finite limits, and asymptotic variances are positive. For instance,

$$(6) \qquad \frac{1}{n} \sum_{i=1}^{n} a_i \to \langle a \rangle,$$

and

$$(7) \qquad \frac{1}{n} \sum_{i=1}^{n} a_i^2 \to \langle a^2 \rangle, \qquad \frac{1}{n} \sum_{i=1}^{n} a_i b_i \to \langle ab \rangle,$$

with

$$(8) \qquad \langle a^2 \rangle > \langle a \rangle^2;$$

likewise for the other variables and pairs of variables. Here, $\langle a \rangle$ and so forth merely denote finite limits. We take $\langle a^2 \rangle$ and $\langle aa \rangle$ as synonymous. In present notation, $\langle a \rangle$ is the limit of $\bar{a}$, the latter being the average of a over the population of size n: see (1).

Condition #3. We assume groups are of order n in size, i.e.,

$$\tilde{p}_A = n_A/n \to p_A > 0, \quad \tilde{p}_B = n_B/n \to p_B > 0, \text{ and}$$

$$(9) \qquad \tilde{p}_C = n_C/n \to p_C > 0,$$

where $p_A + p_B + p_C = 1$. Notice that $\tilde{p}_A$, for instance, is the fraction of subjects assigned to A at stage n; the limit as n increases is p_A.

Condition #4. The variables a, b, c, z have mean 0:

$$(10) \qquad \frac{1}{n} \sum_{i=1}^{n} x_i = 0, \text{ where } x = a, b, c, z.$$

Condition #4 is a normalization for Theorem 1. Without it, some centering would be needed.

Theorem 1. *The CLT. Under Conditions #1–#4, the joint distribution of the 12-vector*

$$\sqrt{n}\bigl(a_A, a_B, a_C, \ldots, z_C\bigr)$$

is asymptotically normal, with parameters given by the limits below:

(i) $E(\sqrt{n}x_A) = 0$;

(ii) $\mathrm{var}(\sqrt{n}x_A) \to \langle x^2\rangle(1 - p_A)/p_A$;

(iii) $\mathrm{cov}(\sqrt{n}x_A, \sqrt{n}y_A) \to \langle xy\rangle(1 - p_A)/p_A$;

(iv) $\mathrm{cov}(\sqrt{n}x_A, \sqrt{n}y_B) \to -\langle xy\rangle$.

Here, $x, y = a, b, c,$ or z. Likewise, A in (i-ii-iii) may be replaced by B or C. And A, B in (iv) may be replaced by any other distinct pair of sets. The theorem asserts, among other things, that the limiting first- and second-order moments coincide with the moments of the asymptotic distribution, which is safe due to the bound on fourth moments. (As noted before, proofs are deferred to a technical appendix, Section 12.A.)

Example 1. Suppose we wish to estimate the effect of C relative to A, that is, $\bar{c} - \bar{a}$. The ITT estimator is $Y_C - Y_A = c_C - a_A$, where the equality follows from (3). As before,

$$Y_C = \sum_{i \in C} Y_i/n_C = \sum_{i \in C} c_i/n_C.$$

The estimator $Y_C - Y_A$ is unbiased by Proposition 1, and its exact variance is

$$\frac{1}{n-1}\left[\frac{1 - \tilde{p}_A}{\tilde{p}_A}\mathrm{var}(a) + \frac{1 - \tilde{p}_C}{\tilde{p}_C}\mathrm{var}(c) + 2\mathrm{cov}(a, c)\right].$$

By contrast, the multiple-regression estimator would be obtained by fitting (4) to the data, and computing $\hat{\Delta} = \hat{\beta}_3 - \hat{\beta}_1$. The asymptotic bias and variance of this estimator will be determined in Theorem 2 below. The performance of the two estimators will be compared in Theorem 4.

12.2 Asymptotics for multiple-regression estimators

In this section, we state a theorem that describes the asymptotic behavior of the multiple-regression estimator applied to experimental data:

There is a random term of order $1/\sqrt{n}$ and a bias term of order $1/n$. As noted above, we have three treatments and one covariate z. The treatment groups are A, B, C, with dummies U, V, W. The covariate is z. If i is assigned to A, we observe the response a_i whereas b_i, c_i remain unobserved. Likewise for B, C. The covariate z_i is always observed and is unaffected by assignment. The response variable Y is given by (3). In Theorem 1, most of the random variables—like a_B or b_A—are unobservable. That may affect the applications, but not the mathematics. Arguments below involve only observable random variables.

The design matrix for the multiple-regression estimator will have n rows and four columns, namely, U, V, W, z. The estimator is obtained by a regression of Y on U, V, W, z, the first three coefficients estimating the effects of A, B, C, respectively. Let $\hat{\beta}_{MR}$ be the multiple-regression estimator for the effects of A, B, C. Thus, $\hat{\beta}_{MR}$ is a 3×1-vector.

We normalize z to have mean 0 and variance 1:

$$(11) \qquad \frac{1}{n} \sum_{i=1}^{n} z_i = 0, \quad \frac{1}{n} \sum_{i=1}^{n} z_i^2 = 1.$$

The mean-zero condition on z overlaps Condition #4, and is needed for Theorem 2. There is no intercept in our regression model; without the mean-zero condition, the mean of z is liable to confound the effect estimates. The technical appendix (Section 12.A) has details. (In the alternative, we can drop one of the dummies and put an intercept into the regression—although we would now be estimating effect differences rather than effects.) The condition on the mean of z^2 merely sets the scale.

Recall that $\tilde{p}_A$ is the fraction of subjects assigned to treatment A. Let

$$(12) \qquad \tilde{Q} = \tilde{p}_A \overline{az} + \tilde{p}_B \overline{bz} + \tilde{p}_C \overline{cz}$$

and

$$(13) \qquad Q = p_A \langle az \rangle + p_B \langle bz \rangle + p_C \langle cz \rangle.$$

Here, for instance, $\overline{az} = \sum_{i=1}^{n} a_i z_i / n$ is the average over the study population. By Condition #2, as the population size grows,

$$\overline{az} = \sum_{i=1}^{n} a_i z_i / n \to \langle az \rangle.$$

Likewise for b and c. Thus,

$$(14) \qquad \tilde{Q} \to Q.$$

The quantities $\tilde{Q}$ and Q are needed for the next theorem, which demonstrates asymptotic normality and isolates the bias term. To state the theorem, recall that $\hat{\beta}_{MR}$ is the multiple-regression estimator for the three effects. The estimand is

$$(15) \qquad \beta = (\overline{a}, \ \overline{b}, \ \overline{c})',$$

where $\overline{a}, \ \overline{b}, \ \overline{c}$ are defined in (1). Define the 3×3 matrix Σ as follows:

$$(16) \qquad \Sigma_{11} = \frac{1 - p_A}{p_A} \lim \operatorname{var}(a - Qz),$$

$$\Sigma_{12} = -\lim \operatorname{cov}(a - Qz, \ b - Qz),$$

and so forth. The limits are taken as the population size $n \to \infty$, and exist by Condition #2. Let

$$(17) \qquad \zeta_n = \sqrt{n}\big(a_A - \tilde{Q}z_A, \ b_B - \tilde{Q}z_B, \ c_C - \tilde{Q}z_C\big)'.$$

This turns out to be the lead random element in $\hat{\beta}_{MR} - \beta$. The asymptotic variance-covariance matrix of ζ_n is Σ, by (14) and Theorem 1. For the bias term, let

$$(18) \quad K_A = \operatorname{cov}(az, z) - \tilde{p}_A\operatorname{cov}(az, z) - \tilde{p}_B\operatorname{cov}(bz, z) - \tilde{p}_C\operatorname{cov}(cz, z),$$

and likewise for K_B, K_C.

Theorem 2. *Assume Conditions #1–#3, not #4, and (11). Define ζ_n by (17), and K_S by (18) for $S = A, B, C$. Then $E(\zeta_n) = 0$ and ζ_n is asymptotically $N(0, \Sigma)$. Moreover,*

$$(19) \qquad \hat{\beta}_{MR} - \beta = \zeta_n/\sqrt{n} - K/n + \rho_n,$$

where $K = (K_A, K_B, K_C)'$ and $\rho_n = O(1/n^{3/2})$ in probability.

Remarks. (i) If $K = 0$, the bias term will be $O(1/n^{3/2})$ or smaller.

(ii) What are the implications for practice? In the usual linear model, $\hat{\beta}$ is unbiased given X. With experimental data and the Neyman model, given the assignment, results are deterministic. At best, we will get unbiasedness on average, over all assignments. Under special circumstances (Theorems 5 and 6 below), that happens. Generally, however, the multiple-regression estimator will be biased. See Example 5. The bias decreases as sample size increases.

(iii) Turn now to random error in $\hat{\beta}$. This is of order $1/\sqrt{n}$, both for the ITT estimator and for the multiple-regression estimator. However, the asymptotic variances differ. The multiple-regression estimator can be more efficient than the ITT estimator—or less efficient—and the difference persists even for large samples. See Examples 3 and 4 below.

12.3 Asymptotic nominal variances

"Nominal" variances are computed by the usual regression formulae, but are likely to be wrong since the usual assumptions do not hold. We sketch the asymptotics here, under the conditions of Theorem 2. Recall that the design matrix X is $n \times 4$, the columns being U, V, W, z. The response variable is Y. The nominal covariance matrix is then

$$(20) \qquad \Sigma_{\text{nom}} = \hat{\sigma}^2 (X'X)^{-1},$$

where $\hat{\sigma}^2$ is the sum of the squared residuals, normalized by the degrees of freedom $(n - 4)$. Recall Q from (13). Let

$$(21) \qquad \sigma^2 = \lim_{n \to \infty} \left[\tilde{p}_A \text{var}(a) + \tilde{p}_B \text{var}(b) + \tilde{p}_C \text{var}(c) \right] - Q^2,$$

where the limit exists by Conditions #2 and #3. Let

$$(22) \qquad D = \begin{pmatrix} p_A & 0 & 0 & 0 \\ 0 & p_B & 0 & 0 \\ 0 & 0 & p_C & 0 \\ 0 & 0 & 0 & 1 \end{pmatrix}.$$

Theorem 3. *Assume Conditions #1–#3, not #4, and* (11). *Define* σ^2 *by* (21) *and* D *by* (22). *In probability,*

(i) $X'X/n \to D$,

(ii) $\hat{\sigma}^2 \to \sigma^2$,

(iii) $n\Sigma_{\text{nom}} \to \sigma^2 D^{-1}$.

What are the implications for practice? The upper left 3×3 block of $\sigma^2 D^{-1}$ will generally differ from Σ in Theorem 2, so the usual regression standard errors—computed for experimental data—can be quite misleading. This difficulty does not go away for large samples. What explains the breakdown? In brief, the multiple regression assumes (i) the expectation of the response given the assignment variables and the co-variates is linear, with coefficients that are constant across subjects; and (ii) the conditional variance of the response is constant across subjects. In the Neyman model, (i) is wrong as noted earlier. Moreover, given the assignments, there is no variance left in the responses.

More technically, variances in the Neyman model are (necessarily) computed across the assignments, for it is the assignments that are the random elements in the model. With regression, variances are computed conditionally on the assignments, from an error term assumed to be IID across subjects, and independent of the assignment variables as well as the covariates. These assumptions do not follow from the randomization, explaining why the usual formulas break down. For additional discussion, see Freedman (2008a).

An example may clarify the issues. Write cov_∞ for limiting covariances, e.g.,

$$\text{cov}_\infty(a, z) = \lim \text{cov}(a, z) = \langle az \rangle - \langle a \rangle \langle z \rangle = \langle az \rangle$$

because $\langle z \rangle = 0$ by (11); similarly for variances. See Condition #2.

Example 2. Consider estimating the effect of C relative to A, so the parameter of interest is $\bar{c} - \bar{a}$. By way of simplification, suppose $Q = 0$. Let $\hat{\Delta}$ be the multiple-regression estimator for the effect difference. By Theorem 3, the nominal variance of $\hat{\Delta}$ is essentially $1/n$ times

$$\left(1 + \frac{p_A}{p_C}\right)\text{var}_\infty(a) + \left(1 + \frac{p_C}{p_A}\right)\text{var}_\infty(c) + \left(\frac{1}{p_A} + \frac{1}{p_C}\right)p_B\text{var}_\infty(b).$$

By Theorem 2, however, the true asymptotic variance of $\hat{\Delta}$ is $1/n$ times

$$\left(\frac{1}{p_A} - 1\right)\text{var}_\infty(a) + \left(\frac{1}{p_C} - 1\right)\text{var}_\infty(c) + 2\text{cov}_\infty(a, c).$$

For instance, we can take the asymptotic variance-covariance matrix of a, b, c, z to be the 4×4 identity matrix, with $p_A = p_C = 1/4$ so $p_B = 1/2$. The true asymptotic variance of $\hat{\Delta}$ is $6/n$. The nominal asymptotic variance is $8/n$ and is too big. On the other hand, if we change $\text{var}_\infty(b)$ to 1/4, the true asymptotic variance is still $6/n$; the nominal asymptotic variance drops to $5/n$ and is too small.

12.4 The gain from adjustment

Does adjustment improve precision? The answer is, sometimes.

Theorem 4. *Assume Conditions #1–#3, not #4, and* (11). *Consider estimating the effect of C relative to A, so the parameter of interest is $\bar{c} - \bar{a}$. If we compare the multiple-regression estimator to the ITT estimator, the asymptotic gain in variance is $\Gamma/(np_A p_C)$, where*

$$(23) \qquad \Gamma = 2Q\big[p_C\langle az\rangle + p_A\langle cz\rangle\big] - Q^2\big[p_A + p_C\big],$$

with Q defined by (13). *Adjustment therefore helps asymptotic precision if $\Gamma > 0$ but hurts if $\Gamma < 0$.*

The next two examples are set up like Example 2, with cov_∞ for limiting covariances. We say the design is *balanced* if n is a multiple of 3 and $n_A = n_B = n_C = n/3$. We say that effects are *additive* if $b_i - a_i$ is constant over i and likewise for $c_i - a_i$. With additive effects, $\mathrm{var}_\infty(a) = \mathrm{var}_\infty(b) = \mathrm{var}_\infty(c)$; write v for the common value. Similarly, $\mathrm{cov}_\infty(a, z) = \mathrm{cov}_\infty(b, z) = \mathrm{cov}_\infty(c, z) = Q = \rho\sqrt{v}$, where ρ is the asymptotic correlation between a and z, or b and z, or c and z.

Example 3. Suppose effects are additive. Then

$$\mathrm{cov}_\infty(a, z) = \mathrm{cov}_\infty(b, z) = \mathrm{cov}_\infty(c, z) = Q$$

and

$$\Gamma = Q^2(p_A + p_C) \geq 0.$$

The asymptotic gain from adjustment will be positive if $\mathrm{cov}_\infty(a, z) \neq 0$.

Example 4. Suppose the design is balanced, so $p_A = p_B = p_C = 1/3$. Then

$$3Q = \mathrm{cov}_\infty(a, z) + \mathrm{cov}_\infty(b, z) + \mathrm{cov}_\infty(c, z).$$

Consequently,

$$3\Gamma/2 = Q[2Q - \mathrm{cov}_\infty(b, z)].$$

Let $z = a + b + c$. Choose a, b, c so that $\mathrm{var}_\infty(z) = 1$ and

$$\mathrm{cov}_\infty(a, b) = \mathrm{cov}_\infty(a, c) = \mathrm{cov}_\infty(b, c) = 0.$$

In particular, $Q = 1/3$. Now

$$2Q - \mathrm{cov}_\infty(b, z) = 2/3 - \mathrm{var}_\infty(b).$$

The asymptotic gain from adjustment will be negative if $\mathrm{var}_\infty(b) > 2/3$.

Example 3 indicates one motivation for adjustment: If effects are nearly additive, adjustment is likely to help. However, Example 4 shows that even in a balanced design, the "gain" from adjustment can be negative (if there are subject-by-treatment interactions). More complicated and realistic examples can no doubt be constructed.

12.5 Finite-sample results

This section gives conditions under which the multiple-regression estimator will be exactly unbiased in finite samples. Arguments are from symmetry. As before, the design is *balanced* if n is a multiple of 3 and $n_A = n_B = n_C = n/3$; effects are *additive* if $b_i - a_i$ is constant over i and likewise for $c_i - a_i$. Then $a_i - \bar{a} = b_i - \bar{b} = c_i - \bar{c} = \delta_i$, say, for all i. Note that $\sum_i \delta_i = 0$.

Theorem 5. *If* (11) *holds, the design is balanced, and effects are additive, then the multiple-regression estimator is unbiased.*

Examples show that the balance condition is needed in Theorem 5: Additivity is not enough. Likewise, if the balance condition holds but there is non-additivity, the multiple-regression estimator will usually be biased. We illustrate the first point.

Example 5. Consider a miniature trial with six subjects. Responses a, b, c to treatments A, B, C are shown in Table 12.1, along with the covariate z. Notice that $b - a = 1$ and $c - a = 2$. Thus, effects are additive.

We assign one subject at random to A, one to B, and the remaining four to C. There are $6 \times 5/2 = 15$ assignments. For each assignment, we build up the 6×4 design matrix (one column for each treatment dummy and one column for z); we compute the response variable from Table 12.1, and then the multiple-regression estimator. Finally, we average the results

Table 12.1 Parameter values

a	b	c	z
0	1	2	0
0	1	2	0
0	1	2	0
2	3	4	-2
2	3	4	-2
4	5	6	4

across the 15 assignments, as shown in Table 12.2. The average gives the expected value of the multiple-regression estimator, because the average is taken across all possible designs. "Truth" is determined from the parameters in Table 12.1. Calculations are exact, within the limits of rounding error; no simulations are involved.

For instance, the average coefficient for the A dummy is 3.3825. However, from Table 12.1, the average effect of A is $\bar{a} = 1.3333$. The difference is bias. Consider next the differential effect of B versus A. On average, this is estimated by multiple regression as $1.9965 - 3.3825 = -1.3860$. From Table 12.1, truth is $+1$. Again, this reflects bias in the multiple-regression estimator. With a larger trial, of course, the bias would be smaller: see Theorem 2. Theorem 5 does not apply because the design is unbalanced.

For the next theorem, consider the possible values v of z. Let n_v be the number of i with $z_i = v$. The average of a_i given $z_i = v$ is

$$\frac{1}{n_v} \sum_{\{i:z_i=v\}} a_i.$$

Suppose this is constant across v's, as is

$$\sum_{\{i:z_i=v\}} b_i/n_v, \quad \sum_{\{i:z_i=v\}} c_i/n_v.$$

The common values must be $\bar{a}, \bar{b}, \bar{c}$, respectively. We call this *conditional constancy*. No condition is imposed on z, and the design need not be balanced. (Conditional constancy is violated in Example 5, as one sees by looking at the parameter values in Table 12.1.)

Theorem 6. *With conditional constancy, the multiple-regression estimator is unbiased.*

Table 12.2 Average multiple-regression estimates versus truth

	Ave MR	Truth
A	3.3825	1.3333
B	1.9965	2.3333
C	2.9053	3.3333
z	−0.0105	

Remarks. (i) In the usual regression model, $Y = X\beta + \epsilon$ with $E(\epsilon|X) = 0$. The multiple-regression estimator is then conditionally unbiased. In Theorems 5 and 6, the estimator is conditionally biased, although the bias averages out to 0 across permutations. In Theorem 5, for instance, the conditional bias is $(X'X)^{-1}X'\delta$. Across permutations, the bias averages out to 0. The proof is a little tricky (see Section 12.A). The δ is fixed, as explained before the theorem; it is X that varies from one permutation to another; the conditional bias is a nonlinear function of X. This is all quite different from the usual regression arguments.

(ii) Kempthorne (1952) points to the difference between permutation models and the usual linear regression model: see chapters 7–8, especially section 8.7. Also see *Biometrics* (1957) 13(3). Cox (1956) cites Kempthorne, but appears to contradict Theorem 5 above. I am indebted to Joel Middleton for the reference to Cox.

(iii) When specialized to two-group experiments, the formulas in this chapter (for example, asymptotic variances) differ in appearance but not in substance from those previously reported (Freedman 2008a).

(iv) Although details have not been checked, the results (and the arguments) in this chapter seem to extend easily to any fixed number of treatments, and any fixed number of covariates. Treatment-by-covariate interactions can probably be accommodated too.

(v) In this chapter, treatments have two levels: low or high. If a treatment has several levels—e.g., low, medium, high—and linearity is assumed in a regression model, inconsistency is likely to be a consequence. Likewise, we view treatments as mutually exclusive: If subject i is assigned to group A, then i cannot also turn up in group B. If multiple treatments are applied to the same subject in order to determine joint effects, and a regression model assumes additive or multiplicative effects, inconsistency is again likely.

(vi) The theory developed here applies equally well to 0–1 valued responses. With 0–1 variables, it may seem more natural to use logit or probit models to adjust the data. However, such models are not justified by randomization—any more than the linear model. Preliminary calculations suggest that if adjustments are to be made, linear regression may be a safer choice. For instance, the conventional logit estimator for the odds ratio may be severely biased. On the other hand, a consistent estimator can be based on estimated probabilities in the logit model. For discussion, see Freedman (2008c [Chapter 13]).

(vii) The theory developed here can probably be extended to more complex designs (like blocking) and more complex estimators (like two-stage least squares), but the work remains to be done.

(viii) Victora, Habicht, and Bryce (2004) favor adjustment. However, they do not address the sort of issues raised here, nor are they entirely clear about whether inferences are to be made on average across assignments, or conditional on assignment. In the latter case, inferences might be strongly model-dependent.

(ix) Models are used to adjust data from large randomized controlled experiments in, for example, Cook et al. (2007), Gertler (2004), Chattopadhyay and Duflo (2004), and Rossouw et al. (2002). Cook et al. report on long-term followup of subjects in experiments where salt intake was restricted [see also Chapter 9]; conclusions are dependent on the models used to analyze the data. By contrast, the results in Rossouw et al. for hormone replacement therapy do not depend very much on the modeling [see also Chapter 11].

12.6 Recommendations for practice

Altman et al. (2001) document persistent failures in the reporting of data from clinical trials and make detailed proposals for improvement. The following recommendations are complementary.

(i) Measures of balance between the assigned-to-treatment group and the assigned-to-control group should be reported (this is standard practice).

(ii) After that should come a simple intention-to-treat analysis, comparing rates (or averages and standard deviations) of outcomes among those assigned to treatment and those assigned to the control group.

(iii) Crossover should be discussed as well as deviations from protocol.

(iv) Subgroup analyses should be reported, and corrections for crossover if that is to be attempted. Analysis by treatment received requires special justification, and so does per protocol analysis. (The first compares those who receive treatment with those who do not, regardless of assignment; the second censors subjects who cross over from one arm of the trial to the other, e.g., they are assigned to control but insist on treatment.) Complications are discussed in Freedman (2006b).

(v) Regression estimates (including logistic regression and proportional hazards) should be deferred until rates and averages have been presented. If regression estimates differ from simple intention-to-treat results, and reliance is placed on the models, that needs to be explained. As indicated above, the usual models are not justified by randomization, and simpler estimators may be more robust.

12.A Technical appendix

This section provides technical underpinnings for the theorems discussed above.

Proof of Proposition 1. We prove only claim (iv). If $i = j$, it is clear that $E(U_i V_j) = 0$, because i cannot be assigned both to A and to B. Furthermore,

$$E(U_i V_j) = P(U_i = 1 \,\&\, V_j = 1) = \frac{n_A}{n} \frac{n_B}{n-1}$$

if $i \neq j$. This is clear if $i = 1$ and $j = 2$; but permuting indices will not change the joint distribution of assignment dummies. We may assume without loss of generality that $\bar{x} = \bar{y} = 0$. Now

$$\operatorname{cov}(x_A, y_B) = \frac{1}{n_A} \frac{1}{n_B} \sum_{i \neq j} E(U_i V_j x_i y_j)$$

$$= \frac{1}{n(n-1)} \sum_{i \neq j} x_i y_j$$

$$= \frac{1}{n(n-1)} \left(\sum_i x_i \sum_j y_j - \sum_i x_i y_i \right)$$

$$= -\frac{1}{n(n-1)} \sum_i x_i y_i = -\frac{1}{n-1} \operatorname{cov}(x, y)$$

as required, where $i, j = 1, \ldots, n$.   QED

Proof of Theorem 1. The theorem can be proved by appealing to Höglund (1978) and computing conditional distributions. Another starting point is Hoeffding (1951), with suitable choices for the matrix from which summands are drawn. With either approach, the usual linear-combinations trick can be used to reduce dimensionality. In view of (9), the limiting distribution satisfies three linear constraints.

A formal proof is omitted, but we sketch the argument for one case, starting from Theorem 3 in Hoeffding (1951). Let α, β, γ be three constants. Let M be an $n \times n$ matrix, with

$$M_{ij} = \alpha a_j \text{ for } i = 1, \ldots, n_A,$$
$$= \beta b_j \text{ for } i = n_A + 1, \ldots, n_A + n_B,$$
$$= \gamma c_j \text{ for } i = n_A + n_B + 1, \ldots, n.$$

Pick one j at random from each row, without replacement (interpretation: If j is picked from row $i = 1, \ldots, n_A$, subject j goes into treatment group A). According to Hoeffding's theorem, the sum of the corresponding matrix entries will be approximately normal. So the law of $\sqrt{n}(a_A, b_B, c_C)$ tends to multivariate normal. Theorem 1 in Hoeffding's paper will help get the regularity conditions in his Theorem 3 from #1–#4 above.

Let X be an $n \times p$ matrix of rank $p \leq n$. Let Y be an $n \times 1$ vector. The multiple-regression estimator computed from Y is $\hat{\beta}_Y = (X'X)^{-1}X'Y$. Let θ be a $p \times 1$ vector. The "invariance lemma" is a purely arithmetic result; the well-known proof is omitted.

Lemma 1. *The invariance lemma.* $\hat{\beta}_{Y+X\theta} = \hat{\beta}_Y + \theta$.

The multiple-regression estimator for Theorem 2 may be computed as follows. Recall from (2) that Y_A is the average of Y over A, i.e., $\sum_{i \in A} Y_i / n_A$; likewise for B, C. Let

$$(A1) \qquad e_i = Y_i - Y_A U_i - Y_B V_i - Y_C W_i,$$

which is the residual when Y is regressed on the first three columns of the design matrix. Let

$$(A2) \qquad f_i = z_i - z_A U_i - z_B V_i - z_C W_i,$$

which is the residual when z is regressed on those columns. Let $\hat{Q}$ be the slope when e is regressed on f:

$$(A3) \qquad \hat{Q} = e \cdot f / |f|^2.$$

The next result is standard.

Lemma 2. *The multiple-regression estimator for the effect of A, i.e., the first element in $(X'X)^{-1}X'Y$, is*

$$(A4) \qquad Y_A - \hat{Q}z_A$$

and likewise for B, C. The coefficient of z in the regression of Y on U, V, W, z is $\hat{Q}$.

We turn now to $\hat{Q}$; this is the key technical quantity in the chapter, and we develop a more explicit formula for it. Notice that the dummy variables U, V, W are mutually orthogonal. By the usual regression arguments,

$$(A5) \qquad |f|^2 = |z|^2 - n_A z_A^2 - n_B z_B^2 - n_C z_C^2,$$

where $|f|^2 = \sum_{i=1}^n f_i^2$. Recall (3). Check that $Y_A = a_A$ where $a_A = \sum_{i \in A} a_i / n_A$; likewise for B, C. Hence,

$$(A6) \qquad e_i = (a_i - a_A)U_i + (b_i - b_B)V_i + (c_i - c_C)W_i,$$

where the residual e_i was defined in (A1). Likewise,

$$(A7) \qquad f_i = (z_i - z_A)U_i + (z_i - z_B)V_i + (z_i - z_C)W_i,$$

where the residual f_i was defined in (A2). Now

$$(A8) \qquad e_i f_i = (a_i - a_A)(z_i - z_A)U_i + (b_i - b_B)(z_i - z_B)V_i$$
$$+ (c_i - c_C)(z_i - z_C)W_i,$$

and

$$(A9) \qquad \sum_{i=1}^n e_i f_i = n_A[(az)_A - a_A z_A] + n_B[(bz)_B - b_B z_B]$$
$$+ n_C[(cz)_C - c_C z_C],$$

where, for instance, $(az)_A = \sum_{i \in A} a_i z_i / n_A$.

Recall that $\tilde{p}_A = n_A / n$ is the fraction of subjects assigned to treatment A; likewise for B and C. These fractions are deterministic, not random. We can now give a more explicit formula for the $\hat{Q}$ defined in (A3), dividing numerator and denominator by n. By (A5) and (A9),

$$(A10) \qquad \hat{Q} = N/D, \text{ where}$$
$$N = \tilde{p}_A[(az)_A - a_A z_A]$$
$$+ \tilde{p}_B[(bz)_B - b_B z_B]$$
$$+ \tilde{p}_C[(cz)_C - c_C z_C],$$
$$D = 1 - \tilde{p}_A(z_A)^2 - \tilde{p}_B(z_B)^2 - \tilde{p}_C(z_C)^2.$$

In the formula for D, we used (11) to replace $|z|^2/n$ by 1.

The reason $\hat{Q}$ matters is that it relates the multiple-regression estimator to the ITT estimator in a fairly simple way. Indeed, by (3) and Lemma 2,

$$(A11) \qquad \hat{\beta}_{MR} = (Y_A - \hat{Q}z_A, \ Y_B - \hat{Q}z_B, \ Y_C - \hat{Q}z_C)'$$
$$= (a_A - \hat{Q}z_A, \ a_B - \hat{Q}z_B, \ a_C - \hat{Q}z_C)'.$$

We must now estimate $\hat{Q}$. In view of (11), Theorem 1 shows that

$$(A12) \qquad (z_A, z_B, z_C) = O(1/\sqrt{n}).$$

(All O's are in probability.) Consequently,

$$(A13) \qquad \text{the denominator } D \text{ of } \hat{Q} \text{ in } (A10) \text{ is } 1 + O(1/n).$$

Two deterministic approximations to the numerator N were presented in (12–13).

Proof of Theorem 2. By Lemma 1, we may assume $\overline{a} = \overline{b} = \overline{c} = 0$. To see this more sharply, recall (3). Let $\hat{\beta}$ be the result of regressing Y on U, V, W, z. Furthermore, let

$$(A14) \qquad Y_i^* = (a_i + a^*)U_i + (b_i + b^*)V_i + (c_i + c^*)W_i.$$

The result of regressing Y^* on U, V, W, z is just $\hat{\beta} + (a^*, b^*, c^*, 0)'$. So the general case of Theorem 2 would follow from the special case. That is why we can, without loss of generality, assume Condition #4. Now

$$(A15) \qquad (a_A, b_B, c_C) = O(1/\sqrt{n}).$$

We use $(A10)$ to evaluate $(A11)$. The denominator of $\hat{Q}$ is essentially 1, i.e., the departure from 1 can be swept into the error term ρ_n, because the departure from 1 gets multiplied by $(z_A, z_B, z_C)' = O(1/\sqrt{n})$. This is a little delicate as we are estimating down to order $1/n^{3/2}$. The departure of the denominator from 1 is multiplied by N, but terms like $a_{AZ}z_A$ are $O(1/n)$ and immaterial, while terms like $(az)_A$ are $O(1)$ by Condition #1 and Proposition 1 (or see the discussion of Proposition 2 below).

For the numerator of $\hat{Q}$, terms like $a_{AZ}z_A$ go into ρ_n: After multiplication by $(z_A, z_B, z_C)'$, they are $O(1/n^{3/2})$. Recall that $\overline{az} = \sum_{i=1}^{n} a_i z_i / n$. What's left of the numerator is $\tilde{Q} + \check{Q}$, where

$$(A16) \qquad \check{Q} = \tilde{p}_A(az - \overline{az})_A + \tilde{p}_B(bz - \overline{bz})_B + \tilde{p}_C(cz - \overline{cz})_C.$$

The term $\tilde{Q}(z_A, z_B, z_C)'$ goes into ζ_n: see (17). The rest of ζ_n comes from (a_A, b_B, c_C) in $(A11)$. The bias in estimating the effects is therefore

$$(A17) \qquad -E\left\{ \check{Q} \begin{pmatrix} z_A \\ z_B \\ z_C \end{pmatrix} \right\}.$$

This can be evaluated by Proposition 1, the relevant variables being az, bz, cz, z.   QED

Additional detail for Theorem 2. We need to show, for instance,

$$\hat{Q}z_A = \tilde{Q}z_A + \check{Q}z_A + O\left(\frac{1}{n^{3/2}}\right).$$

This can be done in three easy steps.

Step 1. $\dfrac{N}{D}z_A = Nz_A + O\left(\dfrac{1}{n^{3/2}}\right).$

Indeed, $N = O(1)$, $D = 1 + O\left(\dfrac{1}{n}\right)$, and $z_A = O\left(\dfrac{1}{\sqrt{n}}\right).$

Step 2. $N = \tilde{Q} + \check{Q} - R,$

where $R = \tilde{p}_A a_A z_A + \tilde{p}_B b_B z_B + \tilde{p}_C c_C z_C$. This is because $(\overline{az})_A = \overline{az}$ and so forth.

Step 3. $R = O\left(\dfrac{1}{n}\right)$ so $Rz_A = O\left(\dfrac{1}{n^{3/2}}\right).$

Remarks. (i) As a matter of notation, $\tilde{Q}$ is deterministic but $\check{Q}$ is random. Both are scalar: compare (12) and (A16). The source of the bias is the covariance between $\check{Q}$ and z_A, z_B, z_C.

(ii) Suppose we add a constant k to z. Instead of (11), we get $\bar{z} = k$ and $\overline{z^2} = 1 + k^2$. Because z_A and so forth are all shifted by the same amount k, the shift does not affect e, f, or $\hat{Q}$: see (A1–3). The multiple-regression estimator for the effect of A is therefore shifted by $\hat{Q}k$; likewise for B, C. This bias does not tend to 0 when sample size grows, but does cancel when estimating differences in effects.

(iii) In applications, we cannot assume the parameters $\bar{a}, \bar{b}, \bar{c}$ are 0—the whole point is to estimate them. The invariance lemma, however, reduces the general case to the more manageable special case, where $\bar{a} = \bar{b} = \bar{c} = 0$, as in the proof of Theorem 2.

(iv) In (19), $K = O(1)$. Indeed, $\bar{z} = 0$, so $\text{cov}(az, z) = \overline{(az)z} = \overline{az^2}$. Now

$$\left|\frac{1}{n}\sum_{i=1}^{n} a_i z_i^2\right| \le \left(\frac{1}{n}\sum_{i=1}^{n} |a_i|^3\right)^{1/3}\left(\frac{1}{n}\sum_{i=1}^{n} |z_i|^3\right)^{2/3}$$

by Hölder's inequality applied to a and z^2. Finally, use Condition #1. The same argument can be used for $\mathrm{cov}(bz, z)$ and $\mathrm{cov}(cz, z)$.

Define $\hat{Q}$ as in (A3); recall (A1–2). The residuals from the multiple regression are $e - \hat{Q}f$ by Lemma 2; according to usual procedures,

$$(A18) \qquad\qquad \hat{\sigma}^2 = |e - \hat{Q}f|^2/(n - 4).$$

Recall f from (A2), and $\hat{Q}$, Q from (A3) and (13).

Lemma 3. *Assume Conditions #1–#3, not #4, and* (11). *Then* $|f|^2/n \to 1$ *and* $\hat{Q} \to Q$. *Convergence is in probability.*

Proof. The first claim follows from (A5) and (A12); the second, from (A10) and Theorem 1.   QED

Proof of Theorem 3. Let M be the 4×4 matrix whose diagonal is $\tilde{p}_A, \tilde{p}_B, \tilde{p}_C, 1$; the last row of M is $(z_A, z_B, z_C, 1)$; the last column of M is $(z_A, z_B, z_C, 1)'$. Pad out M with 0's. Plainly, $X'X/n = M$. As before, $\tilde{p}_A = n_A/n$ is deterministic, and $\tilde{p}_A \to p_A$ by (9). But $z_A = O(1/\sqrt{n})$; likewise for B, C. This proves (i).

For (ii), $e = e - \hat{Q}f + \hat{Q}f$. But $e - \hat{Q}f \perp f$. So $|e - \hat{Q}f|^2 = |e|^2 - \hat{Q}^2|f|^2$. Then

$$
\begin{aligned}
\frac{n-4}{n}\hat{\sigma}^2 &= \frac{|e - \hat{Q}f|^2}{n} \\
&= \frac{|e|^2 - \hat{Q}^2|f|^2}{n} \\
&= \frac{|Y|^2}{n} - \tilde{p}_A(Y_A)^2 - \tilde{p}_B(Y_B)^2 - \tilde{p}_C(Y_C)^2 - \hat{Q}^2\frac{|f|^2}{n} \\
&= \frac{|Y|^2}{n} - \tilde{p}_A(a_A)^2 - \tilde{p}_B(b_B)^2 - \tilde{p}_C(c_C)^2 - \hat{Q}^2\frac{|f|^2}{n}
\end{aligned}
$$

by (A1) and (3). Using (3) again, we get

$$(A19) \qquad\qquad \frac{|Y|^2}{n} = \tilde{p}_A(a^2)_A + \tilde{p}_B(b^2)_B + \tilde{p}_C(c^2)_C.$$

Remember, the dummy variables are orthogonal; as a matter of notation, $(a^2)_A$ is the average of a_i^2 over $i \in A$, and similarly for the other terms.

So

$$(A20) \qquad \frac{n-4}{n}\hat{\sigma}^2 = \tilde{p}_A[(a^2)_A - (a_A)^2]$$
$$+ \tilde{p}_B[(b^2)_B - (b_B)^2]$$
$$+ \tilde{p}_C[(c^2)_C - (c_C)^2] - \hat{Q}^2\frac{|f|^2}{n}.$$

To evaluate $\lim \hat{\sigma}^2$, we may without loss of generality assume Condition #4, by the invariance lemma. Now $a_A = O(1/\sqrt{n})$ and likewise for B, C by $(A15)$. The terms in $(A20)$ involving $(a_A)^2$, $(b_B)^2$, $(c_C)^2$ can therefore be dropped, being $O(1/n)$. Furthermore, $|f|^2/n \to 1$ and $\hat{Q} \to Q$ by Lemma 3. To complete the proof of (ii), we must show that in probability,

$$(A21) \qquad (a^2)_A \to \langle a^2 \rangle, \quad (b^2)_B \to \langle b^2 \rangle, \quad (c^2)_C \to \langle c^2 \rangle.$$

This follows from Condition #1 and Proposition 1. Given (i) and (ii), claim (iii) is immediate.   QED

Proof of Theorem 4. The asymptotic variance of the multiple-regression estimator is given by Theorem 2. The variance of the ITT estimator $Y_C - Y_A$ can be worked out exactly from Proposition 1 (see Example 1). A bit of algebra will now prove Theorem 4.   QED

Proof of Theorem 5. By the invariance lemma, we may as well assume that $\bar{a} = \bar{b} = \bar{c} = 0$. The ITT estimator is unbiased. By Lemma 2, the multiple-regression estimator differs from the ITT estimator by $\hat{Q}z_A$, $\hat{Q}z_B$, $\hat{Q}z_C$. These three random variables sum to 0 by (11) and the balance condition. So their expectations sum to 0. Moreover, the three random variables are exchangeable, so their expectations must be equal. To see the exchangeability more sharply, recall $(A1\text{--}3)$. Because there are no interactions, $Y_i = \delta_i$. So

$$(A22) \qquad e = \delta - \delta_A U - \delta_B V - \delta_C W$$

by $(A1)$, and

$$(A23) \qquad f = z - z_A U - z_B V - z_C W$$

by $(A2)$. These are random n-vectors. The joint distribution of

$$(A24) \qquad e, f, \hat{Q}, z_A, z_B, z_C$$

does not depend on the labels A, B, C: The pairs (δ_i, z_i) are just being divided into three random groups of equal size.   QED

The same argument shows that the multiple-regression estimator for an effect difference (like $\bar{a} - \bar{c}$) is symmetrically distributed around the true value.

Proof of Theorem 6. By Lemma 1, we may assume without loss of generality that $\bar{a} = \bar{b} = \bar{c} = 0$. We can assign subjects to A, B, C by randomly permuting $\{1, 2, \ldots, n\}$: The first n_A subjects go into A, the next n_B into B, and the last n_C into C. Freeze the number of A's, B's— and hence C's—within each level of z. Consider only the corresponding permutations. Over those permutations, z_A is frozen; likewise for B, C. So the denominator of $\hat{Q}$ is frozen: Without condition (11), the denominator must be computed from (A5). In the numerator, z_A, z_B, z_C are frozen, while a_A averages out to zero over the permutations of interest; so do b_B and c_C. With a little more effort, one also sees that $(az)_A$ averages out to zero, as do $(bz)_B$, $(cz)_C$. In consequence, $\hat{Q}z_A$ has expectation 0, and likewise for B, C. Lemma 2 completes the argument.   QED

Remarks. (i) What if $|f| = 0$ in (A2–3)? Then z is a linear combination of the treatment dummies U, V, W; the design matrix $(UVWz)$ is singular, and the multiple-regression estimator is ill-defined. This is not a problem for Theorems 2 or 3, being a low-probability event. But it is a problem for Theorems 4 and 5. The easiest course is to assume the problem away, for instance, requiring

(A25) z is linearly independent of the treatment dummies for every permutation of $\{1, 2, \ldots, n\}$.

Another solution is more interesting: Exclude the permutations where $|f| = 0$, and show the multiple-regression estimator is conditionally unbiased, i.e., has the right average over the remaining permutations.

(ii) All that is needed for Theorems 2–4 is an a priori bound on absolute third moments in Condition #1, rather than fourth moments; third moments are used for the CLT by Höglund (1978). The new awkwardness is in proving results like (A21), but this can be done by familiar truncation arguments. More explicitly, let $x_1, \ldots, x_n$ be real numbers, with

(A26)
$$\frac{1}{n} \sum_{i=1}^{n} |x_i|^\alpha < L.$$

Here, $1 < \alpha < \infty$ and $0 < L < \infty$. As will be seen below, $\alpha = 3/2$ is the relevant case. In principle, the x's can be doubly subscripted, for instance, x_1 can change with n. We draw m times at random without replacement from $\{x_1, \ldots, x_n\}$, generating random variables $X_1, \ldots, X_m$.

Proposition 2. *Under condition (A26), as $n \to \infty$, if m/n converges to a positive limit that is less than 1, then*

$$\frac{1}{m}(X_1 + \cdots + X_m) - E(X_i)$$

converges in probability to 0.

Proof. Assume without loss of generality that $E(X_i) = 0$. Let M be a positive number. Let $U_i = X_i$ when $|X_i| < M$; else, let $U_i = 0$. Let $V_i = X_i$ when $|X_i| \geq M$; else, let $V_i = 0$. Thus, $U_i + V_i = X_i$. Let $\mu = E(U_i)$, so $E(V_i) = -\mu$. Now $\frac{1}{m}(U_1 + \cdots + U_m) - \mu \to 0$. Convergence is almost sure, and rates can be given; see, for instance, Hoeffding (1963).

Consider next $\frac{1}{m}(W_1 + \cdots + W_m)$, where $W_i = V_i + \mu$. The W_i are exchangeable. Fix β with $1 < \beta < \alpha$. By Minkowski's inequality,

$$(A27) \qquad \left[E\left(\left| \frac{W_1 + \cdots + W_m}{m} \right|^\beta \right) \right]^{1/\beta} \leq [E(|W_i|^\beta)]^{1/\beta}.$$

When M is large, the right hand side of $(A27)$ is uniformly small, by a standard argument starting from $(A26)$. In essence,

$$\int_{|X_i|>M} |X_i|^\beta < M^{\beta-\alpha} \int_{|X_i|>M} |X_i|^\alpha < L/M^{\alpha-\beta}. \qquad \text{QED}$$

In proving Theorem 2, we needed $(az)_A = O(1)$. If there is an a priori bound on the absolute third moments of a and z, then $(A26)$ will hold for $x_i = a_i z_i$ and $\alpha = 3/2$ by the Cauchy-Schwarz inequality. On the other hand, a bound on the second moments would suffice by Chebychev's inequality. To get $(A21)$ from third moments, we would for instance set $x_i = a_i^2$; again, $\alpha = 3/2$.

Acknowledgments

Donald Green generated a string of examples where the regression estimator was unbiased in finite samples; ad hoc explanations for the findings gradually evolved into Theorems 5 and 6. Sandrine Dudoit, Winston Lim, Michael Newton, Terry Speed, and Peter Westfall made useful suggestions, as did an anonymous associate editor.

13

Randomization Does Not Justify Logistic Regression

ABSTRACT. *The logit model is often used to analyze experimental data. However, randomization does not justify the model, so the usual estimators can be inconsistent. A consistent estimator is proposed. Neyman's non-parametric setup is used as a benchmark. In this setup, each subject has two potential responses, one if treated and the other if untreated; only one of the two responses can be observed. Beside the mathematics, there are simulation results, a brief review of the literature, and some recommendations for practice.*

13.1 Introduction

The logit model is often fitted to experimental data. As explained below, randomization does not justify the assumptions behind the model. Thus, the conventional estimator of log odds is difficult to interpret; an alternative will be suggested. Neyman's setup is used to define parameters and prove results. (Grammatical niceties apart, the terms "logit model" and "logistic regression" are used interchangeably.)

After explaining the models and estimators, we present simulations to illustrate the findings. A brief review of the literature describes the history and current usage. Some practical recommendations are derived from the theory. Analytic proofs are sketched at the end of the chapter.

Statistical Science (2008) 23: 237–50.

13.2 Neyman

There is a study population with n subjects indexed by $i = 1, \ldots, n$. Fix π_T with $0 < \pi_T < 1$. Choose $n\pi_T$ subjects at random and assign them to the treatment condition. The remaining $n\pi_C$ subjects are assigned to a control condition, where $\pi_C = 1 - \pi_T$. According to Neyman (1923), each subject has two responses: $Y_i{}^T$ if assigned to treatment, and $Y_i{}^C$ if assigned to control. The responses are 1 or 0, where 1 is "success" and 0 is "failure." Responses are fixed, that is, not random.

If i is assigned to treatment (T), then $Y_i{}^T$ is observed. Conversely, if i is assigned to control (C), then $Y_i{}^C$ is observed. Either one of the responses may be observed, but not both. Thus, responses are subject-level parameters. Even so, responses are estimable (Section 13.9). Each subject has a covariate Z_i, unaffected by assignment; Z_i is observable. In this setup, the only stochastic element is the randomization: Conditional on the assignment variable X_i, the observed response

$$Y_i = X_i Y_i{}^T + (1 - X_i) Y_i{}^C$$

is deterministic.

Population-level ITT (intention-to-treat) parameters are defined by taking averages over all n subjects in the study population:

$$(1) \qquad \alpha^T = \frac{1}{n} \sum Y_i^T, \quad \alpha^C = \frac{1}{n} \sum Y_i^C.$$

For example, α^T is the fraction of successes if all subjects are assigned to T; similarly for α^C. A parameter of considerable interest is the differential log odds of success,

$$(2) \qquad \Delta = \log \frac{\alpha^T}{1 - \alpha^T} - \log \frac{\alpha^C}{1 - \alpha^C}.$$

The logit model is all about log odds (more on this below). The parameter Δ defined by (2) may therefore be what investigators think is estimated by running logistic regressions on experimental data, although that idea is seldom explicit.

13.2.1 The intention-to-treat principle

The intention-to-treat principle, which goes back to Bradford Hill (1961, p. 259), is to make comparisons based on treatment assigned rather than treatment received. Such comparisons take full advantage of the randomization, thereby avoiding biases due to self-selection. For example, the unbiased estimators for the parameters in (1) are the fraction of successes in the treatment group and the control group, respectively. Below, these will be called *ITT estimators*. ITT estimators measure the effect of assignment rather than treatment. With crossover, the distinction matters. For additional discussion, see Freedman (2006b).

13.3 The logit model

To set up the logit model, we consider a study population of n subjects, indexed by $i = 1, \ldots, n$. Each subject has three observable random variables: Y_i, X_i, and Z_i. Here, Y_i is the response, which is 0 or 1. The primary interest is the "effect" of X_i on Y_i, and Z_i is a covariate.

For our purposes, the best way to formulate the model involves a latent (unobservable) random variable U_i for each subject. These are assumed to be independent across subjects, with a common *logistic distribution*: for $-\infty < u < \infty$,

$$(3) \qquad P(U_i < u) = \frac{\exp(u)}{[1 + \exp(u)]},$$

where $\exp(u) = e^u$. The model assumes that X and Z are *exogenous*, that is, independent of U. More formally, $\{X_i, Z_i : i = 1, \ldots, n\}$ is assumed to independent of $\{U_i : i = 1, \ldots, n\}$. Finally, the model assumes that $Y_i = 1$ if

$$\beta_1 + \beta_2 X_i + \beta_3 Z_i + U_i > 0;$$

else, $Y_i = 0$.

Given X and Z, it follows that responses are independent across subjects, the conditional probability that $Y_i = 1$ being $p(\beta, X_i, Z_i)$, where

$$(4) \qquad p(\beta, x, z) = \frac{\exp(\beta_1 + \beta_2 x + \beta_3 z)}{1 + \exp(\beta_1 + \beta_2 x + \beta_3 z)}.$$

(To verify this, check first that $-U_i$ is distributed like $+U_i$.) The parameter vector $\beta = (\beta_1, \beta_2, \beta_3)$ is usually estimated by maximum likelihood. We denote the MLE by $\hat{\beta}$.

13.3.1 Interpreting the coefficients in the model

In the case of primary interest, X_i is 1 or 0. Consider the log odds λ_i^T of success when $X_i = 1$, as well as the log odds λ_i^C when $X_i = 0$.

In view of (4),

(5)
$$\lambda_i^T = \log \frac{p(\beta, 1, Z_i)}{1 - p(\beta, 1, Z_i)}$$
$$= \beta_1 + \beta_2 + \beta_3 Z_i,$$
$$\lambda_i^C = \log \frac{p(\beta, 0, Z_i)}{1 - p(\beta, 0, Z_i)}$$
$$= \beta_1 + \beta_3 Z_i.$$

In particular, $\lambda_i^T - \lambda_i^C = \beta_2$ for all i, whatever the value of Z_i may be. Thus, according to the model, $X_i = 1$ adds β_2 to the log odds of success.

13.3.2 Application to experimental data

To apply the model to experimental data, define $X_i = 1$ if i is assigned to T, while $X_i = 0$ if i assigned to C. Notice that the model is not justified by randomization. Why would the logit specification be correct rather than the probit—or anything else? What justifies the choice of covariates? Why are they exogenous? If the model is wrong, what is $\hat{\beta}_2$ supposed to be estimating? The last rhetorical question may have an answer: The parameter Δ in (2) seems like a natural choice, as indicated above.

More technically, from Neyman's perspective, given the assignment variables $\{X_i\}$, the responses are deterministic: $Y_i = Y_i^T$ if $X_i = 1$, while $Y_i = Y_i^C$ if $X_i = 0$. The logit model, on the other hand, views the responses $\{Y_i\}$ as random—with a specified distribution—given the assignment variables and covariates.

The contrast is therefore between two styles of inference.

- Randomization provides a known distribution for the assignment variables; statistical inferences are based on this distribution.

- Modeling assumes a distribution for the latent variables; statistical inferences are based on that assumption. Furthermore, model-based inferences are conditional on the assignment variables and covariates.

A similar contrast will be found in other areas too, including sample surveys. See Koch and Gillings (2005) for a review and pointers to the literature.

13.3.3 What if the logit model is right?

Suppose the model is right, and there is a causal interpretation. We can intervene and set X_i to 1 without changing the Z's or U's, so $Y_i = 1$ if and only if $\beta_1 + \beta_2 + \beta_3 Z_i + U_i > 0$. Similarly, we can set X_i to 0 without changing anything else, and then $Y_i = 1$ if and only if $\beta_1 + \beta_3 Z_i + U_i > 0$. Notice that β_2 appears when X_i is set to 1, but disappears when X_i is set to 0.

On this basis, for each subject, whatever the value of Z_i may be, setting X_i to 1 rather than 0 adds β_2 to the log odds of success. If the model is right, β_2 is a very useful parameter, which is well estimated by the MLE provided n is large. For additional detail on causal modeling and estimation, see Freedman (2009).

Even if the model is right and n is large, β_2 differs from Δ in (2). For instance, α^T will be nearly equal to $\frac{1}{n} \sum_{i=1}^{n} p(\beta, 1, Z_i)$. So $\log \alpha^T - \log(1 - \alpha^T)$ will be nearly equal to

$$(6) \qquad \log \left(\frac{1}{n} \sum_{i=1}^{n} p(\beta, 1, Z_i) \right)$$
$$- \log \left(\frac{1}{n} \sum_{i=1}^{n} \left[1 - p(\beta, 1, Z_i) \right] \right).$$

Likewise, $\log \alpha^C - \log(1 - \alpha^C)$ will be nearly equal to

$$(7) \qquad \log \left(\frac{1}{n} \sum_{i=1}^{n} p(\beta, 0, Z_i) \right)$$
$$- \log \left(\frac{1}{n} \sum_{i=1}^{n} \left[1 - p(\beta, 0, Z_i) \right] \right).$$

Taking the log of an average, however, is quite different from taking the average of the logs. The former is relevant for Δ in (2), as shown by (6–7); the latter for computing

$$(8) \qquad \frac{1}{n} \sum_{i=1}^{n} \left(\lambda_i^T - \lambda_i^C \right) = \beta_2,$$

where the log odds of success λ_i^T and λ_i^C were computed in (5).

The difference between averaging inside and outside the logs may be surprising at first, but in the end, that difference is why you should put confounders like Z into the equation — if you believe the model. Section 13.9 gives further detail and an inequality relating β_2 to Δ.

13.3.4 From Neyman to logits

How could we get from Neyman to the logit model? To begin with, we would allow Y_i^T and Y_i^C to be 0–1 valued random variables; the Z_i can be random too. To define the parameters in (1) and (2), we would replace Y_i^T and Y_i^C by their expectations. None of this is problematic, and the Neyman model is now extremely general and flexible. Randomization makes the assignment variables $\{X_i\}$ independent of the potential responses Y_i^T, Y_i^C.

To get the logit model, however, we would need to specialize this setup considerably, assuming the existence of IID logistic random variables U_i, independent of the covariates Z_i, with

(9) $\quad Y_i^T = 1$ if and only if $\beta_1 + \beta_2 + \beta_3 Z_i + U_i > 0,$

$\quad\quad Y_i^C = 1$ if and only if $\beta_1 + \beta_3 Z_i + U_i > 0.$

Besides (9), the restrictive assumptions are the following:

(i) The U_i are independent of the Z_i.

(ii) The U_i are independent across subjects i.

(iii) The U_i have a common logistic distribution.

If you are willing to make these assumptions, what randomization contributes is a guarantee that the assignment variables $\{X_i\}$ are independent of the latent variables $\{U_i\}$. Randomization does not guarantee the existence of the U_i, or the truth of (9), or the validity of (i)-(ii)-(iii).

13.4 A plug-in estimator for the log odds

If a logit model is fitted to experimental data, average predicted probabilities are computed by plugging $\hat\beta$ into (4):

(10a) $$\tilde\alpha^T = \frac{1}{n}\sum_{i=1}^n p(\hat\beta, 1, Z_i),$$

$$\tilde\alpha^C = \frac{1}{n}\sum_{i=1}^n p(\hat\beta, 0, Z_i).$$

(The tilde notation is needed; $\hat\alpha^T$ and $\hat\alpha^C$ will make their appearances momentarily.) Then the differential log odds in (2) can be estimated by plugging into the formula for Δ:

(10b) $$\tilde\Delta = \log\frac{\tilde\alpha^T}{1-\tilde\alpha^T} - \log\frac{\tilde\alpha^C}{1-\tilde\alpha^C}.$$

As will be seen below, $\tilde{\Delta}$ is consistent. The ITT estimators are defined as follows:

(11a) $$\hat{\alpha}^T = \frac{1}{n_T} \sum_{i \in T} Y_i, \quad \hat{\alpha}^C = \frac{1}{n_C} \sum_{i \in C} Y_i,$$

where $n_T = n\pi_T$ is the number of subjects in T and $n_C = n\pi_C$ is the number of subjects in C. Then

(11b) $$\hat{\Delta} = \log \frac{\hat{\alpha}^T}{1 - \hat{\alpha}^T} - \log \frac{\hat{\alpha}^C}{1 - \hat{\alpha}^C}.$$

The ITT estimators are consistent too, with asymptotics discussed in Freedman (2008a,b [Chapter 12]). The intuition: $\hat{\alpha}^T$ is the average success rate in the treatment group, and the sample average is a good estimator for the population average. The same reasoning applies to $\hat{\alpha}^C$.

13.5 Simulations

The simulations in this section are designed to show what happens when the logit model is fitted to experimental data. The data generating mechanism is not the logit, so the simulations illustrate the consequences of specification error. The stochastic element is the randomization, as in Section 13.2. (Some auxiliary randomness is introduced to construct the individual-level parameters, but that gets conditioned away.) Let $n = 100, 500, 1000, 5000$. For $i = 1, \ldots, n$,

   let U_i, V_i be IID uniform random variables,
   let $Z_i = V_i$,
   let $Y_i^C = 1$ if $U_i > 1/2$, else $Y_i^C = 0$, and
   let $Y_i^T = 1$ if $U_i + V_i > 3/4$, else $Y_i^T = 0$.

Suppose n is very large. The mean response in the control condition is around $P(U_i > 1/2) = 1/2$, so the odds of success in the control condition are around 1. (The qualifiers are needed because the U_i are chosen at random.) The mean response in the treatment condition is around $23/32$, because

$$P(U_i + V_i < 3/4) = (1/2) \times (3/4)^2 = 9/32.$$

So the odds of success in the treatment condition are around

$$(23/32)/(9/32).$$

The parameter Δ in (2) will therefore be around

$$\log \frac{23/32}{9/32} - \log 1 = \log \frac{23}{9} = 0.938.$$

Even for moderately large n, nonlinearity in (2) is an issue, and the approximation given for Δ is unsatisfactory.

The construction produces individual-level variation: A majority of subjects are unaffected by treatment, about 1/4 are helped, about 1/32 are harmed. The covariate is reasonably informative about the effect of treatment—if Z_i is big, treatment is likely to help.

Having constructed Z_i, Y_i^C, and Y_i^T for $i = 1, \ldots, n$, we freeze them, and simulate 1000 randomized controlled experiments, where 25% of the subjects are assigned to C and 75% to T. We fit a logit model to the data generated by each experiment, computing the MLE $\hat{\beta}$ and the plug-in estimator $\tilde{\Delta}$ defined by (10b). The average of the 1000 $\hat{\beta}$'s and $\tilde{\Delta}$'s is shown in Table 13.1, along with the true value of the differential log odds, namely, Δ in (2). We distinguish between the standard deviation and the standard error. Below each average, the table shows the corresponding standard deviation.

For example, with $n = 100$, the average of the 1000 $\hat{\beta}_2$'s is 1.344. The standard deviation is 0.540. The Monte Carlo standard error in the average is therefore

$$0.540/\sqrt{1000} = 0.017.$$

The average of the 1000 plug-in estimates is 1.248, and the true Δ is 1.245. When $n = 5000$, the bias in $\hat{\beta}_2$ as an estimator of Δ is $1.134 - 0.939 = 0.195$, with a Monte Carlo standard error of $0.076/\sqrt{1000} = 0.002$. There is a confusion to avoid: n is the number of subjects in the

Table 13.1 Simulations for $n = 100, 500, 1000, 5000$. Twenty-five percent of the subjects are assigned at random to C, the rest to T. Averages and standard deviations are shown for the MLE $\hat{\beta}$ and the plug-in estimator $\tilde{\Delta}$, as well as the true value of the differential log odds Δ defined in (2). There are 1000 simulated experiments for each n.

n	$\hat{\beta}_1$	$\hat{\beta}_2$	$\hat{\beta}_3$	Plug-in	Truth
100	−0.699	1.344	2.327	1.248	1.245
	0.457	0.540	0.621	0.499	
500	−1.750	1.263	3.318	1.053	1.053
	0.214	0.234	0.227	0.194	
1000	−1.568	1.046	3.173	0.885	0.883
	0.155	0.169	0.154	0.142	
5000	−1.676	1.134	3.333	0.937	0.939
	0.071	0.076	0.072	0.062	

study population, varying from 100 to 5000, but the number of simulated experiments is fixed at 1000. (The *Monte Carlo standard error* measures the impact of randomness in the simulation, which is based on a sample of "only" 1000 observations.)

The plug-in estimator is essentially unbiased and less variable than $\hat{\beta}_2$. The true value of Δ changes from one n to the next, since values of $Y_i{}^C$, $Y_i{}^T$ are generated by Monte Carlo for each n. Even with $n = 5000$, the true value of Δ would change from one run to another, the standard deviation across runs being about 0.03 (not shown in the table).

Parameter choices—for instance, the joint distribution of (U_i, V_i)— were somewhat arbitrary. Surprisingly, bias depends on the fraction of subjects assigned to T. On the other hand, changing the cutpoints used to define $Y_i{}^C$ and $Y_i{}^T$ from 1/2 and 3/4 to 0.95 and 1.5 makes little differ- ence to the performance of $\hat{\beta}_2$ and the plug-in estimator. In these exam- ples, the plug-in estimator and the ITT estimators are essentially unbiased; the latter has slightly smaller variance.

The bias in $\hat{\beta}_2$ depends very much on the covariate. For instance, if the covariate is $U_i + V_i$ rather than V_i, then $\hat{\beta}_2$ hovers around 3. Truth remains in the vicinity of 1, so the bias in $\hat{\beta}_2$ is huge. The plug-in and ITT estimators remain essentially unbiased, with variances much smaller than $\hat{\beta}_2$; the ITT estimator has higher variance than the plug-in estimator (data not shown for variations on the basic setup, or ITT estimators).

The Monte Carlo results suggest the following:

   (i) As n gets large, the MLE $\hat{\beta}$ stabilizes.
  (ii) The plug-in estimator $\tilde{\Delta}$ is a good estimator of the differential log odds Δ.
 (iii) $\hat{\beta}_2$ tends to over-estimate $\Delta > 0$.

These points will be verified analytically below.

13.6 Extensions and implications

Suppose the differential log odds of success is the parameter to be estimated. Then $\hat{\beta}_2$ is generally the wrong estimator to use—whether the logit model is right or the logit model is wrong (Section 13.9 has a mathematical proof). It is better to use the plug-in estimator (10) or the ITT estimator (11). These estimators are nearly unbiased, and in many examples have smaller variances too.

Although details remain to be checked, the convergence arguments in Section 13.8 seem to extend to probits, the parameter corresponding to (2) being

$$\Phi^{-1}(\alpha^T) - \Phi^{-1}(\alpha^C),$$

where Φ is the standard normal distribution function. On the other hand, with the probit, the plug-in estimators are unlikely to be consistent, since the analogs of the likelihood equations (16–18) below involve weighted averages rather than simple averages.

In simulation studies (not reported here), the probit behaves very much like the logit, with the usual difference in scale: Probit coefficients are about 5/8 of their logit counterparts (Amemiya 1981, p. 1487). Numerical calculations also confirm inconsistency of the plug-in estimators, although the asymptotic bias is small.

According to the logit and probit models, if treatment improves the chances of success, it does so for all subjects. In reality, of course, treatment may help some subgroups and hurt others. Subgroup analysis can therefore be a useful check on the models. Consistency of the plug-in estimators—as defined here—does not preclude subgroup effects.

Logit models, probit models, and their ilk are not justified by randomization. This has implications for practice. Rates and averages for the treatment and control groups should be compared before the modeling starts. If the models change the substantive results, that raises questions that need to be addressed.

There may be an objection that models take advantage of additional information. The objection has some merit *if* the models are right or nearly right. On the other hand, if the models cannot be validated, conclusions drawn from them must be shaky. "Cross-tabulation before regression" is a slogan to be considered.

13.7 Literature review

Logit and probit models are often used to analyze experimental data. See Pate and Hamilton (1992), Gilens (2001), Hu (2003), Duch and Palmer (2004), Frey and Meier (2004), and Gertler (2004). The plug-in estimator discussed here is similar to the "average treatment effect" sometimes reported in the literature; see, for example, Evans and Schwab (1995). For additional discussion, see Lane and Nelder (1982) and Brant (1996).

Lim (1999) conjectured that plug-in estimators based on the logit model would be consistent, with an informal argument based on the likelihood equation. He also conjectured inconsistency for the probit. Middleton (2007) discusses inconsistent logit estimators.

The logistic distribution may first have been used to model population growth. See Verhulst (1845) and Yule (1925). Later, the distribution was used to model dose-response in bioassays (Berkson 1944). An early biomedical application to causal inference is Truett, Cornfield, and Kannel (1967). The history is considered further in Freedman (2005). The present chapterextends previous results on linear regression (Freedman 2008a,b [Chapter 12]).

Statistical models for causation go back to Jerzy Neyman's work on agricultural experiments in the early part of the twentieth century. The key paper, Neyman (1923), was in Polish. There was an extended discussion by Scheffé (1956), and an English translation by Dabrowska and Speed (1990). The model was covered in elementary textbooks in the 1960's; see, for instance, Hodges and Lehmann (1964, section 9.4). The setup is often called "Rubin's model," due in part to Holland (1986); that mistakes the history.

Neyman, Kolodziejczyk, and Iwaszkiewicz (1935) develop models with subject-specific random effects that depend on assignment, the objective being to estimate average expected values under various circumstances. This is discussed in section 4 of Scheffé (1956).

Heckman (2000) explains the role of potential outcomes in econometrics. In epidemiology, a good source is Robins (1999). Rosenbaum (2002) proposes using models and permutation tests as devices for hypothesis testing. This avoids difficulties outlined here: (i) if treatment has no effect, then $Y_i^T = Y_i^C = Y_i$ for all i; and (ii) randomization makes all permutations of i equally likely—which is just what permutation tests need.

Rosenblum and van der Laan (2009) suggest that, at least for purposes of hypothesis testing, robust standard errors will fix problems created by specification error. Such optimism is unwarranted. Under the alternative hypothesis, the robust standard error is unsatisfactory because it ignores bias (Freedman 2006a [Chapter 17]).

Under the null hypothesis, the robust standard error may be asymptotically correct, but using it can reduce power (Freedman, 2008a,b [Chapter 12]). In any event, if the null hypothesis is to be tested using model-based adjustments, exact P-values can be computed by permutation methods, as suggested by Rosenbaum (2002).

Models are often deployed to infer causation from association. For a discussion from various perspectives, see Berk (2004), Brady and Collier (2004), and Freedman (2005). The last summarizes a cross-section of the literature on this topic (pp. 192–200).

Consider a logit model like the one in Section 13.3. Omitting the covariate Z from the equation is called *marginalizing* over Z. The model is *collapsible* if the marginal model is again logit with the same β_2. In other words, given the X's, the Y's are conditionally independent, and

$$P(Y_i = 1|X_i) = \frac{\exp(\beta_1 + \beta_2 X_i)}{1 + \exp(\beta_1 + \beta_2 X_i)}.$$

Guo and Geng (1995) give conditions for collapsibility; also see Ducharme and Lepage (1986). Gail (1986, 1988) discusses collapsing when a design is balanced. Robinson and Jewell (1991) show that collapsing will usually decrease variance: Logit models differ from linear models. Aris et al. (2000) review the literature and consider modeling strategies to compensate for non-collapsibility.

13.8 Sketch of proofs

We are fitting the logit model, which is incorrect, to data from an experiment. As before, let X_i be the assignment variable, so $X_i = 1$ if $i \in T$ and $X_i = 0$ if $i \in C$. Let Y_i be the observed response, so $Y_i = X_i Y_i^T + (1 - X_i) Y_i^C$. Let $L_n(\beta)$ be the "log-likelihood function" to be maximized. The quote marks are there because the model is wrong; L_n is therefore only a pseudo-log-likelihood function. Abbreviate $p_i(\beta)$ for $p(\beta, X_i, Z_i)$ in (4). The formula for $L_n(\beta)$ is this:

(12a) $$L_n(\beta) = \sum_{i=1}^{n} T_i,$$

where

(12b) $$T_i = \log[1 - p_i(\beta)] + (\beta_1 + \beta_2 X_i + \beta_3 Z_i) Y_i.$$

(The T is for term, not treatment.) It takes a moment to verify (12), starting from the equation

(13) $$T_i = Y_i \log(p_i) + (1 - Y_i) \log(1 - p_i).$$

Each T_i is negative. The function $\beta \to L_n(\beta)$ is strictly concave, as one sees by proving that L_n'' is a negative definite matrix. Consequently, there is a unique maximum at the MLE $\hat{\beta}_n$. We write $\hat{\beta}_n$ to show dependence on the size n of the study population, although that creates a conflict in the notation. If pressed, we could write $\hat{\beta}_{n,j}$ for the jth component of the MLE.

The ith row of the "design matrix" is $(1, X_i, Z_i)$. Tacitly, we are assuming this matrix is nonsingular. For large n, the assumption will follow from regularity conditions to be imposed. The concavity of L_n is well known. See, for instance, pp. 122–23 in Freedman (2005) or p. 273 in Amemiya (1985). Pratt (1981) discusses the history and proves a more general result.

For reference, we record one variation on these ideas. Let M be an $n \times p$ matrix of rank p; write M_i for the ith row of M. Let y be an $n \times 1$ vector of 0's and 1's. Let β be a $p \times 1$ vector. Let $w_i > 0$ for $i = 1, \ldots, n$. Consider M and y as fixed, β as variable. Define $L(\beta)$ as

$$\sum_{i=1}^{n} w_i \{ - \log[1 + \exp(M_i \cdot \beta)] + (M_i \cdot \beta) y_i \}.$$

Proposition 1. *The function* $\beta \to L(\beta)$ *is strictly concave.*

One objective in the rest of this section is showing that

(14) $\qquad \beta_n$ converges to a limit β_∞ as $n \to \infty$.

A second objective is showing that

(15) $\qquad$ the plug-in estimator $\tilde{\Delta}$ is consistent.

The argument actually shows a little more. The plug-in estimator $\tilde{\alpha}^T$, the ITT estimator $\hat{\alpha}^T$, and the parameter α^T become indistinguishable as the size n of the study population grows; likewise for $\tilde{\alpha}^C$, $\hat{\alpha}^C$, and α^C.

The ITT estimators $\hat{\alpha}^T$, $\hat{\alpha}^C$ were defined in (11). Recall too that $n_T = n\pi_T$ and $n_C = n\pi_C$ are the numbers of subjects in T and C respectively. The statement of Lemma 1 involves the *empirical distribution* of Z_i for $i \in T$, which assigns mass $1/n_T$ to Z_i for each $i \in T$. Similarly, the empirical distribution of Z_i for $i \in C$ assigns mass $1/n_C$ to Z_i for each $i \in C$.

To prove Lemma 1, we need the likelihood equation $L'_n(\beta) = 0$. This vector equation unpacks to three scalar equations in three unknowns, the components of β that make up $\hat{\beta}_n$:

(16) $\qquad \dfrac{1}{n_T} \sum_{i \in T} p(\hat{\beta}_n, 1, Z_i) = \dfrac{1}{n_T} \sum_{i \in T} Y_i,$

(17) $\qquad \dfrac{1}{n_C} \sum_{i \in C} p(\hat{\beta}_n, 0, Z_i) = \dfrac{1}{n_C} \sum_{i \in C} Y_i,$

(18) $\qquad \dfrac{1}{n} \sum_{i=1}^{n} p(\hat{\beta}_n, X_i, Z_i) Z_i = \dfrac{1}{n} \sum_{i=1}^{n} Y_i Z_i.$

This follows from (12–13) after differentiating with respect to β_1, β_2, and β_3—and then doing a bit of algebra.

Lemma 1. *If the empirical distribution of Z_i for $i \in T$ matches the empirical distribution for $i \in C$ (the first balance condition), then the plug-in estimators $\tilde{\alpha}^T$ and $\tilde{\alpha}^C$ match the ITT estimators. More explicitly,*

$$\frac{1}{n} \sum_{i=1}^{n} p(\hat{\beta}_n, 1, Z_i) = \frac{1}{n_T} \sum_{i \in T} Y_i,$$

$$\frac{1}{n} \sum_{i=1}^{n} p(\hat{\beta}_n, 0, Z_i) = \frac{1}{n_C} \sum_{i \in C} Y_i.$$

Proof. The plug-in estimators $\tilde{\alpha}^T$, $\tilde{\alpha}^C$ were defined in (10); the ITT estimators $\hat{\alpha}^T$, $\hat{\alpha}^C$ in (11). We begin with $\tilde{\alpha}^T$. By (16),

$$\frac{1}{n_T} \sum_{i \in T} p(\hat{\beta}_n, 1, Z_i) = \frac{1}{n_T} \sum_{i \in T} Y_i = \hat{\alpha}^T.$$

By the balance condition,

$$\frac{1}{n_C} \sum_{i \in C} p(\hat{\beta}_n, 1, Z_i) = \frac{1}{n_T} \sum_{i \in T} p(\hat{\beta}_n, 1, Z_i)$$

equals $\hat{\alpha}^T$ too. Finally, the average of $p(\hat{\beta}_n, 1, Z_i)$ over all i is a mixture of the averages over T and C. So $\tilde{\alpha}^T = \hat{\alpha}^T$ as required. The same argument works for $\tilde{\alpha}^C$, using (17).   QED

For the next lemma, recall α^T, α^C from (1). The easy proof is omitted, being very similar to the proof of the previous result.

Lemma 2. *Suppose the empirical distribution of the pairs (Y_i^T, Y_i^C) for $i \in T$ matches the empirical distribution for $i \in C$ (the second balance condition). Then $\hat{\alpha}^T = \alpha^T$ and $\hat{\alpha}^C = \alpha^C$.*

Lemma 3. *Let x be any real number. Then*

$$e^x - \frac{1}{2}e^{2x} < \log(1 + e^x) < e^x,$$

$$x + e^{-x} - \frac{1}{2}e^{-2x} < \log(1 + e^x) < x + e^{-x}.$$

The first bound is useful when x is large and negative; the second, when x is large and positive. To get the second bound from the first, write $1 + e^x = e^x(1 + e^{-x})$, then replace x by $-x$. The first bound will look more familiar on substituting $y = e^x$. The proof is omitted, being "just" calculus.

For the next result, let G be an open, bounded, convex subset of Euclidean space. Let f_n be a strictly concave function on G, converging uniformly to f_∞, which is also strictly concave. Let f_n take its maximum at x_n, while f_∞ takes its maximum at $x_\infty \in G$. Although the lemma is well known, a proof may be helpful. We write $G \setminus H$ for the set of points that are in G but not in H.

Lemma 4. $x_n \to x_\infty$ and $f_n(x_n) \to f_\infty(x_\infty)$.

Proof. Choose a small neighborhood H of $x_\infty = \arg\max f_\infty$. There is a small positive δ with $f_\infty(x) < f_\infty(x_\infty) - \delta$ for $x \in G \setminus H$. For all sufficiently large n, we have $|f_n - f_\infty| < \delta/3$. In particular, $f_n(x_\infty) > f_\infty(x_\infty) - \delta/3$. On the other hand, if $x \in G \setminus H$, then

$$f_n(x) < f_\infty(x) + \delta/3 < f_\infty(x_\infty) - 2\delta/3.$$

Thus, $\arg\max f_n \in H$. And $f_n(x_n) \geq f_n(x_\infty) > f_\infty(x_\infty) - \delta/3$. In the other direction, $f_\infty(x_\infty) \geq f_\infty(x_n) > f_n(x_n) - \delta/3$. So

$$|\max f_n - \max f_\infty| < \delta/3,$$

which completes the proof.   QED

For the final lemma, consider a population consisting of n objects. Suppose r are red, and $r/n \to \rho$ with $0 < \rho < 1$. (The remaining $n - r$ objects are colored black.) Now choose m out of the n objects at random without replacement, where $m/n \to \lambda$ with $0 < \lambda < 1$. Let X_m be the number of red objects that are chosen. So X_m is hypergeometric. The lemma puts no conditions on the joint distribution of the $\{X_m\}$. Only the marginals are relevant.

Lemma 5. $X_m/n \to \lambda\rho$ almost surely as $n \to \infty$.

Proof. Of course, $E(X_m) = rm/n$. The lemma can be proved by using Chebychev's inequality, after showing that

$$E\left[\left(X_m - r\frac{m}{n}\right)^4\right] = O(n^2).$$

Tedious algebra can be reduced by appealing to theorem 4 in Hoeffding (1963). In more detail, let W_i be independent 0–1 variables with $P(W_i = 1) = r/n$. Thus, $\sum_{i=1}^{m} W_i$ is the number of reds in m draws with replacement, while X_m is the number of reds in m draws without replacement. According to Hoeffding's theorem, X_m is more concentrated around the common expected value. In particular,

$$E\left\{\left(X_m - r\frac{m}{n}\right)^4\right\} < E\left\{\left[\sum_{i=1}^{m}\left(W_i - \frac{r}{n}\right)\right]^4\right\}.$$

Expanding $\left[\sum_{i=1}^{m}(W_i - \frac{r}{n})\right]^4$ yields m terms of the form $(W_i - \frac{r}{n})^4$. Each of these terms is bounded above by 1. Next consider terms like $(W_i - \frac{r}{n})^2(W_j - \frac{r}{n})^2$ with $i \neq j$. The number of such terms is of order m^2, and each term is bounded above by 1. All remaining terms have expectation 0. Thus, $E[(X_n - r\frac{m}{n})^4]$ is of order $m^2 < n^2$.   QED

Note. There are m^4 terms in $(a_1 + \cdots + a_m)^4 = \sum_{ijk\ell} a_i a_j a_k a_\ell$. By combinatorial arguments—

(i) m terms are like a_i^4, with one index only.

(ii) $3m(m-1)$ are like $a_i^2 a_j^2$, with two different indices.

(iii) $4m(m-1)$ are like $a_i^3 a_j$, with two different indices.

(iv) $6m(m-1)(m-2)$ are like $a_i^2 a_j a_k$, with three different indices.

(v) $m(m-1)(m-2)(m-3)$ are like $a_i a_j a_k a_\ell$, with four different indices.

The counts can also be derived from the "multinomial theorem," which expands $(a_1 + \cdots + a_m)^N$. For an early—and very clear—textbook exposition, see Chrystal (1889, pp. 14–15). A little care is needed, since our counts do not restrict the order of the indices: $i < j$ and $i > j$ are both allowed. By contrast, in the usual statements of the multinomial theorem, indices are ordered ($i < j$). German scholarship traces the theorem ("der polynomische Lehrsatz") back to correspondence between Leibniz and Johann Bernoulli in 1695; see, for instance, Tauber (1963), Netto (1927, p. 58), and Tropfke (1903, p. 332). On the other hand, de Moivre (1697) surely deserves some credit.

We return now to our main objectives. In outline, we must show that $L_n(\beta)/n$ converges to a limit $L_\infty(\beta)$, uniformly over β in any bounded set; this will follow from Lemma 5. The limiting $L_\infty(\beta)$ is a strictly concave function of β, with a unique maximum at β_∞: see Proposition 1. Furthermore, $\hat{\beta}_n \to \beta_\infty$ by Lemma 4. In principle, randomization ensures that the balance conditions are nearly satisfied, so the plug-in estimator is consistent by Lemmas 1–2. A rigorous argument gets somewhat intricate; one difficulty is showing that remote β's can be ignored, and Lemma 3 helps in this respect.

Some regularity conditions are needed. Technicalities will be minimized if we assume that Z_i takes only a finite number of values; notational overhead is reduced even further if $Z_i = 0, 1,$ or 2. There are now $3 \times 2 \times 2 = 12$ possible values for the triples Z_i, Y_i^C, and Y_i^T. We say that subject i is of *type* (z, c, t) provided

$$Z_i = z, \ Y_i^C = c, \ Y_i^T = t.$$

Let $\theta_{z,c,t}$ be the fraction of subjects that are of type (z, c, t); the number of these subjects is $n\theta_{z,c,t}$.

The θ's are population-level parameters. They are not random. They sum to 1. We assume the θ's are all positive. Recall that π_T is the fraction of subjects assigned to T. This is fixed (not random), and $0 < \pi_T < 1$. The fraction assigned to C is $\pi_C = 1 - \pi_T$. In principle, π_T, π_C, and the $\theta_{z,c,t}$ depend on n. As n increases, we assume these quantities have respective limits λ_T, λ_C, and $\lambda_{z,c,t}$, all positive. Since z takes only finitely many values, $\sum_{z,c,t} \lambda_{z,c,t} = 1$.

When n is large, within type (z, c, t), the fraction of subjects assigned to T is random, but essentially λ_T: Such subjects necessarily have response $Y_i = t$. Likewise, the fraction assigned to C is random, but essentially λ_C: Such subjects necessarily have response $Y_i = c$. In the limit, the Z's are exactly balanced between T and C within each type of subject. That is the essence of the argument; details follow.

Within type (z, c, t), let $n_{z,c,t}^T$ and $n_{z,c,t}^C$ be the number of subjects assigned to T and C, respectively. So

$$n_{z,c,t}^T + n_{z,c,t}^C = n\theta_{z,c,t}.$$

The variables $n_{z,c,t}^T$ are hypergeometric. They are unobservable. This is because type is unobservable: Y_i^C and Y_i^T are not simultaneously observable.

To analyze the log-likelihood function $L_n(\beta)$, recall that

$$Y_i = X_i Y_i^T + (1 - X_i) Y_i^C$$

is the observed response. Let $n_{z,x,y}$ be the number of i with

$$Z_i = z, X_i = x, Y_i = y.$$

Here $z = 0, 1$, or 2; $x = 0$ or 1; and $y = 0$ or 1. The $n_{z,x,y}$ are observable because Y_i is observable. They are random because X_i is random. Also let $n_{z,x} = n_{z,x,0} + n_{z,x,1}$, which is the number of subjects i with $Z_i = z$ and $X_i = x$. Now $L_n(\beta)/n$ in (12) is the sum

$$(19a) \qquad \sum_{z,x} T_{z,x},$$

where

$$(19b) \qquad T_{z,x} = -\frac{n_{z,x}}{n} \log\left[1 + \exp\left(\beta_1 + \beta_2 x + \beta_3 z\right)\right]$$
$$+ \frac{n_{z,x,1}}{n}\left(\beta_1 + \beta_2 x + \beta_3 z\right).$$

(Again, T is for "term," not "treatment.") This can be checked by grouping the terms T_i in (12) according to the possible values of (Z_i, X_i, Y_i). There are six terms $T_{z,x}$ in (19), corresponding to $z = 0, 1$, or 2 and $x = 0$ or 1.

We claim

$$(20) \qquad n_{z,x,y} = \begin{cases} n^T_{z,0,y} + n^T_{z,1,y}, & \text{if } x = 1 \\ n^C_{z,y,0} + n^C_{z,y,1}, & \text{if } x = 0. \end{cases}$$

The trick is seeing through the notation. For instance, take $x = 1$. By definition, $n_{z,1,y}$ is the number of i with $Z_i = z$, $X_i = 1$, $Y_i = y$. The i's with $X_i = 1$ correspond to subjects in the treatment group, so $Y_i = Y_i^T$. Thus, $n_{z,1,y}$ is the number of i with $Z_i = z$, $X_i = 1$, $Y_i^T = y$. Also by definition, $n^T_{z,c,y}$ is the number of subjects with $Z_i = z$, $X_i = 1$, $Y_i^C = c$, $Y_i^T = y$. Now add the numbers for $c = 0, 1$: How these subjects would have responded to the control regime is at this point irrelevant. A similar argument works if $x = 0$, completing the discussion of (20).

Recall that $\theta_{z,c,y} \to \lambda_{z,c,y}$ as $n \to \infty$. Let

$$\theta_z = \sum_{c,y} \theta_{z,c,y} \quad \text{and} \quad \lambda_z = \sum_{c,y} \lambda_{z,c,y}.$$

Thus, θ_z is the fraction of subjects with $Z_i = z$, and $\theta_z \to \lambda_z$ as $n \to \infty$. As $n \to \infty$, we claim that

$$(21) \qquad n_{z,1,y}/n \to \lambda_T(\lambda_{z,0,y} + \lambda_{z,1,y}),$$

$$(22) \qquad\qquad n_{z,1}/n \to \lambda_T \lambda_z,$$

$$(23) \qquad\qquad n_{z,0,y}/n \to \lambda_C(\lambda_{z,y,0} + \lambda_{z,y,1}),$$

$$(24) \qquad\qquad n_{z,0}/n \to \lambda_C \lambda_z,$$

where, for instance, λ_T is the limit of π_T as $n \to \infty$. More specifically, there a set $\mathcal{N}$ of probability 0, and (21–24) hold true outside of $\mathcal{N}$. Indeed, (21) follows from (20) and Lemma 5. Then (22) follows from (21) by addition over $y = 0, 1$. The last two lines are similar to the first two.

A little more detail on (21) may be helpful. What is the connection with Lemma 5? Consider $n_{z,0,y}^T$, which is the number of subjects of type $(z, 0, y)$ that are assigned to T. The "reds" are subjects of type $(z, 0, y)$, so the fraction of reds in the population converges to $\lambda_{z,0,y}$, by assumption. We are drawing m times at random without replacement from the population to get the treatment group, and $m/n \to \lambda_T$, also by assumption. Now X_m is the number of reds in the sample, that is, the number of subjects of type $(z, 0, y)$ assigned to treatment. The lemma tells us that $X_m \to \lambda_T \lambda_{z,0,y}$ almost surely. The same argument works for $n_{z,1,y}^T$. Add to get (21).

Next, fix a positive, finite, real number B. Consider the open, bounded, convex polyhedron G_B defined by the six inequalities

$$(25) \qquad\qquad |\beta_1 + \beta_2 x + \beta_3 z| < B$$

for $x = 0, 1$ and $z = 0, 1, 2$. As $n \to \infty$, we claim that $L_n(\beta)/n \to L_\infty(\beta)$ uniformly over $\beta \in G_B$, where

$$(26a) \qquad\qquad L_\infty(\beta) = \lambda_T \Lambda_T + \lambda_C \Lambda_C,$$

$$(26b) \qquad \Lambda_T = \sum_z \Big(-\lambda_z \log\big[1 + \exp\big(\phi_T(z)\big)\big]$$
$$+ (\lambda_{z,0,1} + \lambda_{z,1,1})\phi_T(z)\Big),$$

$$(26c) \qquad \Lambda_C = \sum_z \Big(-\lambda_z \log\big[1 + \exp\big(\phi_C(z)\big)\big]$$
$$+ (\lambda_{z,1,0} + \lambda_{z,1,1})\phi_C(z)\Big),$$

(26d) $\phi_T(z) = \beta_1 + \beta_2 + \beta_3 z, \quad \phi_C(z) = \beta_1 + \beta_3 z.$

(Recall that λ_T was the limit of π_T as $n \to \infty$, and likewise for λ_C.) This follows from (21–24), on splitting the sum in (19) into two sums, one with terms $(z, 1)$ and the other with terms $(z, 0)$. The $z, 1$ terms give us $\lambda_T \Lambda_T$, and the $z, 0$ terms give us $\lambda_C \Lambda_C$. The conclusion holds outside the null set $\mathcal{N}$ defined for (21–24).

It may be useful to express the limiting distribution of $\{Z, X, Y\}$ in terms of λ_T, λ_C and $\lambda_{z,c,t}$, the latter being the limiting fraction of subjects of type (z, c, t). See Table 13.2. For example, what fraction of subjects have $Z = z, X = 1, Y = 1$ in the limit? The answer is the first row, second column of the table. The other entries can be read in a similar way.

The function $\beta \to L_\infty(\beta)$ is strictly concave, by Proposition 1 with $n = 12$ and $p = 3$. The rows of $(M\,y)$ run through all twelve combinations of $1\,z\,x\,y$ with $z = 1, 2, 3$, and $x = 0, 1$, and $y = 0, 1$. The weights are shown in Table 13.2.

Let β_∞ be the β that maximizes $L_\infty(\beta)$. Choose B in (25) so large that $\beta_\infty \in G_B$. Lemma 4 shows that $\max_{\beta \in G_B} L_n(\beta)/n$ is close to $L_\infty(\beta_\infty)$ for all large n. Outside G_B—if B is large enough—$L_n(\beta)/n$ is too small to matter; additional detail is given below. Thus, $\hat{\beta}_n \in G_B$ for all large n, and converges to β_∞.

This completes the argument for (14), and we turn to proving (15)—the consistency of the plug-in estimators defined by (10). Recall that θ_z is the fraction of i's with $Z_i = z$; and $\theta_z \to \lambda_z$ as $n \to \infty$. Now

$$
\begin{aligned}
\tilde{\alpha}^T &= \frac{1}{n} \sum_{i=1}^{n} p(\hat{\beta}_n, 1, Z_i) \\
&= \sum_z \theta_z p(\hat{\beta}_n, 1, z) \\
&\to \sum_z \lambda_z p(\beta_\infty, 1, z),
\end{aligned}
$$

where the function $p(\beta, x, z)$ was defined in (4). Remember, z takes only finitely many values! A similar argument shows that

$$
\tilde{\alpha}^C \to \sum_z \lambda_z p(\beta_\infty, 0, z).
$$

Table 13.2 Asymptotic distribution of $\{Z, X, Y\}$
expressed in terms of λ_T, λ_C and $\lambda_{z,c,t}$

Value	Weight
$z, 1, 1$	$\lambda_T(\lambda_{z,0,1} + \lambda_{z,1,1})$
$z, 1, 0$	$\lambda_T(\lambda_{z,0,0} + \lambda_{z,1,0})$
$z, 0, 1$	$\lambda_C(\lambda_{z,1,0} + \lambda_{z,1,1})$
$z, 0, 0$	$\lambda_C(\lambda_{z,0,0} + \lambda_{z,0,1})$

The limiting distribution for $\{Z_i, Y_i^C, Y_i^T\}$ is defined by the $\lambda_{z,c,t}$, where $\lambda_{z,c,t}$ is the limiting fraction of subjects of type (z, c, t); recall that $\lambda_z = \sum_{c,t} \lambda_{z,c,t}$. We claim

$$(27) \qquad \sum_z \lambda_z p(\beta_\infty, 1, z) = \sum_{z,c} \lambda_{z,c,1},$$

$$(28) \qquad \sum_z \lambda_z p(\beta_\infty, 0, z) = \sum_{z,t} \lambda_{z,1,t}.$$

Indeed, (22) and (24) show that in the limit, the Z_i are exactly balanced between T and C. Likewise, (21) and (23) show that in the limit, the pairs Y_i^T, Y_i^C are exactly balanced between T and C. Apply Lemmas 1–2. The left hand side of (27) is the plug-in estimator for the limiting α^T. The right hand side is the ITT estimator, as well as truth. The three values coincide by the lemmas. The argument for (28) is the same, completing the discussion of (27–28).

The right hand side of (27) can be recognized as the limit of

$$\frac{1}{n} \sum_{i=1}^{n} Y_i^T = \sum_{z,c} \theta_{z,c,1}.$$

Likewise, the right hand side of (28) is the limit of $\frac{1}{n} \sum_{i=1}^{n} Y_i^C$. This completes the proof of (15). In effect, the argument parlays Fisher consistency into almost-sure consistency, the exceptional null set being the $\mathcal{N}$ where (21–24) fail.

Our results give an indirect characterization of $\lim \beta_n$ as the β at which the limiting log-likelihood function (26) takes on its maximum. Furthermore, asymptotic normality of $\{n_{z,c,t}^T\}$ entails asymptotic normality of $\hat{\beta}_n$ and the plug-in estimators, but that is a topic for another day.

13.8.1 Additional detail on boundedness

Consider a $z, 1$ term in (19). We are going to show that for B large, this term is too small to matter. Fix a small positive ϵ. By (22), for all large n,

$$n_{z,1}/n > (1 - \epsilon)\lambda_T \lambda_z;$$

by (21),

$$n_{z,1,1}/n < (1 + \epsilon)\lambda_T(\lambda_{z,0,1} + \lambda_{z,1,1}).$$

Let $z' = \beta_1 + \beta_2 + \beta_3 z \geq B > 0$. By Lemma 3,

$$\log[1 + \exp(z')] > z' + \exp(-z') - \frac{1}{2}\exp(-2z') > z'$$

because $z' \geq B > 0$. Our $z, 1$ term is therefore bounded above for all large n by

$$\left[- (1 - \epsilon)\lambda_z + (1 + \epsilon)(\lambda_{z,0,1} + \lambda_{z,1,1})\right]\lambda_T z'.$$

The largeness needed in n depends on ϵ not B.

We can choose $\epsilon > 0$ so small that

$$(1 + \epsilon)(\lambda_{z,0,1} + \lambda_{z,1,1}) < (1 - 2\epsilon)\lambda_z,$$

because $\lambda_{z,0,1} + \lambda_{z,1,1} < \lambda_z$. Our $z, 1$ term is therefore bounded above by $-\epsilon\lambda_T\lambda_z B$. For B large enough, this term is so negative as to be irrelevant. The argument works because all $\lambda_{z,c,t}$ are assumed positive, and there are only finitely many of them. A similar argument works for $z' = \beta_1 + \beta_2 + \beta_3 z \leq -B$, and for terms $(z, 0)$ in (19). These arguments go through outside the null set $\mathcal{N}$ defined for (21–24).

13.8.2 Summing up

It may be useful to summarize the results so far. The parameter α^T is defined in terms of the study population, as the fraction of successes that would be obtained if all members of the population were assigned to treatment; likewise for α^C. See (1). The differential log odds Δ of success is defined by (2). There is a covariate taking a finite number of values. A fraction of the subjects are assigned at random to treatment, and the rest to control. We fit a logit model to data from this randomized controlled experiment, although the model is likely false. The MLE is $\hat{\beta}_n$. ITT and plug-in estimators are defined by (10–11).

The size of the population is n. This is increasing to infinity. "Types" of subjects are defined by combinations of possible values for the covariate, the response to control, and the response to treatment. We assume that the fraction of subjects assigned to treatment converges to a positive limit, along with the fraction in each type. The parameters α^T and α^C converge too. This may seem a little odd, but α^T and α^C may depend on the study population, hence on n.

Theorem 1. *Under the conditions of this section, if a logit model is fitted to data from a randomized controlled experiment: (i) the MLE $\hat{\beta}_n$ converges to a limit β_∞; (ii) the plug-in estimator $\tilde{\alpha}^T$, the ITT estimator $\hat{\alpha}^T$, and the parameter α^T have a common limit; (iii) $\tilde{\alpha}^C$, $\hat{\alpha}^C$, and α^C have a common limit; and (iv) $\tilde{\Delta}$, $\hat{\Delta}$, and Δ have a common limit. Convergence of estimators holds almost surely, as the sample size grows.*

13.8.3 Estimating individual-level parameters

At the beginning of the chapter, it was noted that the individual-level parameters $Y_i{}^T$ and $Y_i{}^C$ are estimable. The proof is easy. Recall that $X_i = 1$ if i is assigned to treatment, and $X_i = 0$ otherwise; furthermore, $P(X_i = 1) = \pi_T$ is in $(0, 1)$. Then $Y_i X_i/\pi_T$ is an unbiased estimator for $Y_i{}^T$, and $Y_i(1 - X_i)/(1 - \pi_T)$ is an unbiased estimator for $Y_i{}^C$, where $Y_i = X_i Y_i{}^T + (1 - X_i)Y_i{}^C$ is the observed response.

13.9 An inequality

Let subject i have probability of success p_i if treated, q_i if untreated, with $0 < q_i < 1$ and the q_i not all equal. Suppose

$$\frac{p_i}{1 - p_i} = \lambda \frac{q_i}{1 - q_i}$$

for all i, where $\lambda > 1$. Thus,

$$p_i = \frac{\lambda q_i}{1 + (\lambda - 1)q_i}$$

and $0 < p_i < 1$. Let $\overline{p} = \frac{1}{n}\sum_i p_i$ be the average value of p_i, and likewise for $\overline{q}$. We define the *pooled multiplier* as

$$\frac{\overline{p}/(1 - \overline{p})}{\overline{q}/(1 - \overline{q})}.$$

The log of this quantity is analogous to the differential log odds in (2).

The main object in this section is showing that

(29) λ is strictly larger than the pooled multiplier.

Russ Lyons suggested this elegant proof. Fix $\lambda > 1$. Let $f(x) = x/(1-x)$ for $0 < x < 1$. So f is strictly increasing. Let $h(x) = f^{-1}(\lambda f(x))$, so $p_i = h(q_i)$. Inequality (29) says that $f(\overline{p}) < \lambda f(\overline{q})$, that is, $\overline{p} < h(\overline{q})$. Since $p_i = h(q_i)$, proving (29) comes down to proving that h is strictly concave. But

$$h(x) = \frac{\lambda x}{1 + (\lambda - 1)x}$$

$$= \frac{\lambda}{\lambda - 1}\left(1 - \frac{1}{1 + (\lambda - 1)x}\right),$$

and $y \to 1/y$ is strictly convex for $y > 0$. This completes the proof of (29).

In the other direction,

$$(30) \qquad \frac{\overline{p}}{1 - \overline{p}} - \frac{\overline{q}}{1 - \overline{q}} = \frac{\overline{p} - \overline{q}}{(1 - \overline{p})(1 - \overline{q})} > 0$$

because $p_i > q_i$ for all i. So the pooled multiplier exceeds 1. In short, given the assumptions of this section, pooling moves the multiplier downward towards 1. Of course, if $\lambda < 1$, we could simply interchange p and q. The conclusion: Pooling moves the multiplier toward 1.

In this chapter, we are interested in estimating differential log odds. If the logit model (4) is right, the coefficient β_2 of the treatment indicator is a biased estimator of the differential log odds Δ in (2)—biased away from 0. That is what the inequalities of this section demonstrate, the assumptions being $\beta_3 \neq 0$, Z_i is non-random, and Z_i shows variation across i. (Random Z_i are easily accommodated.)

If the logit model is wrong, the inequalities show that $\hat{\beta}_2 > \hat{\Delta}$ if $\hat{\Delta} > 0$, while $\hat{\beta}_2 < \hat{\Delta}$ if $\hat{\Delta} < 0$. The assumptions are the same, with β_3 replaced by $\hat{\beta}_3$, attention being focused on the limiting values defined in the previous section. Since the plug-in estimator $\hat{\Delta}$ is consistent, $\hat{\beta}_2$ must be inconsistent.

The pooling covered by (29–30) is a little different from the collapsing discussed in Guo and Geng (1995). (i) Pooling does not involve a joint distribution for $\{X_i, Z_i\}$, or a logit model connecting Y_i to X_i and Z_i. (ii) Guo and Geng consider the distribution of one triplet $\{Y_i, X_i, Z_i\}$ only, that is, $n = 1$.

Acknowledgments

Thad Dunning, Winston Lim, Russ Lyons, Philip B. Stark, and Peter Westfall made helpful comments, as did an anonymous editor. Ed George deserves special thanks for helpful comments and moral support.

14

The Grand Leap

With Paul Humphreys

> "The grand leap of the whale up the Fall of Niagara is esteemed,
> by all who have seen it, as one of the finest spectacles in Nature."
> —Benjamin Franklin

ABSTRACT. *A number of algorithms purport to discover causal struc-ture from empirical data with no need for specific subject-matter knowl-edge. Advocates claim that the algorithms are superior to methods al-ready used in the social sciences (regression analysis, path models, factor analysis, hierarchical linear models, and so on). But they have no real success stories to report. The algorithms are computationally impressive and the associated mathematical theory may be of some interest. How-ever, the problem solved is quite removed from the challenge of causal inference from imperfect data. Nor do the methods resolve long-standing philosophical questions about the meaning of causation.*

Causation, Prediction, and Search by Peter Spirtes, Clark Glymour, and Richard Scheines (SGS) is an ambitious book. SGS claim to have

British Journal for the Philosophy of Science (1996) 47: 113–23.

methods for discovering causal relations based only on empirical data, with no need for subject-matter knowledge. These methods—which combine graph theory, statistics, and computer science—are said to allow quick, virtually automated, conversion of association to causation. The algorithms are held out as superior to methods already in use in the social sciences (regression analysis, path models, factor analysis, hierarchical linear models, and so on). According to SGS, researchers who use these other methods are sometimes too timid, sometimes too bold, and sometimes just confused:

> Chapters 5 and 8 illustrate a variety of cases in which features of linear models that have been justified at length on theoretical grounds are produced immediately from empirical covariances by the procedures we describe. We also describe cases in which the algorithms produce plausible alternative models that show various conclusions in the social scientific literature to be unsupported by the data. (p. 14)

> In the absence of very strong prior causal knowledge, multiple regression should not be used to select the variables that influence an outcome or criterion variable in data from uncontrolled studies. So far as we can tell, the popular automatic regression search procedures [like stepwise regression] should not be used at all in contexts where causal inferences are at stake. Such contexts require improved versions of algorithms like those described here to select those variables whose influence on an outcome can be reliably estimated by regression. (p. 257)

SGS are exaggerating more than a little. Indeed, they have no real success stories to report. The algorithms and the associated mathematical theory may be of some interest; computationally, the algorithms are quite impressive. However, in the end, the whole development is only tangentially related to long-standing philosophical questions about the meaning of causation, or to real problems of statistical inference from imperfect data. We will summarize the evidence below.[1]

Statistical relationships are often displayed in graphical form, path models being an early example.[2] Such models represent variables as nodes in a graph; an arrow from X to Y means that X is related to Y, given the prior variables.[3] For instance, take Figure 14.1; the regression equation for Y in terms of U, V, and X should include only X: The only arrow into Y is from X. However, the equation for X in terms of U and V should include both variables: There are arrows into X from U and V.

Figure 14.1 Directed Acyclic Graph.

Starting from the joint distribution of the variables, the "Markov condition," and the so-called "faithfulness assumption," SGS have algorithms for determining the presence or absence of arrows. However, there is no coherent ground for thinking that arrows represent causation. Indeed, the connection between arrows and causes is made on the basis of yet another assumption—the "causal Markov condition." These assumptions will be discussed below.

SGS focus on a special class of graphical models, the Directed Acyclic Graph (DAG). Mathematical properties of these graphs are summarized in chapter 2; the Markov condition and the faithfulness assumption are stated there. The Markov condition says, roughly, that past and future are conditionally independent given the present. Figure 14.1 illustrates the idea: Y is independent of U and V given X. With DAG's, there is mathematical theory that permits conditional independence relations to be read off the graph. And the faithfulness assumption says there are no "accidental" relations: Conditional independence holds according to presence or absence of arrows, not in virtue of specific parameter values. Under such circumstances, the probability distribution is said to be "faithful" to the graph.[4] If the probability distribution is faithful to a graph for which the Markov condition holds, that graph can be inferred (in whole or in part) from the conditional independence relations defined by the distribution. The object of the SGS algorithms is to reconstruct the graph from these statistical relationships.

The causal Markov condition is introduced in chapter 3. The connection with chapter 2 is only the Causal Representation Convention (p. 47), according to which causal graphs are DAG's where arrows represent causation. In other words, the causal Markov condition is just the Markov condition, plus the assumption that arrows represent causation. Thus, causation is not proved into the picture, it is assumed in. To compound the confusion between mathematical objects in the theory and applications to real data, SGS make the convention (p. 56) that the "Markov property" means the "causal Markov property."[5]

Philosophers are nowadays used to a style of formal axiomatization within which uninterpreted logical or mathematical formulae are used as axioms (the syntactic approach) or classes of abstract structures are defined in the axiomatization (the semantic approach).[6] These axiomatic approaches make a clear distinction between a mathematical theory and its interpretation. SGS do not use either of these approaches, and positively invite the confusion that axiomatics are supposed to prevent. SGS themselves seem to have no real interest in interpretative issues:

> Views about the nature of causation divide very roughly into those that analyze causal influence as some sort of probabilistic relation, those that analyze causal influence as some sort of counterfactual relation (sometimes a counterfactual relation having to do with manipulations or interventions), and those that prefer not to talk of causation at all. We advocate no definition of causation, but in this chapter attempt to make our usage systematic, and to make explicit our assumptions connecting causal structure with probability, counterfactuals and manipulations. With suitable metaphysical gyrations the assumptions could be endorsed from any of these points of view, perhaps including even the last. (p. 41)[7]

SGS do not give a reductive definition of "A causes B" in noncausal terms. And their axiomatics require that you already understand what causes are. Indeed, the causal Markov condition and the faithfulness assumption boil down to this: Direct causes can be represented by arrows when the data are faithful to the true causal graph that generates the data. In short, causation is defined in terms of causation.[8] That is why the mathematics in SGS will be of little interest to philosophers seeking to clarify the meaning of causation.

The SGS algorithms for inferring causal relations from data are embodied in a computer program called TETRAD. We give a rough description. The program takes as input the joint distribution of the variables, and it searches over DAG's. In real applications, of course, the full joint distribution is unknown, and must be estimated from sample data. In its present incarnation, TETRAD can handle only two kinds of sample data, governed by conventional and unrealistic textbook models: (i) independent, identically distributed multivariate Gaussian observations, or (ii) independent, identically distributed multinomial observations. These

Figure 14.2 Orienting the edges.

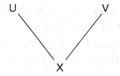

assumptions are not emphasized in SGS, but appear in the computer documentation and the computer output.[9]

In essence, TETRAD begins with a "saturated" graph, where any
pair of nodes are joined by an edge. If the null hypothesis of independence cannot be rejected—at, say, the 5% level, using the t-test—the edge
is deleted. The t-test is relevant only because of the statistical assumptions.
After examining all pairs of nodes, TETRAD moves on to triples, and so
forth. According to the faithfulness assumption, independence cannot be
due to the cancellation of conditional dependencies. That is why an edge,
once deleted, never returns.

TETRAD also orients edges that remain. For example, take the graph
in Figure 14.2. If U and V are conditionally independent given X, the arrows cannot go from U and V to X—that would violate the faithfulness
assumption. However, it is exact independence that is relevant, and exact
independence cannot be determined from any finite amount of sample
data. Consequently, the mathematical demonstrations in SGS (e.g., theorem 5.1 on p. 405) do not cope with the most elementary of statistical
ideas. Even if all the assumptions hold, the t-test makes mistakes. The
test has to make mistakes, because sample data do not determine the joint
distribution. (The problem is compounded when, as here, multiple tests
are made.)

Therefore, the SGS algorithms can be shown to work only when the
exact conditional independencies and dependencies are given. Similarly,
with the faithfulness condition, it is only exact conditional independence
that protects against confounding. As a result, the SGS algorithms must
depend quite sensitively on the data and even on the underlying distribution: Tiny changes in the circumstances of the problem have big impacts
on causal inferences.[10]

Exact conditional independence cannot be verified, even in principle, by mere statisticians using real data. Approximate conditional
independence—which is knowable—has no consequences in the SGS
scheme of things. That is one reason why the SGS theory is unrelated

to the real problems of inference from limited data. The artificiality of the assumptions is the other reason.[11]

Setting theoretical issues to the side, SGS seem also to offer empirical proof for the efficacy of their methods: Their book is studded with examples. However, the proof is illusory. Many of the examples (for instance, the ALARM network, p. 11 and pp. 145ff) turn out to be simulations, where the computer generates the data. The ALARM network is supposed to represent causal relations between variables relevant to hospital emergency rooms, and SGS claim (p. 11) to have discovered almost all of the adjacencies and edge directions "from the sample data." However, these "sample data" are simulated. The hospitals and patients exist only in the minds of the computer programmers. The statistical assumptions made by SGS are all satisfied, having been programmed into the computer. Simulations tell us very little about the likelihood that that SGS's assumptions will be satisfied in real applications. Furthermore, arguments about causation seem out of place in the context of a computer simulation. What can it mean for one computer-generated variable to "cause" another?

SGS use the health effects of smoking as a running example to illustrate their theory (pp. 18, 19, 75ff, 172ff, 179ff). However, that only creates another illusion. The causal diagrams are all hypothetical, no contact is made with data, and no substantive conclusions are drawn. If the diagrams were proposed as real descriptions of causal mechanisms, they would be laughed out of court.

Does smoking cause lung cancer, heart disease, and many other illnesses? SGS appear not to believe the epidemiological evidence. When they get down to arguing their case, they use a rather old-fashioned method—a literature review with arguments in ordinary English (pp. 291–302). Causal models and search algorithms have disappeared. Thus, SGS elected not to use their analytical machinery on one of their leading examples. This is a remarkable omission.

In the end, SGS do not make bottom-line judgments on the effects of smoking. Their principal conclusion is methodological. Nobody besides them understood the issues:

> Neither side understood what uncontrolled studies could and could not determine about causal relations and the effects of interventions. The statisticians pretended to an understanding of causality and correlation they did not have; the epidemiologists resorted to informal and often irrelevant criteria, appeals to plausibility, and in the worst case to ad hominem . . . . While the statisticians didn't get the connection between causality and

probability right, the . . . 'epidemiological criteria for causality' were an intellectual disgrace, and the level of argument . . .
was sometimes more worthy of literary critics than scientists.
(pp. 301–02)

On pp. 132–52 and 243–50, SGS analyze a number of real examples, mainly drawn from the social-science literature. What are the scoring rules? Apparently, SGS count a win if their algorithms more or less
reproduce the original findings (rule #1); but they also count a win if their
algorithms yield different findings (rule #2). This sort of empirical test is
not particularly harsh.[12] Even so, the SGS algorithms reproduce original
findings only if one is very selective in reading the computer output, as
will be seen below.

SGS make strong empirical claims for their methods. To evaluate
those claims, empirical evidence is relevant. We ran TETRAD on the
four most solid-looking examples in SGS. The results were similar; we
report on one example here.[13] Rindfuss et al. (1980) developed a model
to explain the process by which a woman decides how much education
to get, and when to have her first child. The variables in the model are
defined in Table 14.1.

The statistical assumptions made by Rindfuss et al., let alone the
stronger conditions used by SGS, may seem rather implausible if exam-

Table 14.1 Variables in the model[14]

ED	Respondent's education (years of schooling completed at first marriage)
AGE	Respondent's age at first birth
DADSOCC	Respondent's father's occupation
RACE	Race of respondent (Black = 1, other = 0)
NOSIB	Respondent's number of siblings
FARM	Farm background (1 if respondent grew up on a farm, else 0)
REGN	Region where respondent grew up (South = 1, other = 0)
ADOLF	Broken family (0 if both parents were present at age fourteen, else 1)
REL	Religion (Catholic = 1, other = 0)
YCIG	Smoking (1 if respondent smoked before age sixteen, else 0)
FEC	Fecundability (1 if respondent had a miscarriage before first birth; else 0)

ined at all closely. For now, we set such questions aside, and focus on the results of the data analysis. SGS report only a graphical version of their model:

> Given the prior information that ED and AGE are not causes of the other variables, the PC algorithm (using the .05 significance level for tests) directly finds the model [in the left hand panel of Figure 14.3] where connections among the regressors are not pictured. (p. 139)

Apparently, the left hand panel in Figure 14.3 is close to the model in Rindfuss et al., and SGS claim a victory under their scoring rule #1. However, the graph published in *Causation, Prediction, and Search* (p. 140) is only a subset of the one actually produced by TETRAD. The whole graph—which SGS do not report—is shown in the right hand panel of Figure 14.3. This graph says, for instance, that race and religion cause region of residence. Comments on the sociology may be unnecessary, but consider the arithmetic. REGN takes only two values (Table 14.1), so it cannot be presented as a linear combination of prior variables with an additive Gaussian error, as required by TETRAD's statistical assumptions. FARM creates a similar problem. So does NOSIB. In short, the SGS algorithms have produced a model that fails the most basic test—internal consistency. Even by the fairly relaxed standards of the social science literature, Figure 14.3 is a minor disaster.

> Figure 14.3 The left hand panel shows the model reported by SGS (p.140). The right hand panel shows the whole graph produced by the SGS search program TETRAD.[15]

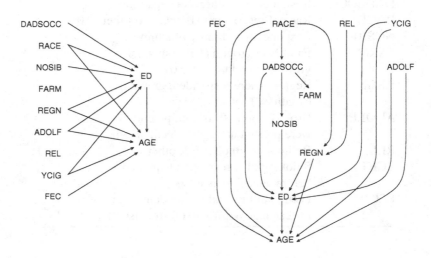

SGS seem to buy the Automation Principle: The only worthwhile knowledge is the knowledge that can be taught to a computer. This principle is perverse. Despite SGS's agnosticism, the epidemiologists discovered an important truth: Smoking is bad for you.[16] The epidemiologists made this discovery by looking at the data and using their brains—two skills that are not readily automated. SGS, on the other hand, taught their computer to discover Figure 14.3. The examples in SGS count against the Automation Principle, not for it.

Researchers in the field of Artificial Intelligence are seldom accused of false modesty, while causal models in the social sciences often promise more than they deliver. TETRAD is an AI package that generates causal models. The rest is just show business.

Notes

Springer-Verlag, New York, published *Causation, Prediction, and Search* in 1993.

1. For more details, see Freedman (1997) and Humphreys (1997).

2. Path models originate in the work of Sewell Wright (1921). Perhaps the first application to social-science data was Blau and Duncan (1967). For discussion, see the *Journal of Educational Statistics*, Summer (1987).

3. In this review, we try to give the intuition not the rigor; mathematical definitions are only sketched.

4. The Markov condition for DAG's was developed by Kiiveri and Speed (1982). Faithfulness was introduced by Pearl (1988). Verma and Pearl (1990) proved the deep connection between the graph theory and conditional independence; also see Geiger, Verma, and Pearl (1990). The Markov condition must hold for the original variables to which the algorithms will be applied; it is not enough if the condition holds for recoded variables. SGS state the Markov condition as follows (p. 33): "A directed acyclic graph G over [a vertex set] V and a probability distribution $P(V)$ satisfy the Markov Condition if and only if for every W in V, W is [statistically] independent of $V \backslash (\textbf{Descendants}(W) \cup \textbf{Parents}(W))$ given **Parents**(W)." Here, a "parent" of W is any vertex immediately preceding W in the graph, and a "descendant" of W is any vertex with a path from W to that vertex. Our informal definition of faithfulness is paraphrased from SGS (p. 35); also see note 5 below.

5. SGS state the causal Markov condition as follows: "Let G be a causal graph with vertex set V and P be a probability distribution over the ver-

tices in V generated by the causal structure represented by G. G and P satisfy the Causal Markov Condition if and only if for every W in V, W is [statistically] independent of $V \backslash (\textbf{Descendants}(W) \cup (\textbf{Parents}(W))$ given $\textbf{Parents}(W)$" (p. 54). SGS state the faithfulness condition as follows: "Let G be a causal graph and P a probability distribution generated by G. $<G, P>$ satisfies the Faithfulness Condition if and only if every conditional independence relation true in P is entailed by the Causal Markov Condition applied to G" (p. 56). For causal inference, it is not enough that the distribution be faithful to *some* graph; the distribution must be faithful to the *true* causal graph that generates the data, the latter being a somewhat informal idea in SGS's framework. See Freedman (1997, section 12.3).

6. The statistical literature does offer some formal treatments of causation, in the sense of effects of hypothetical interventions. See, for example, Neyman (1923), Robins (1986, 1987a,b), Holland (1988), and Pearl (1995).

7. SGS justify their lack of an explicit definition by noting that probability theory has made progress despite notorious difficulties of interpretation— perhaps the first innocence-by-association argument in causal modeling. On the other hand, lack of clarity in the foundations of statistics may be one source of difficulties in applying the techniques. For discussion, see *Sociological Methodology* (1991) and *Foundations of Science*, Winter (1995).

8. The Causal Representation Convention says: "A directed graph $G = <V, E>$ represents a causally sufficient structure C for a population of units when the vertices of G denote the variables in C, and there is a directed edge from A to B in G if and only if A is a direct cause of B relative to V" (p. 47, footnote omitted). Following the chain of definitions, we have that "A set V of variables is **causally sufficient** for a population if and only if in the population every common cause of any two or more variables in V is in V or has the same value for all units in the population" (p. 45, footnote omitted). What constitutes a direct cause? "C is a **direct cause** of A relative to V just in case C is a member of some set C included in $V \backslash \{A\}$ such that (i) the events in C are causes of A, (ii) the events in C, were they to occur, would cause A no matter whether the events in $V \backslash (\{A\} \cup C)$ were or were not to occur, and (iii) no proper subset of C satisfies (i) and (ii)" (p. 43). This is perhaps intelligible if you already know what causation means; a non-starter otherwise.

9. The most interesting examples are based on the assumption of a multivariate Gaussian distribution, and we focus on those examples. The

documentation for TETRAD is Spirtes, Scheines, Glymour, and Meek (1993); point 2 on p. 71 gives the statistical assumptions, which also appear on the computer printout. The algorithms are discussed in SGS pp. 112ff, 165ff, and 183ff: These include the "PC" and "FCI" algorithms used in TETRAD.

10. Thus, a correlation that equals 0.000 precludes certain kinds of confounding and permits causal inference; a correlation that equals 0.001 has no such consequences. For examples and discussion, see Freedman (1997, section 12.1), which develops work by James Robins.

11. The statistical assumptions (i.e., conditions on the joint distribution) include the Markov property and faithfulness. For the algorithms to work efficiently and give meaningful output, the graph must be sparse, i.e., relatively few pairs of nodes are joined by arrows. Observations are assumed independent and identically distributed; the common distribution is multivariate Gaussian or multinomial (note 9). There is the further, non-statistical, assumption that arrows represent direct causes (notes 5 and 8). This non-statistical assumption may be the most problematic: see the Summer (1987) issue of the *Journal of Educational Statistics* or the Winter (1995) issue of *Foundations of Science*.

12. SGS eventually do acknowledge some drawbacks to their rules: "With simulated data the examples illustrate the properties of the algorithms on samples of realistic sizes. In the empirical cases we often do not know whether an algorithm produces the truth" (pp. 132–33).

13. Our discussion is largely based on the references in note 1 above. Rindfuss et al. is discussed by SGS on pp. 139ff; the other examples are AFQT [the Armed Forces Qualification Test] and Spartina [a salt-tolerant marsh grass] (see also Freedman and Humphreys 1999), and Timberlake and Williams (1984). See pp. 243–50 in SGS.

14. The data are from a probability sample of 1766 women thirty-five to forty-four years of age residing in the continental United States; the sample was restricted to ever-married women with at least one child. DADSOCC was measured on Duncan's scale, combining information on education and income; missing values were imputed at the overall mean. SGS give the wrong definitions for NOSIB and ADOLF; the covariance matrix they report has incorrect entries (p. 139).

15. The right hand panel is computed using the BUILD module in TETRAD. BUILD asks whether it should assume "causal sufficiency." Without this assumption (note 8), the program output is uninformative; therefore, we told BUILD to make the assumption. Apparently, that is

what SGS did for the Rindfuss example. Also see Spirtes et al. (1993, pp. 13–15). Data are from Rindfuss et al. (1980), not SGS; with the SGS covariance matrix, FARM "causes" REGN and YCIG "causes" ADOLF.

16. See Cornfield et al. (1959), International Agency for Research on Cancer (1986), and U.S. Department of Health and Human Services (1990).

15

On Specifying Graphical Models for Causation, and the Identification Problem

ABSTRACT. *Graphical models for causation can be set up using fewer hypothetical counterfactuals than are commonly employed. Invariance of error distributions may be essential for causal inference, but the errors themselves need not be invariant. Graphs can be interpreted using conditional distributions so that one can better address connections between the mathematical framework and causality in the world. The identification problem is posed in terms of conditionals. As will be seen, causal relationships cannot be inferred from a data set by running regressions unless there is substantial prior knowledge about the mechanisms that generated the data. There are few successful applications of graphical models, mainly because few causal pathways can be excluded on a priori grounds. The invariance conditions themselves remain to be assessed.*

In this chapter, I review the logical basis for inferring causation from regression equations, proceeding by example. The starting point is a simple regression, next is a path model, and then simultaneous equations (for supply and demand). After that come nonlinear graphical models.

Evaluation Review (2004) 28: 267–93.

The key to making a causal inference from nonexperimental data by regression is some kind of invariance, exogeneity being a further issue. Parameters need to be invariant to interventions. This well-known condition will be stated here with a little more precision than is customary. Invariance is also needed for errors or error distributions, a topic that has attracted less attention. Invariance for distributions is a weaker assumption than invariance for errors. I will focus on invariance of error distributions in stochastic models for individual behavior, eliminating the need to assume sampling from an ill-defined super-population.

With graphical models, the essential mathematical features can be formulated in terms of conditional distributions ("Markov kernels"). To make causal inferences from nonexperimental data using such techniques, the kernels need to be invariant to intervention. The number of plausible examples is at best quite limited, in part because of sampling error, in part because of measurement error, but more fundamentally because few causal pathways can be excluded on a priori grounds. The invariance condition itself remains to be assessed.

Many readers will "know" that causal mechanisms can be inferred from nonexperimental data by running regressions. I ask from such readers an unusual boon—the suspension of belief. (Suspension of disbelief is all too readily at hand, but that is another topic.) There is a complex chain of assumptions and reasoning that leads from the data via regression to causation. One objective in the present essay is to explicate this logic. Please bear with me: What seems obvious at first may become less obvious on closer consideration, and properly so.

15.1 A first example: Simple regression

Figure 15.1 is the easiest place to start. In order to make causal inferences from simple regression, it is now conventional (at least for a small group of mathematical modelers) to assume something like the setup in equation (1). I will try to explain the key features in the formalism, and then offer an alternative. As will become clear, the equation makes very strong invariance assumptions, which cannot be tested from data on X and Y.

$$(1) \qquad\qquad Y_{i,x} = a + bx + \delta_i.$$

The subscript i indexes the individuals in a study, or the occasions in a repeated-measures design, and so forth. A treatment may be applied at

Figure 15.1 Linear regression.

$$X \longrightarrow Y$$

various levels x. The expected response is $a + bx$. By assumption, this is linear in x, with intercept a and slope b. The parameters a and b are the same, again by assumption, for all subjects and all levels of treatment.

When treatment at level x is applied to subject i, the response $Y_{i,x}$ deviates from the expected by a "random error" or "disturbance" δ_i. This presumably reflects the impact of chance. For some readers, it may be more natural to think of $a + \delta_i$ in (1) as a random intercept. Others may classify $Y_{i,x}$ as a "potential outcome": More about that later.

In this chapter, as is commonplace among statisticians, random variables like δ_i are functions on a probability space Ω. Informally, chance comes in when Nature chooses a point at random from Ω, which fixes the value of δ_i. The choice is made once and once only: Nature does not re-randomize if x is changed in (1). More technically, $Y_{i,x}$ is a function of x and δ_i, but δ_i does not vary with x. (The formalism is compact, which has certain advantages; on the other hand, it is easy to lose track of the ideas.)

The δ_i are assumed to be independent and identically distributed. The common "error distribution" $\mathcal{D}$ is unknown but its mean is assumed to be 0. Nothing in (1) is observable. To generate the data, Nature is assumed to choose $\{X_i : i = 1, \ldots, n\}$ independently of $\{\delta_i : i = 1, \ldots, n\}$, showing us

$$(X_i, Y_i),$$

where

$$Y_i = Y_{i,X_i} = a + bX_i + \delta_i$$

for $i = 1, \ldots, n$.

Notice that x in (1) could have been anything. The model features multiple parallel universes, all of which remain counterfactual hypotheticals—because, of course, we did no intervening at all. Instead, we passively observed X_i and Y_i. (If we had done the experiment, none of these interesting issues would be worth discussing.) Nature obligingly randomizes for us. She chooses X_i at random from some distribution, independently of δ_i, and then sets $Y_i = a + bX_i + \delta_i$ as required by (1).

"Exogeneity" is the assumed independence between the X_i and the errors δ_i. Almost as a bookkeeping matter, your response Y_i is computed from your X_i and error term δ_i. Nobody else's X and δ get into the act, precluding interactions across subjects. According to the model, δ_i exists—incorruptible and unchanging—in all the multiple unrealized counterfactual hypothetical universes, as well as in the one real factual observed universe. This is a remarkably strong assumption. All is flux, except a, b, and δ_i.

An alternative setup will be presented next, more like standard regression, to weaken the invariance assumption. We start with unknown parameters a and b and an error distribution $\mathcal{D}$. The last is unknown, but has mean 0. Nature chooses $\{X_i : i = 1, \ldots, n\}$ at random from some n-dimensional distribution. Given the X's, the Y's are assumed to be conditionally independent, and the random errors

$$Y_i - a - bX_i$$

are assumed to have common distribution $\mathcal{D}$. In other words, the Y's are built up from the X's as follows. Nature computes the linear function $a + bX_i$, then adds some noise drawn at random from $\mathcal{D}$ to get Y_i. We get to see the pairs (X_i, Y_i) for $i = 1, \ldots, n$.

In this alternative formulation, there is a fixed error distribution $\mathcal{D}$ but there are no context-free random errors. Indeed, errors may be functions of treatment levels among other things. The alternative has both a causal and an associational interpretation: (i) assuming invariance of error, distributions to interventions leads to the causal interpretation; and (ii) mere insensitivity to x when we condition on $X_i = x$ gives the associational interpretation—the probability distribution of $Y_i - a - bX_i$ given $X_i = x$ is the same for all x. This can at least in principle be tested against the data. Invariance to interventions cannot, unless interventions are part of the design.

The key difference between equation (1) and the alternative is this: In (1), the errors themselves are invariant; in the alternative formulation, only the error distribution is invariant. In (1), inference is to the *numerical value* that Y_i would have had, if X_i had been set to x. In the alternative formulation, causal inference can only be to the *probability distribution* that Y_i would have had. With either setup, the inference is about specific individuals, indexed by i. Inference at the level of individuals is possible because—by assumption—parameters a and b are the same for all individuals. The two formulations of invariance, with the restrictions on the X's, express different ideas of exogeneity. The second set of assumptions is weaker than the first and seems generally more plausible.

An example to consider is Hooke's law. The stretch of a spring is proportional to the load: a is length under no load and b is stretchiness. The disturbance term would represent measurement error. We could run an experiment to determine a and b. Or we could passively observe the behavior of springs and weights. If heavier weights are attracted to bigger errors, there are problems. Otherwise, passive observation might give the right answer. Moreover, we can with more or less power test the hypothesis that the random errors $Y_i - a - bX_i$ are independent and identically distributed. By way of contrast, consider the hypothesis that $Y_i - a - bX_i$ itself would have been the same if X_i had been seven rather than three. Even in an experiment, testing that seems distinctly unpromising.

What happens without invariance? The answer will be obvious. If intervention changes the intercept a, the slope b, or the mean of the error distribution, the impact of the intervention becomes difficult to determine. If the variance of the error term is changed, the usual confidence intervals lose their meaning.

How would any of this be possible? Suppose, for instance, that—unbeknownst to the statistician—X and Y are both the effects of a common cause operating through linear statistical laws like (1). Suppose errors are independent and normal, while Nature randomizes the common cause to have a normal distribution. The scatter diagram will look lovely, a regression line is easily fitted, and the straightforward causal interpretation will be wrong.

15.2 Conditionals

Let us assume (informally) that the regression in Figure 15.1 is causal. What the Y_i's would have been if we had intervened and set X_i to x_i—this too isn't quite mathematics, but does correspond to either of two formal systems, which involve two sets of objects. The first set of objects is generated by equation (1): the random variables $Y_i = a + bx_i + \delta_i$ for $i = 1, \ldots, n$. The second set of objects is this: n independent Y's, the ith being distributed as $a + bx_i$ plus a random draw from the error distribution $\mathcal{D}$. One system is defined in terms of random variables; the other, in terms of conditional distributions. There is a similar choice for the examples presented below.

So far, I have been discussing linear statistical laws. In Figure 15.1, for example, suppose we set $X = x$. Conditionally, Y will be distributed like $a + bx$ plus random noise with distribution $\mathcal{D}$. Call this conditional distribution $K_x(dy)$. On the one hand, K_x may just represent the conditional distribution of Y given $X = x$, a rather dry statistical idea. On the other hand, K_x may represent the result of a hypothetical intervention: the distribution that Y would have had if only we had intervened and set X to x. This is the more exciting causal interpretation.

Data analysis on X and Y cannot decide whether the causal inter-pretation is viable. Instead, to make causal inferences from a system of regression equations, causation is assumed from the beginning. As Cartwright (1989) says, "No causes in, no causes out." This view con-trasts rather sharply with rhetoric that one finds elsewhere.

Of course, solid arguments for causation have been made from ob-servational data, but fitting regressions is only one aspect of the activity (Freedman 1999). Replication seems to be critical, with good study de-signs and many different kinds of evidence. Also see Freedman (1997), noting the difference between conditional probabilities that arise from se-lection of subjects with $X = x$, and conditional probabilities arising from an intervention that sets X to x. The data structures may look the same, but the implications can be worlds apart.

15.3 Two linear regressions

The discussion can now be extended to path diagrams, with similar conclusions. Figure 15.2 involves three variables and is a cameo version of applied statistics. If we are interested in the effect of Y on Z, then X confounds the relationship. Some adjustment is needed to avoid biased estimates, and regression is often used. The diagram unpacks into two response schedules:

(2a) $$Y_{i,x} = a + bx + \delta_i,$$
(2b) $$Z_{i,x,y} = c + dx + ey + \epsilon_i.$$

We assume that $\delta_1, \ldots, \delta_n$ and $\epsilon_1, \ldots, \epsilon_n$ are all independent. The δ's have a common distribution $\mathcal{D}$. The ϵ's have another common distribu-

Figure 15.2 A path model with three variables.

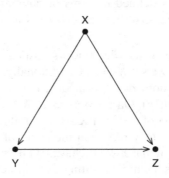

tion $\mathcal{F}$. These two distributions are unknown, but are assumed to have mean 0. Again, nothing in (2) is observable. To generate the data, Nature chooses $\{X_i : i = 1, \ldots, n\}$ independently of $\{\delta_i, \epsilon_i : i = 1, \ldots, n\}$. We observe

$$(X_i, Y_i, Z_i)$$

for $i = 1, \ldots, n$, where

$$Y_i = Y_{i,X_i} = a + bX_i + \delta_i,$$
$$Z_i = Z_{i,X_i,Y_i} = c + dX_i + eY_i + \epsilon_i.$$

Basically, this is a recursive system with two equations. The X's are "exogenous," that is, independent of the δ's and ϵ's. According to the model, Nature plugs the X's into (2a) to compute the Y's. In turn, those very X's and Y's get plugged into (2b) to generate the Z's. That is the recursive step.

In other words, Y_i is computed as a linear function of X_i, with intercept a and slope b, plus the error term δ_i. Then Z_i is computed as a linear function of X_i and Y_i. The intercept is c, the coefficient on X_i is d, and the coefficient on Y_i is e. At the end, the error ϵ_i is tagged on. Again, the δ's and ϵ's remain the same no matter what x's and y's go into (2). So do the parameters a, b, c, d, and e. (Interactions across subjects are precluded because, for instance, subject i's response Y_i is computed from X_i and δ_i rather than X_j and δ_j.)

The proposed alternative involves not random errors but their distributions $\mathcal{D}$ and $\mathcal{F}$. These distributions are unknown but have mean 0. We still have the parameters a, b, c, d, and e. To generate the data, we assume that Nature chooses $X_1, \ldots, X_n$ at random from some n-dimensional distribution. Given the X's, the Y's are assumed to be conditionally independent: Y_i is generated by computing $a + bX_i$, then adding some independent noise distributed according to $\mathcal{D}$. Given the X's and Y's, the Z's are assumed to be conditionally independent: Z_i is generated as $c + dX_i + eY_i$, with independent additive noise distributed according to $\mathcal{F}$. The exogeneity assumption is the independence between the X's and the errors.

As before, the second setup assumes less invariance than the first. It is error distributions that are invariant, not error terms. The inference is to distributions rather than specific numerical values. Either way, there are unbiased estimates for the parameters a, b, c, d, and e. The error distributions $\mathcal{D}$ and $\mathcal{F}$ are identifiable. Parameters and error distributions are constant in both formulations. As before, the second setup may be used to describe conditional distributions of random variables. If those conditional distributions admit a causal interpretation, then causal inferences can made from observational data. In other words, regression succeeds in determining the effect of Y on Z if we know that X is the confounder and that the statistical relationships are linear and causal.

What can go wrong? Omitted variables are a problem, as discussed before. Assuming the wrong causal order is another issue. For example, suppose equation (2) is correct. The errors are independent and normally distributed. Moreover, the exogenous variable X has been randomized to have a normal distribution. However, the unfortunate statistician regresses Y on Z, then X on Y and Z. Diagnostics will indicate success: The distribution of residuals will not depend on the explanatory variables. But causal inferences will be all wrong. The list of problem areas can easily be extended to include functional form, stochastic specification, measurement, ...

The issue boils down to this. Does the conditional distribution of Y given X represent mere association, or does it represent the distribution Y would have had if we had intervened and set the values of X? There is a similar question for the distribution of Z given X and Y. These questions cannot be answered just by fitting the equations and doing data analysis on X, Y, and Z. Additional information is needed. From this perspective, the equations are "structural" if the conditional distributions inferred from the equations tell us the likely impact of interventions, thereby allowing a causal rather than an associational interpretation. The take-home message will be clear: You cannot infer a causal relationship from a data set by running regressions—unless there is substantial prior knowledge about the mechanisms that generated the data.

15.4 Simultaneous equations

Similar considerations apply to models with simultaneous equations. The invariance assumptions will be familiar to many readers. Changing pace, I will discuss hypothetical supply and demand equations for butter in the state of Wisconsin. The endogenous variables are Q and P, the quantity and price of butter. The exogenous variables in the supply equation are the agricultural wage rate W and the price H of hay. The exogenous variables in the demand equation are the prices M of margarine and B of bread (substitutes and complements). For the moment, "exogeneity" just means "externally determined." Annual data for the previous twenty years are available on the exogeneous variables, and on the quantity of Wisconsin butter sold each year as well as its price. Linearity is assumed, with the usual stochastics.

The model can be set up formally with two linear equations in two unknowns, Q and P:

(3a) Supply $\quad Q = a_0 + a_1 P + a_2 W + a_3 H + \delta_t,$

(3b) Demand $\quad Q = b_0 + b_1 P + b_2 M + b_3 B + \epsilon_t.$

On the right hand sides, there are parameters (the a's and b's). There are also error terms (δ_t, ϵ_t), which are assumed to be independent and identically distributed for $t = 1, \ldots, 20$. The common two-dimensional "error distribution" $\mathcal{C}$ for (δ_t, ϵ_t) is unknown but is assumed to have mean 0.

Each equation describes a thought experiment. In the first, we set $P, W, H, M,$ and B and observe how much butter comes to market. By assumption, M and B have no effect on supply, while $P, W,$ and H have additive linear effects. In the second we set $P, W, H, M,$ and B and observe how much butter is sold: W and H have no effect on demand, while $P, M,$ and B have additive linear effects. In short, we have linear supply and demand schedules. Again, the error terms themselves are invariant to all interventions, as are the parameters. Since this is a hypothetical, there is no need to worry about the EEC [European Economic Community, now the European Community], NAFTA [the North American Free Trade Agreement], or the economics.

A third gedanken experiment is described by taking equations (3a) and (3b) together. Any values of the exogenous variables $W, H, M,$ and B—perhaps within certain ranges—can be substituted in on the right, and the two equations solved together for the two unknowns Q and P, giving us the transacted quantity and price in a free market, denoted

(4) $\qquad\qquad Q_{W,H,M,B}$ and $P_{W,H,M,B}.$

Since δ and ϵ turn up in the formulas for both Q and P, the random variables in (4) are correlated—barring some rare parameter combinations—with the error terms. The correlation is "simultaneity."

So far, we have three thought experiments expressing various assumptions, but no data: None of the structure of the equation, including the error distribution, is observable. We assume that Nature generates data for us by choosing $W_t, H_t, M_t,$ and B_t for $t = 1, \ldots, 20$, at random from some high-dimensional distribution, independently of the δ's and ϵ's. This independence is the exogeneity assumption, which gives the concept a more technical shape. For each t, we get to see the values of the exogenous variables

$$W_t, \ H_t, \ M_t, \ B_t,$$

and the corresponding endogenous variables computed by solving (3a,b) together, namely,

$$Q_t = Q_{W_t, H_t, M_t, B_t} \quad \text{and} \quad P_t = P_{W_t, H_t, M_t, B_t}.$$

Of course, we do not get to see the parameters or the disturbance terms. A regression of Q_t on P_t and the exogenous variables leads to "simultaneity bias," because P_t is correlated with the error term; hence two-stage least squares and related techniques. With such estimators, enough data, and the assumptions detailed above, we can (almost) recover the supply and demand schedules (3a,b) from the free market data—using the exogenous variables supplied by Nature.

The other approach, sketched above for Figures 15.2 and 15.3, suggests that we start from the parameters and the error distribution $\mathcal{C}$. If we were to set P, W, H, M, and B, then Nature would be assumed to choose the errors in (3) from $\mathcal{C}$: Farmers would respond according to the supply equation (3a) and consumers according to the demand equation (3b). If we were to set only W, H, M, and B and allow the free market to operate, then quantity and price would in this parable be computed by solving the pair of equations (3a,b).

The notation for the error terms in (3) is a bit simplistic now, since these terms may be functions of W, H, M, and B. Allowing the errors to be functions of P may make sense if (3a) and (3b) are considered in isolation. But if the two equations are considered together, this extra generality would lead to a morass. We therefore allow errors to be functions of W, H, M, and B but not P. To generate data, we assume that Nature chooses the exogenous variables at random from some multi-dimensional distribution. The market quantities and prices are still computed by solving the pair of equations (3a,b) for Q and P, with independent additive errors for each period drawn from $\mathcal{C}$; the usual statistical computations can still be carried out.

In this setup, it is not the error terms that are invariant but their distribution. Of course, parameters are taken to be invariant. The exogeneity assumption is the independence of $\{W_t, H_t, M_t, B_t : t = 1, 2 \ldots\}$ and the error terms. The inference is, for instance, to the probability distribution of butter supply, if we were to intervene in the market by setting price as well as the exogenous variables. By contrast, with assumed invariance for the error terms themselves, the inference is to the numerical quantity of butter that would be supplied.

I have presented the second approach with a causal interpretation. An associational interpretation is also possible, although less interesting. The exposition may seem heavy-handed, because I have tried to underline the critical invariance assumptions that need to be made in order to draw causal conclusions from nonexperimental data: Parameters are invariant to interventions, and so are errors or their distributions. Exogeneity is another concern. In a real example, as opposed to a butter hypothetical, real questions would have to be asked about these assumptions. Why are the equations "structural," in the sense that the required invariance assumptions hold true?

Obviously, there is some tension here. We want to use regression to draw causal inferences from nonexperimental data. To do that, we need to know that certain parameters and certain distributions would remain invariant if we were to intervene. That invariance can seldom if ever be demonstrated by intervention. What, then, is the source of the knowledge? "Economic theory" seems like a natural answer, but an incomplete one. Theory has to be anchored in reality. Sooner or later, invariance needs empirical demonstration, which is easier said than done.

15.5 Nonlinear models: Figure 15.1 revisited

Graphical models can be set up with nonlinear versions of equation (1), as in Pearl (1995, 2000). The specification would be something like $Y_{i,x} = f(x, \delta_i)$, where f is a fairly general (unknown) function. The interpretation is this: If the treatment level were set to x, the response by subject i would be $Y_{i,x}$. The same questions about interventions and counterfactual hypotheticals would then have to be considered.

Instead of rehashing such issues, I will indicate how to formulate the models using conditional distributions ("Markov kernels"), so that the graphs can be interpreted either distributionally or causally. In the nonlinear case, K_x—the conditional distribution of Y given that $X = x$ —depends on x in some fashion more complicated than linearity with additive noise. For example, if X and Y are discrete, then K can be visualized as the matrix of conditional probabilities $P(Y = y | X = x)$. For any particular x, K_x is a row in this matrix.

Inferences will be to conditional distributions, rather than specific numerical values. There will be some interesting new questions about identifiability. And the plausibility of causal interpretations can be assessed separately, as will be shown later. I will organize most of the discussion around two examples used by Pearl (1995); also see Pearl (2000, pp. 66–68, 83–85). But first, consider Figure 15.1. In the nonlinear case, the exogenous variables have to be assumed independent and identically distributed in order to make sense out of the mathematics. Otherwise, there are substantial extra complications, or we have to impose additional smoothness conditions on the kernel.

Assume now that (X_i, Y_i) are independent and distributed like (X, Y) for $i = 1, \ldots, n$; the conditional distribution of Y_i given $X_i = x$ is K_x, where K is an unknown Markov kernel. With a large enough sample, the joint distribution of (X, Y) can be estimated reasonably well; so can K_x, at least for x's that are likely to turn up in the data. If K is only a conditional probability, that is what we obtain from data analysis. If K admits a causal interpretation—by prior knowledge or assumption, not by data analysis on the X's and Y's—then we can make a causal inference: What would the distribution of Y_i have been if we had intervened and set X_i to x? (The answer is K_x.)

15.6 Technical notes

The conditional distribution of Y given X tells you the conditional probability that Y is in one set C or another, given that $X = x$. A Markov kernel K assigns a number $K_x(C)$ to pairs (x, C). The first element x of the pair is a point; the second, C, is a set. With x fixed, K_x is a probability. With C fixed, the function that sends x to $K_x(C)$ should satisfy some minimal regularity condition. Below, I will write $K_x(dy)$ as shorthand for the kernel whose value at (x, C) is $K_x(C)$, where C is any reasonable set of values for Y. Matters will be arranged so that $K_x(C)$ is the conditional probability that $Y \in C$ given $X = x$, and perhaps given additional information. Thus, $K_x(C) = P(Y \in C | X = x \ldots)$.

Without further restrictions, graphical models are non-parametric, because kernels are infinite-dimensional "parameters." Our ability to estimate such things depends on the degree of regularity that is assumed. With minimal assumptions, you may get minimal performance—but that is a topic for another day. Even in the linear case, some of the fine points about estimation have been glossed over. To estimate the model in Figure 15.1, we would need some variation in X and δ. To get standard errors, we would assume finite variances for the error terms. Conditions for identifiability in the simultaneous-equations setup do not need to be rehearsed here, and I have assumed a unique solution for (3). Two-stage least squares will have surprising behavior unless variances are assumed for the errors. Some degree of correlation between the exogenous and endogenous variables would also be needed.

More general specifications can be assumed for the errors. For example, in (1) the δ_i may be assumed to be independent, with common variances and uniformly bounded fourth moments. Then the hypothesis of a common distribution can be dropped. In (3), an ARIMA [autoregressive integrated moving average] model may be assumed. And so forth. The big picture does not change, because questions about invariance remain and even an ARIMA model requires some justification.

15.7 More complicated examples

The story behind Figure 15.3 will be explained below. For the moment, it is an abstract piece of mathematical art. The diagram corresponds to three kernels: $K_x(dy)$, $L_y(dz)$, and $M_{x,z}(dw)$. These kernels describe the joint distribution of the random variables shown in the diagram (X, Y, Z, W).

The conditional distribution of Y given $X = x$ is K_x. The conditional distribution of Z given $X = x$ and $Y = y$ is L_y. There is no subscript x on L because—by assumption—there is no arrow from X to Z in the diagram. The conditional distribution of W given $X = x$, $Y = y$, and $Z = z$ is $M_{x,z}$. There is no subscript y on M because—again by assumption—there is no arrow leading directly from Y to W in the diagram.

You can think of building up the variables X, Y, Z, and W from the kernels and a base distribution μ for X, in a series of steps:

(i) Chose X at random according to $\mu(dx)$.

(ii) Given the value of X from step (i), say $X = x$, choose Y at random from $K_x(dy)$.

(iii) Given $X = x$ and $Y = y$, choose Z at random from $L_y(dz)$.

(iv) Given $X = x$, $Y = y$, and $Z = z$, choose W at random from $M_{x,z}(dw)$.

The recipe is equivalent to the graph.

Figure 15.3 A graphical model with four variables; three are observed.

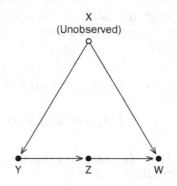

By assumption, the four-tuples (X_i, Y_i, Z_i, W_i) are independent and distributed like (X, Y, Z, W) for $i = 1, \ldots, n$. There is one more wrinkle. The circle marked "X" in the diagram is open, meaning that X is not observed. In other words, Nature hides $X_1, \ldots, X_n$ but shows us

$$Y_1, \ldots, Y_n, \ Z_1, \ldots, Z_n, \ W_1, \ldots, W_n.$$

That is our data set.

The base distribution μ and the kernels K, L, and M are unknown. However, with many observations on independent and identically distributed triplets (Y_i, Z_i, W_i), we can estimate their joint distribution reasonably well. Moreover—and this should be a little surprising—we can compute L_y from that joint distribution, as well as

$$(5a) \qquad \mathcal{M}_z(dw) = \int M_{x,z}(dw)\,\mu(dx),$$

where μ is the distribution of the unobserved confounder X. Hence we can also compute

$$(5b) \qquad \mathcal{L}_y(dw) = \int \mathcal{M}_z(dw)\,L_y(dz).$$

Here is the idea: L is computable because the relationship between Y and Z is not confounded by X. Conditional on Y, the relationship between Z and W is not confounded, so $\mathcal{M}_z$ in (5a) is computable. Then (5b) follows.

More specifically, with "P" for probability, the identity

$$P(Z \in C | Y = y) = P(Z \in C | X = x, Y = y) = L_y(C)$$

can be used to recover L from the joint distribution of Y and Z.

Likewise, we can recover $\mathcal{M}_z$ in (5a) from the joint distribution of Y, Z, and W, although the calculation is a little more intricate. Let $P_{x,y,z} = P(\cdot | X = x, Y = y, Z = z)$ be a regular conditional probability given X, Y, and Z. Then

$$P(W \in D | Y = y, Z = z) = \int P_{x,y,z}(W \in D)\,P(X \in dx | Y = y, Z = z)$$

$$= \int M_{x,z}(D)\,P(X \in dx | Y = y),$$

because $P_{x,y,z}(W \in D) = M_{x,z}(D)$ by construction, and X is independent of Z given Y by a side-calculation.

We have recovered $\int M_{x,z}(D) P(X \in dx | Y = y)$ from the joint distribution of Y, Z, and W. Hence we can recover

$$\int\int M_{x,z}(D)\, P(X \in dx | Y = y) P(Y \in dy) = \int M_{x,z}(D)\, \mu(dx)$$
$$= \mathcal{M}_z(D),$$

although the distribution μ of X remains unknown and so does the kernel M.

These may all just be facts about conditional distributions, in which case (5) is little more than a curiosity. On the other hand, if K, L, and M have causal interpretations, then $\mathcal{M}_z$ in (5a) tells you the effect of setting $Z = z$ on W, averaged over the possible X's in the population. Similarly, $\mathcal{L}_y$ in (5b) tells you the effect of Y on W. If you intervene and set Y to y, then the distribution of W will be $\mathcal{L}_y$, on the average over all X and Z in the population. (There may be exceptional null sets, which are being ignored.) How to estimate $\mathcal{M}$ and $\mathcal{L}$ in a finite sample is another question, which will not be discussed here.

The next example (Figure 15.4) is a little more complicated. (Again, the story behind the figure is deferred.) There are two unobserved variables, A and B. The setup involves six kernels, which characterize the

Figure 15.4 A graphical model with seven variables; five are observed.

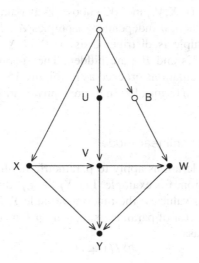

joint distribution of the random variables (A, B, U, X, V, W, Y) in the diagram:

$$K_a(db) = P(B \in db | A = a),$$

$$L_a(du) = P(U \in du | A = a),$$

$$M_a(dx) = P(X \in dx | A = a),$$

$$N_{u,x}(dv) = P(V \in dv | A = a, B = b, U = u, X = x),$$

$$Q_{b,v}(dw) = P(W \in dw | A = a, B = b, U = u, X = x, V = v),$$

$$R_{x,v,w}(dy) =$$
$$P(Y \in dy | A = a, B = fb, U = u, X = x, V = v, W = w).$$

Here, P represents "probability"; it seemed more tasteful not to have kernels labeled O or P. There is no a, b, or u among the subscripts on R because there are no arrows going directly from A, B, or U to Y in the diagram; similarly for the other kernels. The issue is to determine the effect of X on Y, integrating over the unobserved confounders A and B. This is feasible, because conditional on the observed U, V, and W, the relationship between X and Y is not confounded. (If the kernels have causal interpretations, "effect" is meant literally; if not, figuratively.)

To fix ideas, we can go through the construction of the random variables. There is a base probability μ for A. First, choose A at random from μ. Given A, choose B, U, and X independently at random from K_A, L_A, and M_A, respectively. Given A, B, U, and X, choose V at random from $N_{U,X}$. Given A, B, U, X, and V, choose W at random from $Q_{B,V}$. Finally, given A, B, U, X, V, and W, choose Y at random from $R_{X,V,W}$. The data set consists of n independent septuples A_i, B_i, U_i, X_i, V_i, W_i, and Y_i. Each septuple is distributed as A, B, U, X, V, W, and Y. The kicker is that the A's and B's are hidden. The "parameters" are μ and the six kernels. Calculations proceed as for Figure 15.3. Again, the graph and the description in terms of kernels are equivalent. Details are (mercifully?) omitted.

15.8 Parametric nonlinear models

Similar considerations apply to parametric nonlinear models. Take the logit specification, for example. Let X_i be a p-dimensional random vector, with typical value x_i; the random variable Y_i is 0 or 1. Let β be a p-dimensional vector of parameters. For the p-dimensional data vector x, let K_x assign mass

$$e^{\beta x} / (1 + e^{\beta x})$$

to 1, and the remaining mass to 0. Given $X_1, \ldots, X_n$, each being a p-vector, suppose the Y_i are conditionally independent, and

(6) $$P(Y_i = 1 | X_1 = x_1, \ldots, X_n = x_n) = K_{x_i}.$$

On the right hand side of (6), the subscript on K is x_i. The conditional distribution of Y for a subject depends only on that subject's x. If the $x_1, \ldots, x_n$ are reasonably spread out, we can estimate β by maximum likelihood. (With a smooth, finite-dimensional parameterization, we do not need the X_i to be independent and identically distributed.)

Of course, this model could be set up in a more strongly invariant form, like (1). Let U_i be independent (unobservable) random variables with a common logistic distribution: $P(U_i < u) = e^u / (1 + e^u)$. Then

(7) $$Y_{i,x} = 1 \iff U_i < \beta x.$$

The exogeneity assumption would make the X's independent of the U's, and the observable Y_i would be Y_{i,X_i}. That is, $Y_i = 1$ if $U_i < \beta X_i$, else $Y_i = 0$.

This is all familiar territory, except perhaps for (7); so familiar that the critical question may get lost. Does K_x merely represent the conditional probability that $P(Y_i = 1 | X_i = x)$, as in (6)? Or does K_x tell us what the law of Y_i would have been, if we had intervened and set X_i to x? Where would the U_i come from, and why would they be invariant if we were to intervene and manipulate x? Nothing in the mysteries of Euclidean geometry and likelihood statistics can possibly answer this sort of question. Other kinds of information are needed.

15.9 Concomitants

Some variables are potentially manipulable; others ("concomitants") are not. For example, education and income may be manipulable; age, sex, race, personality, . . . , are concomitants. So far, we have ignored this distinction, which is less problematic for kernels, but a difficulty for the kind of strong invariance in equation (1). If Y depends on a manipulable X and a concomitant W through a linear causal law with additive error, we can rewrite (1) as

(8) $$Y_{i,x} = a + bx + cW_i + \delta_i.$$

In addition to the usual assumptions on the δ's, we would have to assume independence between the δ's and the W's. Similar comments apply when there are several manipulable variables, or logits, probits, and so forth. In applications, defining and isolating the intervention may not be so easy, but that is a topic for another day. Also see Robins (1986, 1987a,b).

15.10 The story behind Figures 15.3 and 15.4

When some variables are unobserved, Pearl (1995) develops an interesting calculus to define confounding and decide which kernels or composites—see (5) for example—can be recovered from the joint distribution of the observed variables. That is a solution to the identification problem for such diagrams. He uses Figure 15.3 to illustrate his "front-door criterion." The unobserved variable X is genotype. The observed variables Y, Z, and W represent smoking, tar deposits in the lung, and lung cancer, respectively (Figure 15.5). The objective is to determine the effect of smoking on lung cancer, via (5).

Data in this example would consist of a long series of independent triplets (Y_i, Z_i, W_i), each distributed like (Y, Z, W). Pearl interprets the graph causally. The timeworn idea that subjects in a study form a random sample from some hypothetical super-population still deserves a moment of respectful silence. Moreover, there are three special assumptions in Figure 15.5:

(i) Genotype has no direct effect on tar deposits.

(ii) Smoking has no direct effect on lung cancer.

(iii) Tar deposits can be measured with reasonable accuracy.

There is no support for these ideas in the literature. (i) The lung has a mechanism—"the mucociliary escalator"—for eliminating foreign matter, including tar. This mechanism seems to be under genetic control. (Of

Figure 15.5 A graphical model for smoking and lung cancer.

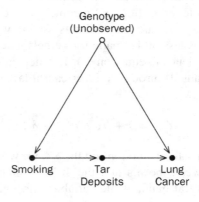

Genotype
(Unobserved)

Smoking   Tar   Lung
             Deposits   Cancer

course, clearance mechanisms can be overwhelmed by smoking.) The forbidden arrow from genotype to tar deposits may have a more solid empirical basis than the permitted arrows from genotype to smoking and lung cancer. Assumption (ii) is just that—an assumption. And (iii) is clearly wrong. The consequences are severe. If arrows are permitted from genotype to tar deposits or from smoking to lung cancer, or if measurements of tar are subject to error, then formula (5) does not apply. Graphical models cannot solve the problem created by an unmeasured confounder without introducing strong and artificial assumptions.

The intellectual history is worth mentioning. Fisher's "constitutional hypothesis" explained the association between smoking and disease on the basis of a gene that caused both. This idea is refuted not by making assumptions but by doing some empirical work. For example, Kaprio and Koskenvuo (1989) present data from their twin study. The idea is to find pairs of identical twins where one smokes and one does not. That sets up a race: Who will die first, the smoker or the non-smoker? The smokers win hands down, for total mortality or death from heart disease. The genetic hypothesis is incompatible with these data.

For lung cancer, the smokers win two out of the two races that have been run. (Why only two? Smoking-discordant twin pairs are unusual, lung cancer is a rare disease, and the population of Scandinavia is small.) Carmelli and Page (1996) have a similar analysis with a larger cohort of twins. Do not bet on Fisher. International Agency for Research on Cancer (1986) reviews the health effects of smoking and indicates the difficulties in measuring tar deposits (pp. 179–98). Nakachi et al. (1993) and Shields et al. (1993) illustrate conflicts on the genetics of smoking and lung cancer. Also see Miller et al. (2003). The lesson: Finding the mathematical consequences of assumptions matters, but connecting assumptions to reality matters even more.

Pearl uses Figure 15.4 to illustrate his "back-door criterion," calling the figure a "classical example due to Cochran," with a cite to Wainer (1989). Pearl's vision is that soil fumigants X are used to kill eelworms and improve crop yields Y for oats. The decision to apply fumigants is affected by the worm population A before the study begins, hence the arrow from A to X. The worm population is measured at baseline, after fumigation, and later in the season: The three measurements are U, V, and W. The unobserved B represents "birds and other predators."

This vision is whimsical. The example originates with Cochran (1957, p. 266) who had several fumigants applied under experimental control, with measurements of worm cysts and crop yield. Pearl converts this to an observational study with birds, bees, and so forth—entertaining, a teaching tool, but unreal. It might be rude to ask too many questions about Figure 15.4, but surely crops attract predators? Don't birds eat oat seeds? If early birds get the worms, what stops them from eating worms at baseline? In short, where have all the arrows gone?

15.11 Models and kernels revisited

Graphical models may lead to some interesting mathematical developments. The number of successful applications, however, is at best quite limited. The examples discussed here are not atypical. Given that the arrows and kernels represent causation, while variables are independent and identically distributed, we can use Pearl's framework to determine from the diagram which effects are estimable. This is a step forward. However, we cannot use the framework to answer the more basic question: Does the diagram represent the causal structure? As everyone knows, there are no formal algorithmic procedures for inferring causation from association; everyone is right.

Pearl (1995) considers only models with a causal interpretation, the latter being partly formalized; and there is new terminology that some readers may find discouraging. On the other hand, he draws a clear distinction between averaging Y's when the corresponding X is
- set to x, and
- observed to be x in the data.

That is a great advantage of his formalism.

The approach sketched here would divide the identification problem in two: (i) reconstructing kernels, viewed as ordinary conditional distributions, from partial information about joint distributions, and (ii) deciding whether these kernels bear a causal interpretation. Problem (i) can be handled entirely within the conventional probability calculus. Problem (ii) is one of the basic problems in applied statistics. Of course, kernels—especially mixtures like (5)—may not be interesting without a causal interpretation.

In sum, graphical models can be formulated using conditional distributions ("Markov kernels"), without invariance assumptions. Thus, the graphs can be interpreted either distributionally or causally. The theory governing recovery of kernels and their mixtures can be pushed through with just the distributional interpretation. That frees us to consider whether or not the kernels admit a causal interpretation.

So far, the graphical modelers have few if any examples where the causal interpretation can be defended. Pearl generally agrees with this discussion (personal communication):

Causal analysis with graphical models does not deal with defending modeling assumptions, in much the same way that differential calculus does not deal with defending the physical validity of a differential equation that a physicist chooses to use. In fact no analysis void of experimental data can possibly defend modeling assumptions. Instead, causal analysis deals with the conclusions that logically follow from the combination of data and a given set of assumptions, just in case one is prepared to accept the latter. Thus, all causal inferences are necessarily *conditional*. These limitations are not unique to graphical models. In complex fields like the social sciences and epidemiology, there are only a few (if any) real life situations where we can make enough compelling assumptions that would lead to identification of causal effects.

15.12 Literature review

The model in (1) was proposed by Neyman (1923). It has been rediscovered many times since; see, for instance, Hodges and Lehmann (1964, section 9.4). The setup is often called "Rubin's model," but this simply mistakes the history. See Dabrowska and Speed (1990), with a comment by Rubin. Compare Rubin (1974) and Holland (1986). Holland (1986, 1988) explains the setup with a super-population model to account for the randomness, rather than individualized error terms. These error terms are often described as the overall effects of factors omitted from the equation. But this description introduces difficulties of its own, as shown by Pratt and Schlaifer (1984, 1988). Stone (1993) presents a clear super-population model with some observed covariates and some unobserved.

Dawid (2000) objects to counterfactual inference. Counterfactual distributions may be essential to any account of causal inference by regression methods. On the other hand, as the present chapter tries to show, invariant counterfactual random variables—like δ_i in equation (1)—are dispensable. In particular, with kernels, there is no need to specify the joint distribution of random variables across inconsistent hypotheticals.

There is by now an extended critical literature on statistical modeling, starting perhaps with the exchange between Keynes (1939, 1940) and Tinbergen (1940). Other familiar citations in the economics literature include Liu (1960), Lucas (1976), and Sims (1980). Manski (1995) returns to the under-identification problem that was posed so sharply by Liu and Sims. In brief, a priori exclusion of variables from causal equations can seldom be justified, so there will typically be more parameters than data.

Manski suggests methods for bounding quantities that cannot be estimated. Sims' idea was to use simple, low-dimensional models for policy analysis, instead of complex, high-dimensional ones. Leamer (1978) discusses the issues created by inferring the specification from the data, as does Hendry (1980). Engle, Hendry, and Richard (1983) distinguish several kinds of exogeneity, with different implications for causal inference.

Heckman (2000) traces the development of econometric thought from Haavelmo and Frisch onwards, stressing the role of "structural" or "invariant" parameters and "potential outcomes"; also see Heckman (2001a,b). According to Heckman (2000, pp. 89–91), the enduring contributions are the insights that—

> . . . causality is a property of a model, that many models may explain the same data and that assumptions must be made to identify causal or structural models . . . recognizing the possibility of interrelationships among causes . . . [clarifying] the conditional nature of causal knowledge and the impossibility of a purely empirical approach to analyzing causal questions . . . . The information in any body of data is usually too weak to eliminate competing causal explanations of the same phenomenon. There is no mechanical algorithm for producing a set of "assumption free" facts or causal estimates based on those facts.

For another discussion of causal models from an econometric perspective see Angrist (2001) or Angrist, Imbens, and Rubin (1996). Angrist and Krueger (2001) provide a nice introduction to instrumental variables; an early application of the technique was to fit supply and demand curves for butter (Wright 1928, p. 316).

One of the drivers for modeling in economics and cognate fields is rational choice theory. Therefore, any discussion of empirical foundations must take into account a remarkable series of papers, initiated by Kahneman and Tversky (1974), that explores the limits of rational choice theory. These papers are collected in Kahneman, Slovic, and Tversky (1982) and in Kahneman and Tversky (2000). The heuristics and biases program has attracted its own critics (Gigerenzer 1996). That critique is interesting and has some merit. In the end, however, the experimental evidence demonstrates severe limits to the descriptive power of choice theory (Kahneman and Tversky 1996).

If people are trying to maximize expected utility, they don't do it very well. Errors are large and repetitive, go in predictable directions, and fall into recognizable categories: These are biases, not random errors. Rather than making decisions by optimization—or bounded rationality, or satisficing—people seem to use plausible heuristics that can be identified. If so, rational choice theory is generally not a good basis for justifying empirical models of behavior. Sen (2002) makes a far-reaching critique of rational choice theory, based in part on the work of Kahneman and Tversky.

Recently, modeling issues have been much canvassed in sociology. Berk (2004) is skeptical about the possibility of inferring causation by modeling, absent a strong theoretical base. Abbott (1997) finds that variables (like income and education) are too abstract to have much explanatory power; also see Abbott (1998). Clogg and Haritou (1997) review various difficulties with regression, noting in particular that you can all too easily include endogenous variables as regressors.

Goldthorpe (1999, 2001) describes several ideas of causation and corresponding methods of statistical proof, with different strengths and weaknesses; he finds rational choice theory to be promising. Hedström and Swedberg (1998) edited a lively collection of essays by a number of sociologists, who turn out to be quite skeptical about regression models; rational choice theory takes its share of criticism. Ní Bhrolcháin (2001) has some particularly forceful examples to illustrate the limits of regression. There is an influential book by Lieberson (1985), with a followup by Lieberson and Lynn (2002). The latest in a series of informative papers is Sobel (2000).

Meehl (1978) reports the views of an empirical psychologist; also see Meehl (1954), with data showing the advantage of using regression to make predictions—rather than experts. Meehl and Waller (2002) discuss the choice between similar path models, viewed as reasonable approximations to some underlying causal structure, but do not reach the critical question—how to assess the adequacy of the approximation. Steiger (2001) has a critical review.

There are well-known books by Cook and Campbell (1979) and by Shadish, Cook, and Campbell (2002). In political science, Brady and Collier (2004) compare regression methods with case studies; invariance is discussed under the rubric of causal homogeneity. Cites from other perspectives include Freedman, Rothenberg, and Sutch (1983), Oakes (1990), as well as Freedman (1985, 1987, 1991 [Chapter 3], 1995 [Chapter 1], 1999).

There is an extended literature on graphical models for causation. Greenland, Pearl, and Robins (1999) give a clear account in the context of epidemiology. Lauritzen (1996, 2001) has a careful treatment of the mathematics. These authors do not recognize the difficulties in applying the methods to real problems.

Equation (5) is a special case of the "g-computation algorithm" due to Robins (1986, 1987a,b); also see Gill and Robins (2004), Pearl (1995, 2000), or Spirtes, Glymour, and Scheines (1993). Robins (1995) explains—all too briefly—how to state Pearl's results as theorems about conditionals.

For critical reviews of graphical models (with responses and further citations), see Freedman (1997), Humphreys (1997), Humphreys and Freedman (1996) [Chapter 14], and Freedman and Humphreys (1999): Among other things, these papers discuss various applications proposed by the modelers. Woodward (1997, 1999) stresses the role of invariance.

Freedman and Stark (1999 [Chapter 10]) show that different models for the correlation of outcomes across counterfactual scenarios can have markedly different consequences in the legal context. Scharfstein, Rotnitzky, and Robins (1999) demonstrate a large range of uncertainty in estimates, due to incomplete specifications; also see Robins (1999).

Acknowledgments

Over the years, I learned a great deal about statistics from Tom Rothenberg; it is a pleasure to acknowledge the debt. I would also like to thank some other friends for many helpful conversations on the topics of this chapter: Dick Berk, Paul Holland, Paul Humphreys, Máire Ní Bhrolcháin, Judea Pearl, Jamie Robins, and Philip B. Stark. At the risk of the obvious, thanking people does not imply they agree with my opinions; nor does this caveat imply disagreement.

16

Weighting Regressions by Propensity Scores

With Richard A. Berk

ABSTRACT. *Regressions can be weighted by propensity scores in order to reduce bias. However, weighting is likely to increase random error in the estimates and to bias the estimated standard errors downward, even when selection mechanisms are well understood. Moreover, in some cases, weighting will increase the bias in estimated causal parameters. If investigators have a good causal model, it seems better just to fit the model without weights. If the causal model is improperly specified, there can be significant problems in retrieving the situation by weighting, although weighting may help under some circumstances.*

Estimating causal effects is often the key to evaluating social programs, but the interventions of interest are seldom assigned at random. Observational data are therefore frequently encountered. In order to estimate causal effects from observational data, some researchers weight regressions using "propensity scores." This simple and ingenious idea is due to Robins and his collaborators. If the conditions are right, propensity scores can be used to advantage when estimating causal effects.

Evaluation Review (2008) 32: 392–409.

However, weighting has been applied in many different contexts. The costs of misapplying the technique, in terms of bias and variance, can be serious. Many users, particularly in the social sciences, seem unaware of the pitfalls. Therefore, it may be useful to explain the idea and the circumstances under which it can go astray.

That is what we try to do here. We illustrate the performance of the technique—and some of the problems that can arise—on simulated data where the causal mechanism and the selection mechanism are both known, which makes it easy to calibrate performance.

We focus on cross-sectional parametric models, of the kind commonly seen in applications. Pooling time-series and cross-sectional variation leads to substantial additional complexity. Thus, we consider linear causal models like

$$(1) \qquad Y = a + bX + c_1 Z_1 + c_2 Z_2 + U,$$

where $X = 1$ for subjects in the treatment group and 0 for those in the control group; Z_1 and Z_2 are confounders, correlated with X. The random error U is independent of X, Z_1, and Z_2.

The "propensity score" $\hat{p}$ is an estimate for $P(X = 1|Z_1, Z_2)$, that is, the conditional probability of finding the subject in the treatment group given the confounders. Subjects with $X = 1$ receive weight $1/\hat{p}$; subjects with $X = 0$ receive weight $1/(1 - \hat{p})$. A "weighted" regression minimizes the weighted sum of squares.

We investigated the operating characteristics of weighting in a dozen simulation models. In these simulations, there were $n = 1000$ independent, identically distributed (IID) subjects. In some cases, we re-ran the simulation with $n = 10,000$ subjects to see the effect of larger n on bias and variance.

Each simulation had two components. The first component was a model that explained selection into the treatment or control condition. The second component was a causal model that determined response to treatment and to confounders. (Responses may be continuous or binary.) Selection was exogenous, that is, independent of the error term in the causal model.

The simulations were all favorable to weighting, in three important ways: (i) subjects were IID; (ii) selection was exogenous; and (iii) the selection equation was properly specified. We report in detail on two simulations that were reasonably typical and mention some others in passing. We write Y for the response, X for treatment status (0 if in control, 1 if in treatment), and Z for the confounder. Generally, Z is multivariate normal.

16.1 Simulation #1

Our first simulation had a continuous linear response and probit selection. The causal model is

$$(2) \qquad Y = a + bX + c_1Z_1 + c_2Z_2 + dU,$$

where U is $N(0, 1)$. The selection model is

$$(3) \qquad X = (e + f_1Z_1 + f_2Z_2 + V > 0),$$

where V is $N(0, 1)$. Here, $a, b, c_1, c_2, d, e, f_1$, and f_2 are parameters. Equation (3) may look a bit cryptic. More explicitly, the equation says that $X = 1$ if $e + f_1Z_1 + f_2Z_2 + V > 0$; otherwise, $X = 0$.

By construction, U, V, and $Z = (Z_1, Z_2)$ are all independent, and Z is bivariate normal. The observables are (X, Z, Y). The variables U and V are not observable. In particular, X follows a probit model. To construct the weights, we fit this probit model to the data on (X, Z).

Let $\hat{p}$ be the estimated probability that $X = 1$ given Z. Subjects with $X = 1$ get weight $w = 1/\hat{p}$. Subjects with $X = 0$ get weight $w = 1/(1 - \hat{p})$. Notice that $\hat{p}$ depends on Z, so w depends on X and Z. Notice too that the selection equation is correctly specified.

For simplicity, we put $a = b = c_1 = d = 1$ and $c_2 = 2$ in equation (2). To keep variability in the weights within bounds, we make $e = .5, f_1 = .25$, and $f_2 = .75$ in equation (3). We set $\text{var}(Z_1) = 2$, $\text{var}(Z_2) = 1, \text{cov}(Z_1, Z_2) = 1, E(Z_1) = .5$, and $E(Z_2) = 1$.

We run regressions of Y on X and Z, unweighted and weighted, getting estimates for $a, b, \ldots$, and their nominal standard errors. ("Nominal" standard errors are computed from the usual regression formulae.) We also run a regression of Y on X and Z_1. Finally, we run a simple regression of Y on X.

Without the weights, the latter two regressions are misspecified: There is omitted-variables bias. The point of the weighting, as in most of the social-science literature we reviewed, is to correct omitted-variables bias. In the simulations, truth is known, so we can evaluate the extent to which the correction succeeds.

We repeat the process 250 times, getting the mean of the estimates, the standard deviation of the estimates, and the root mean square of the nominal standard errors. We abbreviate SD for standard deviation, SE for standard error, and RMS for root mean square. The SD measures the likely size of the random error in the estimates.

If Z_1 and Z_2 are both included in the regression, the weighted multiple regression estimates are essentially unbiased. However, the SD of the $\hat{b}$'s is about double the SD in the unweighted regression. Furthermore, the nominal SE's are too small by a factor of three (Table 16.1, first two blocks). When all the covariates are included, weighting the regression is therefore counter-productive. There is no bias to reduce, there is an increase in variance, and the nominal SE's become difficult to interpret.

Next, suppose Z_2 is omitted from the regression. The unweighted regression of Y on X and Z_1 then gives a biased estimate for b. The weighted regression of Y on X and Z_1 is still somewhat biased for b and is quite biased for a and c_1. The bias in $\hat{b}$ is "small-sample bias." The other biases will not disappear with larger samples. The SD's in the weighted regression are rather large, and the nominal SE's are too small (Table 16.1, middle two blocks).

Finally, suppose Z_1 and Z_2 are both omitted from the regression. The bias in the weighted regression is even worse. By comparison, an unweighted simple regression does better at estimating a, worse at estimating b (Table 16.1, last two blocks). Again, the bias in the weighted regression estimate for b is a small-sample bias: With an n of 10,000, this bias will largely disappear.

The bias in $\hat{a}$ comes about because $E(Z) \neq 0$. This bias remains, no matter how large the sample may be. If we wish to estimate the causal effects of the treatment and control regimes separately, conditional on the covariates, this bias cannot be ignored. (It does cancel if we estimate differential effects.)

Some of the trouble is due to variability in the weights. We did the simulation over again, truncating the weights at twenty: In other words, when the weight is above twenty, we replace it by twenty. Qualitatively, results are similar. Quantitatively, there is a noticeable reduction in variance—even though we only trim six weights per 1000 subjects. However, there is some increase in bias. We also tried filtering out subjects with large weights. This was worse than truncation. Variability in the weights is a difficulty that is frequently encountered in applications.

The unweighted simple regression of Y on X has substantial bias, and the nominal SE's are far too optimistic. Why? The error term in this regression is $c_1 Z_1 + c_2 Z_2$. Some of this will be picked up in the intercept and the coefficient of X, explaining the bias. The remainder is heteroscedastic, partly because X is a binary variable so (X, Z_1, Z_2) cannot be jointly normal, partly because weighting converts homoscedastic errors to heteroscedastic errors. That helps to explain why the nominal SE's are deficient.

We return to the weighted regressions. It seems natural to try the Huber-White correction [Chapter 17], but this is unlikely to help. With

Table 16.1 Simulation #1: Linear regression with $n = 1000$ independent subjects. "Ave" is the average value of the estimates and "SD" is their standard deviation across 250 replications. "nom SE" is the nominal SE. The table reports the RMS of the nominal SE's.

Parameters	a	b	c_1	c_2
True values	1	1	1	2

Linear regression of Y on X, Z_1, and Z_2, unweighted

Ave	0.9970	1.0101	1.0003	1.9952
SD	0.0802	0.0974	0.0323	0.0468
nom SE	0.0812	0.0967	0.0320	0.0466

Linear regression of Y on X, Z_1, and Z_2, weighted

Ave	1.0007	1.0089	0.9947	1.9978
SD	0.1452	0.2130	0.1010	0.1400
nom SE	0.0562	0.0635	0.0320	0.0459

Linear regression of Y on X and Z_1, unweighted

Ave	1.6207	2.1310	1.8788
SD	0.1325	0.1574	0.0446
nom SE	0.1345	0.1569	0.0415

Linear regression of Y on X and Z_1, weighted

Ave	2.3994	1.1366	1.9432
SD	0.2995	0.3295	0.1202
nom SE	0.0789	0.1082	0.0401

Linear regression of Y on X, unweighted

Ave	0.1547	5.0232
SD	1.1101	1.0830
nom SE	0.2276	0.2495

Linear regression of Y on X, weighted

Ave	3.0665	1.4507
SD	0.7880	0.7765
nom SE	0.1414	0.1972

omitted variables, errors do not have conditional expectation 0 given the included variables, even after we subtract the projection of the error vector onto the regressors. Again, (X, Z_1, Z_2) isn't normal, and the projection operator depends on the weights. The key assumption behind the correction is false. (Outliers are another problem.)

Indeed, the Huber-White correction did not work very well for us, even in the full multivariate regression. The reason for this last failure may be the length of the tail in the distribution of $1/\hat{p}$, which is our next topic.

Recall that the weights w are defined as follows: $w = 1/\hat{p}$ for subjects with $X = 1$ and $w = 1/(1 - \hat{p})$ for subjects with $X = 0$, where $\hat{p}$ is the estimated value for $P(X = 1|Z_1, Z_2)$. A histogram for $\log \log w$ in one replication is shown in Figure 16.1. The top panel shows the histogram for $X = 0$; the bottom panel, for $X = 1$.

The height of each bar shows the number of observations falling in the corresponding class interval; there were 180 observations with $X = 0$ and 820 with $X = 1$. That is why the bottom histogram is bigger. It also has longer tails. The difference in the length of the tails in the two distributions is one of the problems faced by the weighting procedure. (The difference is not due to the difference in sample sizes.)

Figure 16.1. Top panel: Weights for controls. Bottom panel: Weights for treatment group. Log log transformation.

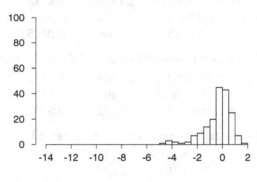

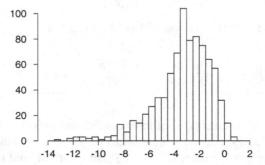

The two logs are needed to get a decent-looking histogram. The low end of the scale corresponds to weights just above 1, that is, $\hat{p}$'s just below 1. The high end of the scale corresponds to weights on the order of 50 to 250 for $X = 0$ and 5 to 15 for $X = 1$, depending on how the random numbers fall. For the particular replication reported here, the maximal weights were about 150 and 7, respectively. However, maxima are notoriously vulnerable to chance fluctuations, and larger weights do occur.

Which way do our assumptions cut? The assumption that subjects are IID is favorable to the modeling enterprise. So is the exogeneity of the selection mechanism. Making V normal is another kindness; without it, the selection equation would be misspecified. Making U normal also seems to be generous, since the response equation is estimated by least squares.

Assuming Z to be normal presents tradeoffs that are more complicated. With shorter-tailed distributions, weighting may work better. With longer-tailed distributions, which seem more common in practice, weighting is likely to do worse.

In our simulations, the exogenous regressors Z_1 and Z_2 are randomized afresh on each of the 250 repetitions. Generating the Z's once and for all at the beginning and reusing the same Z's throughout makes almost no difference to the results. (We tried it.) In principle, the SD's should go down a little, but the difference is too small to see.

16.2 Results for Simulation #2

Simulation #2 is just like Simulation #1, with logit selection and logit response; the parameter values remain the same, along with the joint distribution of (Z_1, Z_2). The causal model is

$$(4) \qquad Y = (a + bX + c_1 Z_1 + c_2 Z_2 + U > 0),$$

and the selection model is

$$(5) \qquad X = (e + f_1 Z_1 + f_2 Z_2 + V > 0),$$

where (Z_1, Z_2), U, and V are independent; U and V follow the standard logistic distribution.

Results are much like those in Simulation #1. See Table 16.2. However, with omitted variables the weighted logistic regression performs very poorly at estimating the coefficient b of the treatment variable.

Table 16.2 Simulation #2: Logistic regression with $n = 1000$ independent subjects. "Ave" is the average value of the estimates and "SD" is their standard deviation across 250 replications. "nom SE" is the nominal SE. The table reports the RMS of the nominal SE's.

Parameters	a	b	c_1	c_2
True values	1	1	1	2

Logistic regression of Y on X, Z_1, and Z_2, unweighted

Ave	1.0100	1.0262	1.0210	2.0170
SD	0.2372	0.2919	0.1611	0.2674
nom SE	0.2296	0.2750	0.1589	0.2525

Logistic regression of Y on X, Z_1, and Z_2, weighted

Ave	1.0178	1.0616	1.0470	2.1018
SD	0.3084	0.3066	0.2593	0.4197
nom SE	0.1286	0.1943	0.0960	0.1453

Logistic regression of Y on X and Z_1, unweighted

Ave	1.5879	1.3711	1.5491
SD	0.2140	0.2543	0.1396
nom SE	0.2027	0.2452	0.1389

Logistic regression of Y on X and Z_1, weighted

Ave	2.5934	0.3214	1.8977
SD	0.3419	0.3218	0.2391
nom SE	0.0977	0.1684	0.0788

Logistic regression of Y on X, unweighted

Ave	0.6779	1.9893
SD	1.1458	1.1778
nom SE	0.1367	0.2016

Logistic regression of Y on X, weighted

Ave	3.9154	−2.1168
SD	0.9632	0.9725
nom SE	0.0729	0.1190

(A "weighted" logistic regression maximizes the weighted log-likelihood function.) When Z_1 and Z_2 are both omitted, the sign of $\hat{b}$ is usually wrong. The unweighted simple logistic regression does substantially better.

The bad behavior of the weighted simple logistic regression is not a small-sample problem. It is quite reproducible. We think it is due to occasional large weights. However, if we truncate the weights above at 20, there is no improvement in the weighted estimator. At 10—and this affects only 65/1000 of the weights—$\hat{b}$ has a fair chance of being positive. In practice, of course, it might be hard to tell how much truncation to do. We return to this point later.

16.3 Covariate balance

Covariate balance in a sample after weighting is sometimes used to justify the results of propensity score weighted regression. We tried Simulation #1 with one covariate instead of two and slightly different values for the parameters $a, b, \ldots$. About 40% of the time, the covariate balanced across treatment and control groups. In these data sets, the simple weighted regression estimator was nearly unbiased for b. But the SD of the $\hat{b}$'s was about double the SD in the unweighted multiple regression, and the nominal SE was much too small. Therefore, covariate balance in the data does not answer our arguments. In our setup, you are better off just running the unweighted multiple regression. Of course, the response equation is correctly specified, which counsels against weighting. The selection equation is correct too, but this counsels in favor of weighting.

16.4 Discussion

When a linear causal model is correctly specified, weighting is usually counter-productive because there is no bias to remove. On the other hand, when the model omits relevant variables, weighting regressions by propensity scores is worth considering. If the propensity scores can be accurately estimated, weighting may lead to a substantial reduction in bias—although, with realistic sample sizes, the bias that remains can be appreciable. The price of bias reduction is an increase in random error, along with a downward bias in the nominal SE's. See Table 16.1.

There are two threshold questions. (i) Were relevant variables omitted from the causal model? (ii) Is there enough information to estimate the propensity scores with good accuracy? If the answer to both questions is "yes," the propensity scores are likely to help reduce bias. However, the conjunction is improbable. If variables are missing from the causal model, variables are likely to be missing from the selection model too. In all our simulation models, the selection model was correctly specified, shifting the balance in favor of weighting.

When the response model is logit, weighting creates substantial bias in coefficient estimates. See Table 16.2. There are parameters that can usefully be estimated in a weighted logit specification, but these are not the usual parameters of interest. Similar comments apply to the probit model and the proportional-hazards model. On the latter, see Hernán, Brumback, and Robins (2001).

In the simulations reported here, as in many social-science papers, weighting is not intended to correct specification errors other than omitted-variables bias. The errors we have in mind include heteroscedasticity, dependence between subjects, endogeneity (selection into treatment correlated with the error term in the causal model), and so forth. In some of our simulations, weighting worsens endogeneity bias in multiple regression but helps in simple regression.

With non-parametric models for response and selection—and this is closer to Robins' original conception—the issues will be different. Still, you need to get at least one of the two models (and preferably both) nearly right in order for weighting to help much. If both models are wrong, weighting could easily be a dead end. There are papers suggesting that under some circumstances, estimating a shaky causal model and a shaky selection model should be doubly robust. Our results indicate that under other circumstances, the technique is doubly frail.

Robins and his collaborators were not estimating structural equations. They were estimating contrasts: What would happen if you put everyone into the treatment condition? the control condition? This is not a suggestion to replace structural equations by non-parametric modeling and contrasts. Our point is that caution is needed when using new techniques. Sometimes you do have to read the fine print. Non-parametric models, Robins' work, and contrasts versus structural equations will be discussed below.

The bottom line for social scientists is this. If you have a causal model that you believe, you should probably just fit it to the data. If there are omitted variables but the propensity scores can somehow be estimated with reasonable accuracy, weighting the regression should reduce bias. If you believe the propensity scores but not the causal model, a good option might be weighted contrasts between the treatment and control groups. On the other hand, weighting is likely to increase random error by a substantial amount, and nominal standard errors (the ones printed out by the software) can be much too small.

If you are going to weight, it rarely makes sense to use the same set of covariates in the response equation and the selection equation. Furthermore, you should always look at the weights. If results are sensitive to a few large weights, it is time to reconsider. Finally, if you go beyond continuous response variables and weighted least squares, each combination of response model and fitting procedure has to be considered separately—to see what the weighted regression is going to estimate.

16.5 Literature review

There have recently been a number of studies that apply propensity score weighting to causal models. Much of the research addresses topics of interest to social scientists. The studies proceed in two steps, which are mimicked by our simulations.

Step 1. A model (typically logit or probit) is used to estimate the probability of selection into the treatment and control groups. The treatment may be an explicit intervention such as hospice care (Gozalo and Miller 2007). Or, it may reflect some feature of an ongoing social process, such as marriage (Sampson et al. 2006). The units of analysis may be individuals (Francesconi and Nicoletti 2006) or larger entities such as neighborhoods (Tita and Ridgeway 2007).

Step 2. Estimated probabilities from the first step are used to construct weights. The weights are then used to fit the causal model of substantive interest. The causal model can take a variety of forms: conventional linear regression (Francisco and Nicoletti 2006), logistic regression (Bluthenthal et al. 2006), Poisson regression (Tita and Ridgeway 2007), hierarchical Poisson regression (Sampson et al. 2006), or proportional hazards (McNiel and Binder 2007).

Sample sizes generally range from several hundred to several thousand. There will typically be several dozen covariates. In one example (Schonlau 2006), there were over 100 possible covariates to choose from, and the sample size was around 650.

Investigators differ on procedures used for choosing regressors in the causal model. Sometimes all available covariates are used (McNiel and Binder 2007). Sometimes there is a screening process, so that only variables identified as important or out of balance are included (Ridgeway et al. 2006). Typically, a multivariate model is used; sometimes, however, there are no covariates (Leslie and Theibaud 2007).

Some investigators use rather elaborate estimation procedures, including the lasso (Ridgeway et al. 2006) and boosting (Schonlau 2006). These estimation procedures, like the variable selection procedures and choice of response model—when combined with weighting—can change the meaning of the parameters that are being estimated. Thus, caution is in order.

Investigators may combine "robust" standard errors and nonlinear response models like hierarchical Poisson regressions (Sampson et al.

2006). The use of robust standard errors implicitly acknowledges that the model has the wrong functional form (Freedman 2006a [Chapter 17]). However, specification error is rarely considered to be a problem.

In this literature, important details of the model specification often remain opaque. See, for instance, pp. 483–89 in Sampson et al. (2006): Although the selection model is clear, the response model remains unclear.

Few authors consider the bias in nominal standard errors or the problems created by large weights. We saw no mention of definitional problems created by nonlinear response models or complex estimation procedures.

Lunceford and Davidian (2004) summarize the theory of weighted regressions with some informative simulations. However, the limitations of the technique are not fully described.

In a biomedical application, Hirano and Imbens (2001) recommend including interactions between the treatment dummy and the covariates. In our simulations, this sometimes reduced bias in the estimated intercept, but usually had little effect.

Two journals have special issues that explore the merits of propensity scores. This includes use of propensity scores in weighted regression and in earlier techniques, such as (i) creating match sets or (ii) computing weighted contrasts between treatment and control groups. See

Review of Economics and Statistics (2004) 86(1)

Journal of Econometrics (2005) 125(1–2).

Other references of interest include Arceneaux, Gerber, and Green (2006); Glazerman, Levy, and Myers (2003); Peikes, Moreno, and Orzol (2008); and Wilde and Hollister (2007). These authors point to serious weaknesses in the propensity-score methods that have been used for program evaluation.

The basic papers on weighted regression include Robins and Rotnitzky (1992, 1995); Robins, Rotnitzky, and Zhao (1994); Rotnitzky, Robins, and Scharfstein (1998); and Bang and Robins (2005). The last describes simulations that show the power of weighted regressions when the assumptions behind the technique are satisfied, even approximately. Kang and Schafer (2007) criticize use of weighted regressions, a central issue being variability in the weights. There is a reply by Robins, Sued, Lei-Gomez, and Rotnitzky (2007). Also see Crump, Hotz, Imbens, and Mitnik (2009) on handling variable weights. Freedman (2008f) describes a measure-theoretic justification for weighting in terms of Radon-Nikodym derivatives.

Weighted regression should be distinguished from the methods suggested by Heckman (1978, 1979). For instance, if U and V in (2)–(3) are correlated, Heckman recommended maximum likelihood, or—in the linear case—including an additional term in the regression to center the errors.

When unbiased estimators do not exist, there are theorems showing that reduction in bias is generally offset by an increase in variance (Doss and Sethuraman, 1989). Evans and Stark (2002) provide a broader context for this discussion.

16.6 Theory

Suppose we have a linear causal model as in Simulation #1,

$$(6) \qquad Y = a + bX + c_1 Z_1 + c_2 Z_2 + dU,$$

where (Z_1, Z_2) is correlated with X. However, we omit Z_1 and Z_2 when we run the regression. Omitted-variables bias is the consequence, and the regression estimator is inconsistent. If we weight the regression using propensity weights, then Z_1 and Z_2 will be asymptotically balanced between treatment $(X = 1)$ and control $(X = 0)$. In other words, after weighting, covariates will be independent of treatment status and hence cannot confound the causal relationship.

From this perspective, what can we say about $\hat{a}$ in a weighted simple regression? (See Table 16.1, last block.) It turns out that $\hat{a}$ estimates, not a itself, but $a + E(c_1 Z_1 + c_2 Z_2)$, which is the average effect of the control condition—averaged across all values of the confounders. Weighting changed the meaning of the estimand. This is often the case.

The discussion here is intended only as a useful heuristic, rather than rigorous mathematics. A rigorous treatment would impose moment conditions on weighted variables, distinguishing between estimated weights and true weights.

Theoretical treatments of weighted regression generally assume that subjects are IID. This is a very strong assumption. By comparison, with structural models the exogenous variables need not be independent or identically distributed across subjects. Instead, it is commonplace to condition on such variables.

The stochastic elements that remain are the latent variables in the selection and response equations. To be sure, if the latents in the two equations fail to be independent within subject, or fail to be IID across subjects, the models will be misspecified. With non-parametric models, the IID assumption may go deeper. That is our next topic.

16.7 Non-parametric estimation

Suppose subject i is observed for time $t = 0, 1, 2, \ldots$. Subjects are assumed to be IID. In period $t > 0$, subject i chooses to be in treatment ($X_{it} = 1$) or control ($X_{it} = 0$). This choice depends on a vector of covariates Z_{it-1} defined in the previous period. There is a response Y_{it} that depends on the choice of regime X_{it} and on the covariates Z_{it-1}. Furthermore, Z_{it} depends on Z_{it-1}, X_{it}, and Y_{it}. The functions f, g, and h determine choice, response, and evolution of covariates respectively. These functions are unknown in form, although subject to a priori smoothness conditions. We do not allow them to depend on i or t. There are unobserved random errors U_{it}, V_{it}, and W_{it}. These are assumed to be independent within subject and IID across subjects, with

(7a) $$X_{it} = f\left(Z_{it-1}, U_{it}\right),$$

(7b) $$Y_{it} = g\left(Z_{it-1}, X_{it}\right) + V_{it},$$

(7c) $$Z_{it} = h\left(Z_{it-1}, X_{it}, Y_{it}\right) + W_{it}.$$

The system is assumed to be complete: Apart from the random errors, there are no unobserved covariates that influence treatment choice or response. (Social-science applications discussed above do not satisfy the completeness assumption—far from it.)

This is a rather complex environment, in which parametric models might not do very well. It is for this sort of environment that Robins and his colleagues developed weighting. The object was to determine what would happen if the choice equation (7a) was no longer operative and various treatment regimes were imposed on the subjects—without changing the response functions g and h or the random errors—a prospect that makes little sense in social-science applications like Sampson et al. (2006) or Schonlau (2006). Sampson et al. at least have the sort of longitudinal data structure where parametric models might run into trouble. Schonlau, among others, uses weights in a cross-sectional data structure.

16.8 Contrasts

Let i index the subjects in the treatment group T and j index the subjects in the control group C, so $w_i = 1/\hat{p}_i$ and $w_j = 1/(1 - \hat{p}_j)$, where p_k is the probability that subject k is in T. Assume that selection into T or C is exogenous and the p_k are well estimated. We would like to know the average response if all study subjects were put into T. A sensible estimator is the weighted average response over the treatment group in the study,

$$(8a) \qquad \sum_{i \in T} Y_i w_i \Big/ \sum_{i \in T} w_i.$$

Likewise, a sensible estimator for the average response if all subjects were put into C is the weighted average over the study's control group,

$$(8b) \qquad \sum_{j \in C} Y_j w_j \Big/ \sum_{j \in C} w_j.$$

These are approximations to the familiar Horvitz-Thompson estimators. The difference between (8a) and (8b) is a weighted contrast.

If selection is endogenous, or the weights are poorly estimated, the estimators in (8) are likely to be unsatisfactory. Even with exogenous selection, a large sample, and good estimates for the weights, variances may be large, and estimated variances may not be satisfactory—if there is a lot of variation in the weights across subjects. For instance, a relatively small number of subjects with large weights can easily determine the outcome, in which case the effective sample size is much reduced.

As a technical matter, the coefficient of the treatment variable in a weighted simple regression coincides with the weighted contrast (although the two procedures are likely to give different nominal variances). Anything distinctive about the weighted regression approach must involve the possibility of multiple regression when estimating the response equation. However, as we suggest above, it may be counter-productive to increase the analytic complexity by introducing multiple regression, variable selection, and the like.

16.9 Contrasts vs structural equations

Linear causal models like (1) are called "response equations" or "structural equations." Implicitly or explicitly, the coefficients are often given causal interpretations. If you switch a subject from control to treatment, all else held constant, X changes from 0 to 1. The response should then increase by the coefficient of X, namely, b. Similarly, if Z_1 is increased by one unit, all else held constant, the response should go up by c_1 units. In the papers by Robins and his school, the focus is quite different. Nothing is held constant. The objective is to estimate the average response—over all values of the confounders—if all subjects are put in treatment or all subjects are put in control. When weights are used, it can take some effort to identify the estimands. For additional discussion of structural equations, see Freedman (2009).

16.10 Conclusions

Investigators who have a causal model that they believe in should probably just fit the equation to the data. If there are omitted variables but the propensity scores can be estimated with reasonable accuracy, weighting the regression should reduce bias.

On the other hand, weighting is likely to increase random error by a substantial amount, the nominal standard errors are often severely biased downward, and substantial bias can still be present in the estimated causal effects. Variation in the weights creates problems; the distribution of the weights should always be examined.

If the causal model is dubious but the selection model is believable, an option to consider is the weighted contrast between the treatment and control groups. However, this analysis may be fragile. Again, random errors can be large, and there can be serious problems in estimating the standard errors.

Going beyond continuous response variables and weighted least squares leads to additional complications. Each combination of response model and fitting procedure has to be considered on its own to see what the weighted regression is going to estimate. Even with weighted least squares, some care is needed to identify estimands.

Acknowledgments

We would like to thank Larry Brown, Rob Gould, Rob Hollister, Guido Imbens, Brian Kriegler, Dan Nagin, Jamie Robins, Paul Rosenbaum, Dylan Small, Mikhail Traskin, David Weisburd, and Peter Westfall for many helpful comments. Any remaining infelicities are the responsibility of the authors. Richard Berk's work on the original paper was funded in part by a grant from the National Science Foundation: SES-0437169, "Ensemble Methods for Data Analysis in the Behavioral, Social and Economic Sciences."

17

On The So-Called "Huber Sandwich Estimator" and "Robust Standard Errors"

ABSTRACT. *The "Huber Sandwich Estimator" can be used to estimate the variance of the MLE when the underlying model is incorrect. If the model is nearly correct, so are the usual standard errors, and robustification is unlikely to help much. On the other hand, if the model is seriously in error, the sandwich may help on the variance side, but the parameters being estimated by the MLE are likely to be meaningless— except perhaps as descriptive statistics.*

17.1 Introduction

This chapter gives an informal account of the so-called "Huber Sandwich Estimator," for which Peter Huber is not to be blamed. We discuss the algorithm and mention some of the ways in which it is applied. Although the chapter is mainly expository, the theoretical framework outlined here may have some elements of novelty. In brief, under rather stringent conditions the algorithm can be used to estimate the variance of the MLE when the underlying model is incorrect. However, the algorithm ignores bias, which may be appreciable. Thus, results are liable to be misleading.

The American Statistician (2006) 60: 299–302. Copyright © 2006 by the American Statistical Association. Reprinted with permission. All rights reserved.

To begin the mathematical exposition, let i index observations whose values are y_i. Let $\theta \in R^p$ be a $p \times 1$ parameter vector. Let $y \to f_i(y|\theta)$ be a positive density. If y_i takes only the values 0 or 1, which is the chief case of interest here, then $f_i(0|\theta) > 0$, $f_i(1|\theta) > 0$, and $f_i(0|\theta) + f_i(1|\theta) = 1$. Some examples involve real- or vector-valued y_i, and the notation is set up in terms of integrals rather than sums. We assume $\theta \to f_i(y|\theta)$ is smooth. (Other regularity conditions are elided.) Let Y_i be independent with density $f_i(\cdot|\theta)$. Notice that the Y_i are not identically distributed: f_i depends on the subscript i. In typical applications, the Y_i cannot be identically distributed, as will be explained below.

The data are modeled as observed values of Y_i for $i = 1, \ldots, n$. The likelihood function is $\prod_{i=1}^{n} f_i(Y_i|\theta)$, viewed as a function of θ. The log-likelihood function is therefore

$$(1) \qquad L(\theta) = \sum_{i=1}^{n} \log f_i(Y_i|\theta).$$

The first and second partial derivatives of L with respect to θ are given by

$$(2) \qquad L'(\theta) = \sum_{i=1}^{n} g_i(Y_i|\theta), \quad L''(\theta) = \sum_{i=1}^{n} h_i(Y_i|\theta).$$

To unpack the notation in (2), let ϕ' denote the derivative of the function ϕ: differentiation is with respect to the parameter vector θ. Then

$$(3) \qquad g_i(y|\theta) = [\log f_i(y|\theta)]' = \frac{\partial}{\partial \theta} \log f_i(y|\theta),$$

a $1 \times p$-vector. Similarly,

$$(4) \qquad h_i(y|\theta) = [\log f_i(y|\theta)]'' = \frac{\partial^2}{\partial \theta^2} \log f_i(y|\theta),$$

a symmetric $p \times p$ matrix. The quantity $-E_\theta h(Y_i|\theta)$ is called the "Fisher information matrix." It may help to note that

$$-E_\theta h_i(Y_i|\theta) = E_\theta \big(g_i(Y_i|\theta)^T g_i(Y_i|\theta)\big) > 0,$$

where T stands for transposition.

Assume for the moment that the model is correct, and θ_0 is the true value of θ. So the Y_i are independent and the density of Y_i is $f_i(\cdot|\theta_0)$. The log-likelihood function can be expanded in a Taylor series around θ_0:

$$(5) \qquad L(\theta) = L(\theta_0) + L'(\theta_0)(\theta - \theta_0)$$
$$+ \frac{1}{2}(\theta - \theta_0)^T L''(\theta_0)(\theta - \theta_0) + \ldots.$$

If we ignore higher-order terms and write $\doteq$ for "nearly equal"—this is an informal exposition—the log-likelihood function is essentially a quadratic, whose maximum can be found by solving the likelihood equation $L'(\theta) = 0$. Essentially, the equation is

$$(6) \qquad L'(\theta_0) + (\theta - \theta_0)^T L''(\theta_0) \doteq 0.$$

So

$$(7) \qquad \hat{\theta} - \theta_0 \doteq [-L''(\theta_0)]^{-1} L'(\theta_0)^T.$$

Then

$$(8) \qquad \text{cov}_{\theta_0} \hat{\theta} \doteq [-L''(\theta_0)]^{-1}[\text{cov}_{\theta_0} L'(\theta_0)][-L''(\theta_0)]^{-1},$$

the covariance being a symmetric $p \times p$ matrix.

In the conventional textbook development, $L''(\theta_0)$ and $\text{cov}_{\theta_0} L'(\theta_0)$ are computed, approximately or exactly, using Fisher information. Thus, $-L''(\theta_0) \doteq -\sum_{i=1}^{n} E_{\theta_0} h_i(Y_i)$. Furthermore,

$$\text{cov}_{\theta_0} L'(\theta_0) = -\sum_{i=1}^{n} E_{\theta_0} h_i(Y_i).$$

The sandwich idea is to estimate $L''(\theta_0)$ directly from the sample data, as $L''(\hat{\theta})$. Similarly, $\text{cov}_{\theta_0} L'(\theta_0)$ is estimated as

$$\sum_{i=1}^{n} g_i(Y_i|\hat{\theta})^T g_i(Y_i|\hat{\theta}).$$

So (8) is estimated as

$$(9a) \qquad \hat{V} = (-A)^{-1} B (-A)^{-1},$$

where

$$(9b) \qquad A = L''(\hat{\theta}) \quad \text{and} \quad B = \sum_{i=1}^{n} g_i(Y_i|\hat{\theta})^T g_i(Y_i|\hat{\theta}).$$

The $\hat{V}$ in (9) is the "Huber sandwich estimator." The square roots of the diagonal elements of $\hat{V}$ are "robust standard errors" or "Huber-White standard errors." The middle factor B in (9) is not centered in any way. No centering is needed, because

$$(10) \qquad E_\theta[g_i(Y_i|\theta)] = 0,$$
$$\mathrm{cov}_\theta\big[g_i(Y_i|\theta)\big] = E_\theta\big[g_i(Y_i|\theta)^T g_i(Y_i|\theta)\big].$$

Indeed,

$$(11) \qquad E_\theta[g_i(Y_i|\theta)] = \int g_i(y|\theta) f_i(y|\theta)\, dy$$
$$= \int \frac{\partial}{\partial\theta} f_i(y|\theta)\, dy$$
$$= \frac{\partial}{\partial\theta} \int f_i(y|\theta)\, dy$$
$$= \frac{\partial}{\partial\theta} 1$$
$$= 0.$$

A derivative was passed through the integral sign in (11). Regularity conditions are needed to justify such maneuvers, but we finesse these mathematical issues.

If the motivation for the middle factor in (9) is still obscure, try this recipe. Let U_i be independent $1 \times p$-vectors, with $E(U_i) = 0$. Now $\mathrm{cov}(\sum U_i) = \sum \mathrm{cov}(U_i) = \sum E(U_i^T U_i)$. Estimate $E(U_i^T U_i)$ by $U_i^T U_i$. Take $U_i = g_i(Y_i|\theta_0)$. Finally, substitute $\hat\theta$ for θ_0.

The middle factor B in (9) is quadratic. It does not vanish, although

$$(12) \qquad \sum_{i=1}^{n} g_i(Y_i|\hat\theta) = 0.$$

Remember, $\hat\theta$ was chosen to solve the likelihood equation

$$L'(\theta) = \sum_{i=1}^{n} g_i(Y_i|\theta) = 0,$$

explaining (12).

In textbook examples, the middle factor B in (9) will be of order n, being the sum of n terms. Similarly, $-L''(\theta_0) = -\sum_{i=1}^{n} h_i(Y_i|\theta_0)$ will be of order n: see (2). Thus, (9) will be of order $1/n$. Under suitable regularity conditions, the strong law of large numbers will apply to $-L''(\theta_0)$, so $-L''(\theta_0)/n$ converges to a positive constant; the central limit theorem will apply to $L'(\theta_0)$, so $\sqrt{n}L'(\theta_0)$ converges in law to a multivariate normal distribution with mean 0. In particular, the randomness in L' is of order $\sqrt{n}$. So is the randomness in $-L''$, but that can safely be ignored when computing the asymptotic distribution of $[-L''(\theta_0)]^{-1}L'(\theta_0)^T$, because $-L''(\theta_0)$ is of order n.

17.2 Robust standard errors

We turn now to the case where the model is wrong. We continue to assume the Y_i are independent. The density of Y_i, however, is φ_i—which is not in our parametric family. In other words, there is specification error in the model, so the likelihood function is in error too. The sandwich estimator (9) is held to provide standard errors that are "robust to specification error." To make sense of the claim, we need the

Key assumption. There is a common θ_0 such that $f_i(\cdot|\theta_0)$ is closest—in the Kullback-Leibler sense of relative entropy, defined in (14) below—to φ_i.

(A possible extension will be mentioned below.) Equation (11) may look questionable in this new context. But

$$(13) \qquad E_0\big[g_i(Y_i|\theta)\big] = \int \left(\frac{\partial}{\partial\theta}f_i(y|\theta)\right)\frac{1}{f_i(y|\theta)}\varphi_i(y)\,dx$$

$$= 0 \quad \text{at} \quad \theta = \theta_0.$$

This is because θ_0 minimizes the Kullback-Leibler relative entropy,

$$(14) \qquad \theta \to \int \log\left[\frac{\varphi_i(y)}{f_i(y|\theta)}\right]\varphi_i(y)\,dy.$$

By the key assumption, we get the same θ_0 for every i.

Under suitable conditions, the MLE will converge to θ_0. Furthermore, $\hat{\theta} - \theta_0$ will be asymptotically normal, with mean 0 and covariance $\hat{V}$ given by (9), that is,

$$(15) \qquad \hat{V}^{-1/2}(\hat{\theta} - \theta_0) \to N(0_p, I_{p\times p}).$$

By definition, $\hat{\theta}$ is the θ that maximizes $\theta \to \prod_i f_i(Y_i|\theta)$—although it is granted that Y_i does not have the density $f_i(\cdot|\theta)$. In short, it is a pseudo-likelihood that is being maximized, not a true likelihood. The asymptotics in (15) therefore describe convergence to parameters of an incorrect model that is fitted to the data.

For some rigorous theory in the independent but not identically distributed case, see Amemiya (1985, section 9.2.2) or Fahrmeir and Kaufmann (1985). For the more familiar IID (independent and identically distributed) case, see Rao (1973, chapter 6) or Lehmann and Casella (2003, chapter 6). Lehmann (1998, chapter 7) and van der Vaart (1998) are less formal, more approachable. These references all use Fisher information rather than (9) and consider true likelihood functions rather than pseudo-likelihoods.

17.3 Why not assume IID variables?

The sandwich estimator is commonly used in logit, probit, or cloglog specifications. See, for instance, Gartner and Segura (2000); Jacobs and Carmichael (2002); Gould, Lavy, and Passerman (2004); Lassen (2005); or Schonlau (2006). Calculations are made conditional on the explanatory variables, which are left implicit here. Different subjects have different values for the explanatory variables. Therefore, the response variables have different conditional distributions. Thus, according to the model specification itself, the Y_i are not IID. If the Y_i are not IID, then θ_0 exists only by virtue of the key assumption.

Even if the key assumption holds, bias should be of greater interest than variance, especially when the sample is large and causal inferences are based on a model that is incorrectly specified. Variances will be small and bias may be large. Specifically, inferences will be based on the incorrect density $f_i(\cdot|\hat{\theta}) \doteq f_i(\cdot|\theta_0)$, rather than the correct density φ_i. Why do we care about $f_i(\cdot|\theta_0)$? If the model were correct, or nearly correct—that is, $f_i(\cdot|\theta_0) = \varphi_i$ or $f_i(\cdot|\theta_0) \doteq \varphi_i$—there would be no reason to use robust standard errors.

17.4 A possible extension

Suppose the Y_i are independent but not identically distributed, and there is no common θ_0 such that $f_i(\cdot|\theta_0)$ is closest to φ_i. One idea is to choose θ_n to minimize the total relative entropy, that is, to minimize

$$(16) \qquad \sum_{i=1}^{n} \int \log\left[\frac{\varphi_i(y)}{f_i(y|\theta)}\right] \varphi_i(y)\, dy.$$

Of course, θ_n would depend on n, and the MLE would have to be viewed as estimating this moving parameter. Many technical details remain to be worked out. For discussion along these lines, see White (1994, pp. 28–30, pp. 192–95).

17.5 Cluster samples

The sandwich estimator is often used for cluster samples. The idea is that clusters are independent, but subjects within a cluster are dependent. The procedure is to group the terms in (9), with one group for each cluster. If we denote cluster j by c_j, the middle factor in (9) would be replaced by

$$(17) \qquad \sum_j \left[\sum_{i \in c_j} g_i(Y_i | \hat{\theta}) \right]^T \left[\sum_{i \in c_j} g_i(Y_i | \hat{\theta}) \right].$$

The two outside factors in (9) would remain the same. The results of the calculation are sometimes called "survey-corrected" variances, or variances "adjusted for clustering."

There is undoubtedly a statistical model for which the calculation gives sensible answers, because the quantity in (17) should estimate the variance of $\sum_j \left[\sum_{i \in c_j} g_i(Y_i | \hat{\theta}) \right]$—if clusters are independent and $\hat{\theta}$ is nearly constant. (Details remain to be elucidated.) It is quite another thing to say what is being estimated by solving the non-likelihood equation $\sum_{i=1}^n g_i(Y_i | \theta) = 0$. This is a non-likelihood equation because $\prod_i f_i(\cdot | \theta)$ does not describe the behavior of the individuals comprising the population. If it did, we would not be bothering with robust standard errors in the first place. The sandwich estimator for cluster samples presents exactly the same conceptual difficulty as before.

17.6 The linear case

The sandwich estimator is often conflated with the correction for heteroscedasticity in White (1980). Suppose $Y = X\beta + \epsilon$. We condition on X, assumed to be of full rank. Suppose the ϵ_i are independent with expectation 0, but not identically distributed. The "OLS estimator" is $\hat{\beta}_{\text{OLS}} = (X'X)^{-1} X'Y$, where OLS means "ordinary least squares." White proposed that the covariance matrix of $\hat{\beta}_{\text{OLS}}$ should be estimated as $(X'X)^{-1} X' \hat{G} X (X'X)^{-1}$, where $e = Y - X\hat{\beta}_{\text{OLS}}$ is the vector of residuals, $\hat{G}_{ij} = e_i^2$ if $i = j$, and $\hat{G}_{ij} = 0$ if $i \neq j$. Similar ideas can be used if the ϵ_i are independent in blocks. White's method often gives good results, although $\hat{G}$ can be so variable that t-statistics are surprisingly non-t-like. Compare Beck, Katz, Alvarez, Garrett, and Lange (1993).

The linear model is much nicer than other models because $\hat{\beta}_{OLS}$ is unbiased even in the case we are considering, although OLS may of course be inefficient, and—more important—the usual standard errors may be wrong. White's correction tries to fix the standard errors.

17.7 An example

Suppose there is one real-valued explanatory variable, x, with values x_i spread fairly uniformly over the interval from zero to ten. Given the x_i, the response variables Y_i are independent, and

$$(18) \qquad \text{logit}\, P(Y_i = 1) = \alpha + \beta x_i + \gamma x_i^2,$$

where logit $p = \log[p/(1 - p)]$. Equation (18) is a logit model with a quadratic response. The sample size is moderately large. However, an unwitting statistician fits a linear logit model,

$$(19) \qquad \text{logit}\, P(Y_i = 1) = a + bx_i.$$

If γ is nearly 0, for example, then $\hat{a} \doteq \alpha$, $\hat{b} \doteq \beta$, and all is well—with or without the robust standard errors. Suppose, however, that $\alpha = 0$, $\beta = -3$, and $\gamma = .5$. (The parameters are chosen so the quadratic has a minimum at 3, and the probabilities spread out through the unit interval.) The unwitting statistician will get $\hat{a} \doteq -5$ and $\hat{b} \doteq 1$, concluding that on the logit scale, a unit increase in x makes the probability that $Y=1$ go up by one, across the whole range of x. The only difference between the usual standard errors and the robust standard errors is the confidence one has in this absurd conclusion.

In truth, for x near zero, a unit increase in x makes the probability of a response go *down* by three (probabilities are measured here on the logit scale). For x near three, increasing x makes no difference. For x near ten, a unit increase in x makes the probability go up by seven.

Could the specification error be detected by some kind of regression diagnostics? Perhaps, especially if we knew what kind of specification errors to look for. Keep in mind, however, that the robust standard errors are designed for use when there is *undetected* specification error.

17.8 What about Huber?

The usual applications of the so-called "Huber sandwich estimator" go far beyond the mathematics in Huber (1967), and our critical comments do not apply to his work. In free translation—this is no substitute for reading the paper—he assumes the Y_i are IID, so $f_i \equiv f$, and $g_i \equiv g$, and $h_i \equiv h$. He considers the asymptotics when the true density is f_0, not in the parametric family. Let $A = \int h(y|\theta_0) f_0(y)\, dy$, and $B = \int g(y|\theta_0)^T g(y|\theta_0) f_0(y)\, dy$. Both are $p \times p$ symmetric matrices. Plainly, $L'(\theta_0) = \frac{1}{n} \sum_{i=1}^{n} g(Y_i|\theta_0)$. Under regularity conditions discussed in the paper,

(i) $\hat{\theta} \to \theta_0$, which minimizes the "distance" between $f(\cdot|\theta)$ and f_0.

(ii) $\frac{1}{n} L''(\theta_0) = \frac{1}{n} \sum_{i=1}^{n} h(X_i|\theta_0) \to A$.

(iii) $n^{1/2} B^{-1/2} L'(\theta_0) \to N(0_p, I_{p \times p})$.

Asymptotic normality of the MLE follows:

(20a) $$C_n^{-1/2}(\hat{\theta} - \theta_0) \to N(0_{p \times 1}, I_{p \times p}),$$

where

(20b) $$C_n = n^{-1}(-A)^{-1} B(-A)^{-1}.$$

Thus, Huber's paper answers a question that (for a mathematical statistician) seems quite natural: What is the asymptotic behavior of the MLE when the model is wrong? Applying the algorithm to data, while ignoring the assumptions of the theorems and the errors in the models—that is not Peter Huber.

17.9 Summary and conclusions

Under stringent regularity conditions, the sandwich algorithm yields variances for the MLE that are asymptotically correct even when the specification—and hence the likelihood function—are incorrect. However, it is quite another thing to ignore bias. It remains unclear why applied workers should care about the variance of an estimator for the wrong parameter.

More particularly, inferences are based on a model that is admittedly incorrect. (If the model were correct, or nearly correct, there would be no need for sandwiches.) The chief issue, then, is the difference between the incorrect model that is fitted to the data and the process that generated the data. This is bias due to specification error. The algorithm does not take bias into account. Applied papers that use sandwiches rarely mention bias. There is room for improvement here.

See Koenker (2005) for additional discussion. On White's correction, see Greene (2007). For a more general discussion of independence assumptions, see Berk and Freedman (2003) [Chapter 2] or Freedman (2009). The latter reference also discusses model-based causal inference in the social sciences.

Acknowledgments

Dick Berk, Paul Ruud, and Peter Westfall made helpful comments.

18

Endogeneity in Probit Response Models

With Jasjeet S. Sekhon

ABSTRACT. *Endogeneity bias is an issue in regression models, including linear and probit models. Conventional methods for removing the bias have their own problems. The usual Heckman two-step procedure should not be used in the probit model: From a theoretical perspective, this procedure is unsatisfactory, and likelihood methods are superior. However, serious numerical problems occur when standard software packages try to maximize the biprobit likelihood function, even if the number of covariates is small. The log-likelihood surface may be nearly flat or may have saddle points with one small positive eigenvalue and several large negative eigenvalues. The conditions under which parameters in the model are identifiable are described; this produces novel results.*

18.1 Introduction

Suppose a linear regression model describes responses to treatment and to covariates. If subjects self-select into treatment, the process being dependent on the error term in the model, endogeneity bias is likely. Similarly, we may have a linear model that is to be estimated on sample data; if subjects self-select into the sample, endogeneity becomes an issue.

A revised version to appear in *Political Analysis*.

Heckman (1978, 1979) suggested a simple and ingenious two-step method for taking care of endogeneity, which works under the conditions described in those papers. This method is widely used. Some researchers have applied the method to probit response models. However, the extension is unsatisfactory. The nonlinearity in the probit model is an essential difficulty for the two-step correction, which will often make bias worse. It is well-known that likelihood techniques are to be preferred—although, as we show here, the numerics are delicate.

In the balance of this article, we define models for (i) self-selection into treatment or control, and (ii) self-selection into the sample, with simulation results to delineate the statistical issues. In the simulations, the models are correct. Thus, anomalies in the behavior of estimators are not to be explained by specification error. Numerical issues are explored. We explain the motivation for the two-step estimator and draw conclusions for statistical practice. We derive the conditions under which parameters in the models are identifiable; we believe these results are new. The literature on models for self-selection is huge, and so is the literature on probits; we conclude with a brief review of a few salient papers.

To define the models and estimation procedures, consider n subjects, indexed by $i = 1, \ldots, n$. Subjects are assumed to be independent and identically distributed. For each subject, there are two manifest variables X_i, Z_i and two latent variables U_i, V_i. Assume that (U_i, V_i) are bivariate normal, with mean 0, variance 1, and correlation ρ. Assume further that (X_i, Z_i) is independent of (U_i, V_i), i.e., the manifest variables are exogenous. For ease of exposition, we take (X_i, Z_i) as bivariate normal, although that is not essential. Until further notice, we set the means to 0, the variances to 1, the correlation between X_i and Z_i to 0.40, and sample size n to 1000.

18.2 A probit response model with an endogenous regressor

There are two equations in the model. The first is the selection equation:

(1) $C_i = 1$ if $a + bX_i + U_i > 0$, else $C_i = 0$.

In application, $C_i = 1$ means that subject i self-selects into treatment. The second equation defines the subject's response to treatment:

(2) $Y_i = 1$ if $c + dZ_i + eC_i + V_i > 0$, else $Y_i = 0$.

Notice that Y_i is binary rather than continuous. The data are the observed values of X_i, Z_i, C_i, Y_i. For example, the treatment variable C_i may indicate whether subject i graduated from college; the response Y_i, whether i has a full-time job.

Endogeneity bias is likely in (2). Indeed, C_i is endogenous due to the correlation ρ between the latent variables U_i and V_i. A two-step correction for endogeneity is sometimes used (although it shouldn't be).

Step 1. Estimate the probit model (1) by likelihood techniques.

Step 2. To estimate (2), fit the expanded probit model

$$(3) \qquad P(Y_i = 1 \,|\, X_i, Z_i, C_i) = \Phi(c + dZ_i + eC_i + fM_i)$$

to the data, where

$$(4) \qquad M_i = C_i \frac{\phi(a + bX_i)}{\Phi(a + bX_i)} - (1 - C_i)\frac{\phi(a + bX_i)}{1 - \Phi(a + bX_i)}.$$

Here, Φ is the standard normal distribution function with density $\phi = \Phi'$. In application, a and b in (4) would be unknown. These parameters are replaced by maximum-likelihood estimates obtained from Step 1. The motivation for M_i is explained in Section 18.6. Identifiability is discussed in Section 18.7: According to Proposition 1, parameters are identifiable unless $b = d = 0$.

The operating characteristics of the two-step correction was determined in a simulation study which draws 500 independent samples of size $n = 1000$. Each sample was constructed as described above. We set $a = 0.50$, $b = 1$, and $\rho = 0.60$. These choices create an environment favorable to correction.

Endogeneity is moderately strong: $\rho = 0.60$. So there should be some advantage to removing endogeneity bias. The dummy variable C_i is 1 with probability about 0.64, so it has appreciable variance. Furthermore, half the variance on the right hand side of (1) can be explained: $\text{var}(bX_i) = \text{var}(U_i)$. The correlation between the regressors is only 0.40: Making that correlation higher exposes the correction to well-known instabilities.

The sample is large: $n = 1000$. Regressors are exogenous by construction. Subjects are independent and identically distributed. Somewhat arbitrarily, we set the true value of c in the response equation (2) to -1, while $d = 0.75$ and $e = 0.50$. As it turned out, these choices were favorable too.

Table 18.1 summarizes results for three kinds of estimates:

    (i) raw (ignoring endogeneity),
    (ii) the two-step correction, and
    (iii) full maximum likelihood.

Table 18.1 Simulation results. Correcting endogeneity bias
when the response is binary probit. There are 500 repetitions.
The sample size is 1000. The correlation between latents is
$\rho = 0.60$. The parameters in the selection equation (1) are set
at $a = 0.50$ and $b = 1$. The parameters in the response equation (2) are set at $c = -1$, $d = 0.75$, and $e = 0.50$. The response equation includes the endogenous dummy C_i defined
by (1). The correlation between the exogenous regressors is
0.40. MLE computed by VGAM 0.7-6.

	c	d	e	ρ
True values				
	−1.0000	0.7500	0.5000	0.6000
Raw estimates				
Mean	−1.5901	0.7234	1.3285	
SD	0.1184	0.0587	0.1276	
Two-step				
Mean	−1.1118	0.8265	0.5432	
SD	0.1581	0.0622	0.2081	
MLE				
Mean	−0.9964	0.7542	0.4964	0.6025
SD	0.161	0.0546	0.1899	0.0900

For each kind of estimate and each parameter, the table reports the mean
of the estimates across the 500 repetitions. Subtracting the true value of
the parameter measures the bias in the estimator. Similarly, the standard
deviation across the repetitions, also shown in the table, measures the
likely size of the random error.

The "raw estimates" in Table 18.1 are obtained by fitting the probit
model

$$P(Y_i = 1 | X_i, Z_i, C_i) = \Phi(c + dZ_i + eC_i)$$

to the data, simply ignoring endogeneity. Bias is quite noticeable.

The two-step estimates are obtained via (3–4), with $\hat{a}$ and $\hat{b}$ obtained
by fitting (1). We focus on d and e, as the parameters in equation (2) that
may be given causal interpretations. Without correction, $\hat{d}$ averages about
0.72; with correction, 0.83. See Table 18.1. Correction doubles the bias.
Without correction, $\hat{e}$ averages 1.33; with correction, 0.54. Correction
helps a great deal, but some bias remains.

With the two-step correction, the standard deviation of $\hat{e}$ is about
0.21. Thus, random error in the estimates is appreciable, even with $n =$

1000. On the other hand, the standard error across the 500 repetitions is $0.21/\sqrt{500} = 0.01$. The bias in $\hat{e}$ cannot be explained in terms of random error in the simulation: Increasing the number of repetitions will not make any appreciable change in the estimated biases.

Heckman (1978) also suggested the possibility of fitting the full model—equations (1) and (2)—by maximum likelihood. The full model is a "bivariate probit" or "biprobit" model. Results are shown in the last two lines of Table 18.1. The MLE is essentially unbiased. The MLE is better than the two-step correction, although random error remains a concern.

We turn to some variations on the setup described in Table 18.1. The simulations reported there generated new versions of the regressors on each repetition. Freezing the regressors makes almost no difference in the results: Standard deviations would be smaller, in the third decimal place.

The results in Table 18.1 depend on ρ, the correlation between the latent variables in the selection equation and the response equation. If ρ is increased from 0.60 to 0.80, say, the performance of the two-step correction is substantially degraded. Likewise, increasing the correlation between the exogenous regressors degrades the performance.

Figure 18.1 The two-step correction. Graph of bias in $\hat{e}$ against ρ, the correlation between the latents. The light lower line sets the correlation between regressors to 0.40; the heavy upper line sets the correlation to 0.60. Other parameters as for Table 18.1. Below 0.35, the lines crisscross.

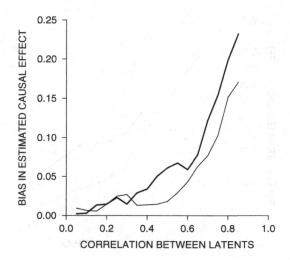

When $\rho = 0.80$ and the correlation between the regressors is 0.60, the bias in the two-step correction (3–4) for $\hat{d}$ is about 0.15; for $\hat{e}$, about 0.20. Figure 18.1 plots the bias in $\hat{e}$ against ρ, with the correlation between regressors set at 0.40 or 0.60, other parameters being fixed at their values for Table 18.1. The wiggles in the graph reflect variance in the Monte Carlo (there are "only" 500 replicates). The MLE is less sensitive to increasing correlations (data not shown).

Results are also sensitive to the distribution of the exogenous regressors. As the variance in the regressors goes down, bias goes up—in the two-step estimates and in the MLE. Furthermore, numerical issues become acute. There is some explanation: Dividing the standard deviation of X by 10, say, is equivalent to dividing b by 10 in equation (1); similarly for Z and d in (2). For small values of b and d, parameters are barely identifiable.

Figure 18.2 plots the bias in $\hat{e}$ against the common standard deviation of X and Z, which is set to values ranging from 0.1 to 1.0. (Other parameters are set as in Table 18.1.) The light line represents the MLE. Some of the "bias" in the MLE is indeed small-sample bias—when the standard deviation is 0.1, a sample with $n = 1000$ is a small sample. Some of the bias, however, reflects a tendency of likelihood maximizers to quit before finding the global maximum.

Figure 18.2 Graph of bias in $\hat{e}$ against the common standard deviation of the regressors X and Z. Other parameters as for Table 18.1. The light line represents the MLE, as computed by VGAM 0.7-6. The heavy line represents the two-step correction.

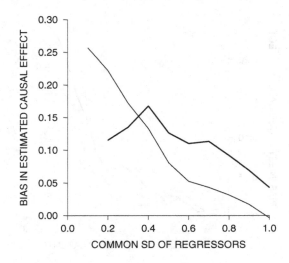

The heavy line represents the two-step correction. (With a standard deviation of 0.1, data for the two-step correction are not shown, because there are huge outliers; even the median bias is quite changeable from one set of 500 repetitions to another, but 0.2 may be a representative figure.) Curiously, the two-step correction is better than the MLE when the standard deviation of the exogenous regressors is set to 0.2 or to 0.3. This is probably due to numerical issues in maximizing the likelihood functions.

We believe the bias in the two-step correction (Figures 18.1 and 18.2) reflects the operating characteristics of the estimator, rather than operating characteristics of the software. Beyond 1.0, the bias in the MLE seems to be negligible. Beyond 1.5, the bias in the two-step estimator for e is minimal, but d continues to be a little problematic.

As noted above, changing the scale of X is equivalent to changing b. Similarly, changing the scale of Z is equivalent to changing d. See equations (1) and (2). Thus, in Figure 18.2, we could leave the standard deviations at 1 and run through a series of (b, d) pairs:

$$(0.1 \times b_0, 0.1 \times d_0), (0.2 \times b_0, 0.2 \times d), \ldots,$$

where $b_0 = 1$ and $d_0 = 0.75$ were the initial choices for Table 18.1.

The number of regressors should also be considered. With a sample size of 1000, practitioners would often use a substantial number of covariates. Increasing the number of regressors is likely to have a negative impact on performance.

18.3 A probit model with endogenous sample selection

Consider next the situation where a probit model is fitted to a sample, but subjects self-select into the sample by an endogenous process. The selection equation is

(5) $\qquad C_i = 1$ if $a + bX_i + U_i > 0$, else $C_i = 0$.

("Selection" means selection into the sample.) The response equation is

(6) $\qquad Y_i = 1$ if $c + dZ_i + V_i > 0$, else $Y_i = 0$.

Equation (6) is the equation of primary interest; however, Y_i and Z_i are observed only when $C_i = 1$. Thus, the data are the observed values of (X_i, C_i) for all i, as well as (Z_i, Y_i) when $C_i = 1$. When $C_i = 0$, however, Z_i and Y_i remain unobserved. Notice that Y_i is binary rather than continuous. Notice too that C_i is omitted from (6); indeed, when (6) can be observed, $C_i \equiv 1$.

Fitting (6) to the observed data raises the question of endogeneity bias. Sample subjects have relatively high values of U_i; hence, high values of V_i. (This assumes $\rho > 0$.) Again, there is a proposed solution that involves two steps.

Step 1. Estimate the probit model (5) by likelihood techniques.

Step 2. Fit the expanded probit model

$$(7) \qquad P(Y_i = 1 | X_i, Z_i) = \Phi(c + dZ_i + fM_i)$$

to the data on subjects i with $C_i = 1$. This time,

$$(8) \qquad M_i = \frac{\phi(a + bX_i)}{\Phi(a + bX_i)}.$$

Parameters in (8) are replaced by the estimates from Step 1. As before, this two-step correction doubles the bias in $\hat{d}$. See Table 18.2. The MLE removes most of the bias. However, as for Table 18.1, the bias in the MLE depends on the standard deviation of the regressors. Bias will be noticeable if the standard deviations are below 0.2. Some of this is small-sample bias in the MLE, and some reflects difficulties in numerical maximization.

Increasing the sample size from 1000 to 5000 in the simulations barely changes the averages, but reduces the standard deviations by a factor of about $\sqrt{5}$, as might be expected. This comment applies both to Table 18.1 and to Table 18.2 (data not shown) but not to the MLE results in Table 18.2. Increasing n would have made the STATA code prohibitively slow to run.

Many applications of Heckman's method feature a continuous response variable rather than a binary variable. Here, the two-step correction is on firmer ground, and parallel simulations (data not shown) indicate that the correction removes most of the endogeneity bias when the parameters are set as in Tables 18.1 and 18.2. However, residual bias is large when the standard deviation of the regressors is set to 0.1 and the sample size is "only" 1000; the issues resolve when $n = 10,000$. The problem with $n = 1000$ is created by (i) large random errors in $\hat{b}$, coupled with (ii) poorly conditioned design matrices. In more complicated situations, there may be additional problems.

Table 18.2 Simulation results. Correcting endogeneity bias in sample selection when the response is binary probit. There are 500 repetitions. The sample size is 1000. The correlation between latents is $\rho = 0.60$. The parameters in the selection equation (5) are set at $a = 0.50$ and $b = 1$. The parameters in the response equation (6) are set at $c = -1$, $d = 0.75$. Response data are observed only when $C_i = 1$, as determined by the selection equation. This will occur for about 64% of the subjects. The correlation between the exogenous regressors is 0.40. MLE computed using STATA 9.2.

	c	d	ρ
True values			
	−1.0000	0.7500	0.6000
Raw estimates			
Mean	−0.7936	0.7299	
SD	0.0620	0.0681	
Two-step			
Mean	−1.0751	0.8160	
SD	0.1151	0.0766	
MLE			
Mean	−0.9997	0.7518	0.5946
SD	0.0757	0.0658	0.1590

18.4 Numerical issues

Exploratory computations were done in several versions of MAT-LAB, R, and STATA. In the end, to avoid confusion and chance capitalization, we redid the computations in a more unified way, with R 2.7 for the raw estimates, the two-step correction; VGAM 0.7-6 for the MLE in (1–2); and STATA 9.2 for the MLE in (5–6). Why do we focus on the behavior of R and STATA? R is widely used in the statistical community, and STATA is almost the lingua franca of quantitative social scientists.

Let b_0 and d_0 be the default values of b and d, namely, 1 and 0.75. As b and d decrease from the defaults, VGAM in R handled the maximization less and less well (Figure 18.2). We believe VGAM had problems computing the Hessian, even for the base case in Table 18.1: Its internally generated standard errors were too small by a factor of about two, for $\hat{c}, \hat{e}, \hat{\rho}$.

By way of counterpoint, STATA did somewhat better when we used it to redo the MLE in (1–2). However, if we multiply the default b_0 and d_0 by 0.3 or 0.4, bias in STATA becomes noticeable. If we multiply by 0.1 or 0.2, many runs fail to converge, and the runs that do converge produce aberrant estimates, particularly for a multiplier of 0.1. For multipliers of 0.2 to 0.4, the bias in $\hat{e}$ is upwards in R but downwards in STATA. In Table 18.2, STATA did well. However, if we scale b_0 and d_0 by 0.1 or 0.2, STATA has problems. In defense of R and STATA, we can say that they produce abundant warning messages when they get into difficulties.

In multi-dimensional problems, even the best numerical analysis routines find spurious maxima for the likelihood function. Our models present three kinds of problems: (i) flat spots on the log-likelihood surface, (ii) ill-conditioned maxima, where the eigenvalues of the Hessian are radically different in size, and (iii) ill-conditioned saddle points with one small positive eigenvalue and several large negative eigenvalues. The maximizers in VGAM and STATA simply give up before finding anything like the maximum of the likelihood surface. This is a major source of the biases reported above.

The model defined by (1–2) is a harder challenge for maximum likelihood than (5–6), due to the extra parameter e. Our computations suggest that most of the difficulty lies in the joint estimation of three parameters, c, e, ρ. Indeed, we can fix a, b, d at the default values for Table 18.1, and maximize the likelihood over the remaining three parameters c, e, ρ. VGAM and STATA still have convergence issues. The problems are the same as with six parameters. For example, we found a troublesome sample where the Hessian of the log-likelihood had eigenvalues 4.7, −1253.6, −2636.9. (We parameterize the correlation between the latents by $\log(1 + \rho) - \log(1 - \rho)$ rather than ρ, since that is how binom2.rho in VGAM does things.)

One of us (JSS) has an improved likelihood maximizer called GENOUD. See http://sekhon.berkeley.edu/genoud/.

GENOUD seems to do much better at the maximization, and its internally generated standard errors are reasonably good. Results for GENOUD and STATA not reported here are available at the URL above, along with the VGAM standard errors.

18.5 Implications for practice

There are two main conclusions from the simulations and the analytic results.

(i) Under ordinary circumstances, the two-step correction should not be used in probit response models. In some cases, the correction will reduce bias, but in many other cases, the correction will increase bias.

(ii) If the bivariate probit model is used, special care should be taken with the numerics. Conventional likelihood maximization algorithms produce estimates that are far away from the MLE. Even if the MLE has good operating characteristics, the "MLE" found by the software package may not. Results from VGAM 0.7-6 should be treated with caution. Results from STATA 9.2 may be questionable for various combinations of parameters.

The models analyzed here are very simple, with one covariate in each of (1–2) and (5–6). In real examples, the number of covariates may be quite large, and numerical behavior will be correspondingly more problematic.

Of course, there is a question more salient than the numerics: What is it that justifies probit models and like as descriptions of behavior? For additional discussion, see Freedman (2009), which has further cites to the literature on this point.

18.6 Motivating the estimator

Consider (1–2). We can represent V_i as $\rho U_i + \sqrt{1 - \rho^2} W_i$, where W_i is an $N(0, 1)$ random variable, independent of U_i. Then

$$
\begin{aligned}
(9) \quad E\{V_i \mid X_i = x, C_i = 1\} \\
= E\{\rho U_i + \sqrt{1 - \rho^2} W_i \mid U_i > -a - bx_i\} \\
= \rho E\{U_i \mid U_i > -a - bx_i\} \\
= \rho \frac{1}{\Phi(a + bx_i)} \int_{-a-bx_i}^{\infty} x\phi(x)\, dx \\
= \rho \frac{\phi(a + bx_i)}{\Phi(a + bx_i)}
\end{aligned}
$$

because $P\{U_i > -a-bx_i\} = P\{U_i < a+bx_i\} = \Phi(a+bx_i)$. Likewise,

$$
(10) \qquad E\{V_i \mid X_i = x, C_i = 0\} = -\rho \frac{\phi(a + bx_i)}{1 - \Phi(a + bx_i)}.
$$

In (2), therefore, $E\{V_i - \rho M_i \mid X_i, C_i\} = 0$. If (2) were a linear regression equation, then OLS estimates would be unbiased, the coefficient of M_i being nearly ρ. (These remarks take a and b as known, with the variance of the error term in the linear regression normalized to 1.)

However, (2) is not a linear regression equation: (2) is a probit model. That is the source of the problem.

18.7 Identifiability

Identifiability means that parameters are determined by the joint distribution of the observables: Parameters that are not identifiable cannot be estimated. In the model defined by (1–2), the parameters are a, b, c, d, e, and the correlation ρ between the latents; the observables are X_i, Z_i, C_i, Y_i. In the model defined by (5–6), the parameters are a, b, c, d, and the correlation ρ between the latents; the observables are $X_i, C_i, \tilde{Z}_i, \tilde{Y}_i$, where $\tilde{Z}_i = Z_i$ and $\tilde{Y}_i = Y_i$ when $C_i = 1$, while $\tilde{Z}_i = \tilde{Y}_i = \mathcal{M}$ when $C_i = 0$. Here, $\mathcal{M}$ is just a special symbol that denotes "missing."

Results are summarized as Propositions 1 and 2. The statements involve the sign of d, which is $+1$ if $d > 0$, 0 if $d = 0$, and -1 if $d < 0$. Since subjects are independent and identically distributed, only $i = 1$ need be considered. The variables (X_1, Z_1) are taken as bivariate normal, with a correlation strictly between -1 and $+1$. This assumption is discussed below.

Proposition 1. *Consider the model defined by (1–2). The parameters a and b in (1) are identifiable, and the sign of d in (2) is identifiable. If $b \neq 0$, the parameters c, d, e, ρ in (2) are identifiable. If $b = 0$ but $d \neq 0$, the parameters c, d, e, ρ are still identifiable. However, if $b = d = 0$, the remaining parameters c, e, ρ are not identifiable.*

Proposition 2. *Consider the model defined by (5–6). The parameters a and b in (5) are identifiable, and the sign of d in (6) is identifiable. If $b \neq 0$, the parameters c, d, ρ in (6) are identifiable. If $b = 0$ but $d \neq 0$, the parameters c, d, ρ are still identifiable. However, if $b = d = 0$, the remaining parameters c, ρ are not identifiable.*

Proof of Proposition 1. Clearly, the joint distribution of C_1 and X_1 determines a and b, so we may consider these as given. The distributions of X_1 and Z_1 are determined (this is not so helpful). We can take the conditional distribution of Y_1 given $X_1 = x$ and $Z_1 = z$ as known. In other words, suppose (U, V) are bivariate normal with mean 0, variance 1 and correlation ρ.

The joint distribution of the observables determines a, b and two functions ψ_0, ψ_1 of x, z:

(11) $\psi_0(x, z) = P(a + bx + U < 0 \ \& \ c + dz + V > 0),$

$\psi_1(x, z) = P(a + bx + U > 0 \ \& \ c + dz + e + V > 0).$

There is no additional information about the parameters.

Fix x at any convenient value, and consider $z > 0$. Then $z \rightarrow \psi_0(x, z)$ is strictly decreasing, constant, or strictly increasing, according as $d < 0$, $d = 0$, or $d > 0$. The sign of d is therefore determined. The rest of the proof, alas, consists of a series of cases.

The case $b \neq 0$ and $d > 0$. Let $u = -a - bx$, $v = -z$, $\xi = U$, and $\zeta = (V + c)/d$. Then (ξ, ζ) are bivariate normal, with unknown correlation ρ. We know ξ has mean 0 and variance 1. The mean and variance of ζ are unknown, being c/d and $1/d^2$, respectively. But

$$(12) \qquad P(\xi < u \ \& \ \zeta > v)$$

is known for all (u, v). Does this determine ρ, c, d? Plainly so, because (12) determines the joint distribution of ξ, ζ. We can then compute ρ, $d = 1/\sqrt{\operatorname{var}(\zeta)}$, and $c = d E(\zeta)$. Finally, ψ_1 in (11) determines e. This completes the argument for the case $b \neq 0$ and $d > 0$.

The case $b \neq 0$ and $d < 0$ is the same, except that $d = -1/\sqrt{\operatorname{var}(\zeta)}$.

The case $b \neq 0$ and $d = 0$. Here, we know

$$(13) \qquad P(U < u \ \& \ c + V > 0) \text{ for all } u.$$

Let $u \rightarrow \infty$: the marginal distribution of V determines c. Furthermore, from (13), we can compute $P(V > -c \mid U = u)$ for all u. Given $U = u$, we know that V is distributed as $\rho u + \sqrt{1 - \rho^2}\, W$, where W is $N(0, 1)$. If $\rho = \pm 1$, then

$$P(V > -c \mid U = u) = 1 \text{ if } \rho u > -c$$
$$= 0 \text{ if } \rho u < -c.$$

If $-1 < \rho < 1$, then

$$(14) \quad P\{V > -c \mid U = u\} = P\left\{W > -\frac{c + \rho u}{\sqrt{1 - \rho^2}}\right\} = \Phi\left(\frac{c + \rho u}{\sqrt{1 - \rho^2}}\right).$$

So we can determine whether $\rho = \pm 1$; and if so, which sign is right. Suppose $-1 < \rho < 1$. Then (14) determines $(c + \rho u)/\sqrt{1 - \rho^2}$. Differentiate with respect u to see that (14) determines $\rho/\sqrt{1 - \rho^2}$. This is a 1–1 function of ρ. Thus, ρ can be determined, and then c; finally, e is obtained from ψ_1 in (11). This completes the argument for the case $b \neq 0$ and $d = 0$.

The case $b = 0$ and $d > 0$. As above, let W be independent of U and $N(0, 1)$; represent V as $\rho U + \sqrt{1 - \rho^2}W$. Let $G = \{U < -a\}$. From ψ_0 and a, we compute

$$(15) \quad P\{V > -c - dz \mid G\} = P\{\rho U + \sqrt{1 - \rho^2}W > -c - dz \mid G\}$$
$$= P\left\{\frac{\rho}{d}U + \frac{\sqrt{1 - \rho^2}}{d}W + \frac{c}{d} > -z \mid G\right\}.$$

Write U_a for U conditioned so that $U < -a$. The right hand side of (15), as a function of z, determines the distribution function of the sum of three terms: two independent random variables, U_a and $\sqrt{1 - \rho^2}W/d$, where W is standard normal, plus the constant c/d. This distribution is therefore known, although it depends on the three unknowns, c, d, ρ.

Write Λ for the log Laplace transform of U_a. This is a known function. Now compute the log Laplace transform of the distribution in (15). This is

$$(16) \qquad\qquad t \to \Lambda\left(\frac{\rho}{d}t\right) + \frac{1 - \rho^2}{d^2}t^2 + \frac{c}{d}t.$$

Again, this function is known, although c, d, ρ are unknown. Consider the expansion of (16) as a power series near 0, of the form $\kappa_1 t + \kappa_2 t^2/2! + \kappa_3 t^3/3! + \cdots$. The κ's are the *cumulants* or *semi-invariants* of the distribution in (15). These are known quantities because the function in (16) is known: κ_1 is the mean of the distribution given by (15), while κ_2 is the variance and κ_3 is the central third moment.

Of course, $\Lambda'(0) = E(U_a) = -\phi(-a)/\Phi(-a)$. Thus, $\kappa_1 = -\phi(-a)/\Phi(-a) + c/d$, which determines c/d. Next, $\Lambda''(0) = \text{var}(U_a)$, so $\kappa_2 = (\rho/d)^2\text{var}(U_a) + (1 - \rho^2)/d^2$ is determined. Finally, $\kappa_3 = \Lambda'''(0)$ is the third central moment of U_a. Since U_a has a skewed distribution, $\Lambda'''(0) \neq 0$. We can compute $(\rho/d)^3$ from κ_3, and then ρ/d. Next, we get $1/d^2$ from κ_2, and then $1/d$. (We are looking at the case $d > 0$.) Finally, c comes from κ_1. Thus, c, d, ρ are determined, and e comes from ψ_1 in (11). This completes the argument for the case $b = 0$ and $d > 0$.

The case $b = 0$ and $d < 0$ follows by the same argument.

The case $b = d = 0$. The three remaining parameters c, e, ρ are not identifiable. For simplicity, take $a = 0$, although this is not essential. Suppose

$$(17) \qquad\qquad P(U < 0 \ \& \ V > -c) = \alpha$$

is given, with $0 < \alpha < 1/2$. Likewise,

$$(18) \qquad\qquad P(U > 0 \ \& \ V > -c - e) = \beta$$

is given, with $0 < \beta < 1/2$. The joint distribution of the observables contains no further information about the remaining parameters c, e, ρ. Choose any particular ρ with $-1 \le \rho \le 1$. Choose c so that (17) holds and e so (18) holds. The upshot: There are infinitely many c, e, ρ triplets yielding the same joint distribution for the observables. This completes the argument for the case $b = d = 0$, and so for Proposition 1.

Proof of Proposition 2. Here, we know the joint distribution of (X_1, C_1), which determines a, b. We also know the joint distribution of (X_1, Z_1, Y_1) given $C_1 = 1$; we do not know this joint distribution given $C_1 = 0$. As in (11), suppose (U, V) are bivariate normal with mean 0, variance 1, and correlation ρ. The joint distributions of the observables determine a, b and the function

$$(19) \qquad \psi_1(x, z) = P(a + bx + U > 0 \;\&\; c + dz + V > 0).$$

There is no other information in the system; in particular, we do not know the analog of ψ_0. Most of the argument is the same as before, or even a little easier. We consider in detail only one case.

The case $b = d = 0$. The two remaining parameters, c, ρ are not identifiable. Again, take $a = 0$. Fix any α with $0 < \alpha < 1/2$. Suppose

$$(20) \qquad P(U > 0 \;\&\; V > -c) = \alpha$$

is given. There is no other information to be had about c, ρ. Fix any ρ with $-1 \le \rho \le 1$ and solve (20) for c. There are infinitely many c, ρ pairs giving the same joint distribution for the observables when $b = d = 0$. This completes our discussion of Proposition 2.

Remarks. (i) The random variable U_a was defined in the course of proving Proposition 1. If desired, the moments of U_a can be obtained explicitly in terms of ϕ and Φ, using repeated integration by parts.

(ii) The Laplace transform of U_a is easily obtained by completing the square, and

$$(21) \qquad t \to \frac{1}{\sqrt{2\pi}} \exp\left(\frac{1}{2}t^2\right) \frac{\Phi(-a-t)}{\Phi(-a)}.$$

The third derivative of the log Laplace transform can be computed from the relationship (21), but it's painful.

(iii) The argument for the case $b = 0$ and $d > 0$ in Proposition 1 is somewhat intricate, but it actually covers all values of b, whether zero or non-zero. The argument shows that for any particular real α, the values of c, d, ρ are determined by the number $P(\alpha + U < 0)$ and the function

$$z \to P(\alpha + U < 0 \ \& \ c + dz + V > 0).$$

(iv) Likewise, the argument for the case $b \neq 0$ and $d = 0$ proves more. If we know $P(U < u)$ and $P(U < u \ \& \ \gamma + V > 0)$ for all real u, that determines γ and ρ.

(v) In (17), for example, if $\alpha = 1/2$, then $\rho = -1$; but c can be anywhere in the range $[0, \infty)$.

(vi) The propositions can easily be extended to cover vector-valued exogenous variables.

(vii) Our proof of the propositions really does depend on the assumption of an imperfect correlation between X_i and Z_i. We hope to consider elsewhere the case where $Z_i \equiv X_i$. The assumption of normality is not material; it is enough if the joint distributions have full support, although positive densities are probably easier to think about.

(viii) The assumption of bivariate normality for the latent variables is critical. If this is wrong, estimates are likely to be inconsistent.

(ix) Suppose (U, V) are bivariate normal with correlation ρ, and $-1 < \rho < 1$. Then

$$\rho \to P(U > 0 \ \& \ V > 0)$$

is strictly monotone. This is Slepian's theorem: see Tong (1980). If the means are 0 and the variances are 1, numerical calculations suggest this function is convex on $(-1, 0)$ and concave on $(0, 1)$.

18.8 Some relevant literature

Cumulants are discussed by Rao (1973, p. 101). The ratio ϕ/Φ in (8) is usually called the "inverse Mills ratio," in reference to Mills (1926)—although Mills tabulates $[1 - \Phi(x)]/\phi(x)$ for $x \geq 0$. Heckman (1978, 1979) proposes the use of M_i to correct for endogeneity and selection bias in the linear case, with a very clear explanation of the issues. He also describes potential use of the MLE. Rivers and Vuong (1988) propose an interesting alternative to the Heckman estimator. Their estimator (perhaps confusingly) is also called a two-step procedure. It seems most relevant when the endogenous variable is continuous; ours is binary.

For other estimation strategies and discussion, see Angrist (2001). Bhattacharya, Goldman, and McCaffrey (2006) discuss several "two-step" algorithms, including a popular IVLS estimator that turns out to be inconsistent; they do not seem to consider the particular two-step estimator of concern in our chapter. Also see Lee (1981) and Rivers and Vuong (1988). Muthen (1979) discusses identifiability in a model with latent causal variables. The VGAM manual (Yee 2007) notes difficulties in computing standard errors. According to Stata (2005), its maximum-likelihood routine "provides consistent, asymptotically efficient estimates for all the parameters in [the] models."

Van de Ven and van Praag (1981) found little difference between the MLE and the two-step correction; the difference doubtless depends on the model under consideration. Instabilities in the two-step correction are described by Winship and Mare (1992), Copas and Li (1997), and Briggs (2004), among others. For additional citations, see Dunning and Freedman (2007). Ono (2007) uses the two-step correction with probit response in a study of the Japanese labor market; X and Z are multi-dimensional. The sample size is 10,000, but only 300 subjects select into the treatment condition. Bushway, Johnson, and Slocum (2007) describe many over-enthusiastic applications of the two-step correction in the criminology literature: Binary response variables are among the least of the sins.

We do not suggest that finding the true maximum of the likelihood function guarantees the goodness of the estimator, because there are situations where the MLE performs rather badly. Freedman (2007) has a brief review of the literature on this topic. However, we would suggest that spurious maxima are apt to perform even less well, particularly with the sort of models considered here.

Acknowledgments

Derek Briggs, Allan Dafoe, Thad Dunning, Joe Eaton, Eric Lawrence, Walter Mebane, Jim Powell, Rocío Titiunik, and Ed Vytlacil made helpful comments. Errors and omissions remain the responsibility of the authors.

19

Diagnostics Cannot Have Much Power Against General Alternatives

ABSTRACT. *Model diagnostics are shown to have little power unless alternative hypotheses can be narrowly defined. For example, independence of observations cannot be tested against general forms of dependence. Thus, the basic assumptions in regression models cannot be inferred from the data. Equally, the proportionality assumption in proportional-hazards models is not testable. Specification error is a primary source of uncertainty in forecasting, and this uncertainty will be difficult to resolve without external calibration. Model-based causal inference is even more problematic.*

19.1. Introduction

The object here is to sketch a demonstration that, unless additional regularity conditions are imposed, model diagnostics have power only against a circumscribed class of alternative hypotheses. The chapter is organized around the familiar requirements of statistical models. Theorems 1 and 2, for example, consider the hypothesis that distributions are continuous and have densities. According to the theorems, such hypotheses cannot be tested without additional structure.

International Journal of Forecasting, in press.

Let us agree, then, that distributions are smooth. Can we test independence? Theorems 3 and 4 indicate the difficulty. Next, we grant independence and consider tests that distinguish between (i) independent and identically distributed random variables on the one hand, and (ii) independent but differently distributed variables on the other. Theorem 5 shows that, in general, power is lacking.

For ease of exposition, we present results for the unit interval; transformation to the positive half-line or the whole real line is easy. At the end of the chapter, we specialize to more concrete situations, including regression and proportional-hazards models. We consider the implications for forecasting, mention some pertinent literature, and make some recommendations.

Definitions. A randomized test function is a measurable function ϕ with $0 \leq \phi(x) \leq 1$ for all x. A non-randomized test function ϕ has $\phi(x) = 0$ or 1. The size of ϕ is the supremum of $\int \phi \, d\mu$ over μ that satisfy the null hypothesis, a set of probabilities that will be specified in Theorems 1–5 below. The power of ϕ at a particular μ satisfying the alternative hypothesis is $\int \phi \, d\mu$. A simple hypothesis describes just one μ; otherwise, the hypothesis is composite. Write λ for Lebesgue measure on the Borel subsets of $[0, 1]$.

Interpretation. Given a test ϕ and data x, we reject the null with probability $\phi(x)$. Size is the maximal probability of rejection at μ that satisfy the null. Power at μ is the probability of rejection, defined for μ that satisfy the alternative.

Theorem 1. *Consider probabilities μ on the Borel unit interval. Consider testing the simple null hypothesis*

   N: $\mu = \lambda$

against the composite alternative

   A: μ *is a point mass at some (unspecified) point.*

Under these circumstances, any test of size α has power at most α against some alternatives.

Proof. Let ϕ be a randomized test function. If $\phi(x) > \alpha$ for all $x \in [0, 1]$, then $\int \phi(x) \, dx > \alpha$. We conclude that $\phi(x) \leq \alpha$ for some x, indeed, for a set of x's of positive Lebesgue measure.   QED

Remarks. (i) If we restrict ϕ to be non-randomized, then $\phi(x) = 0$ for some x. In other words, power would be 0 rather than α.

(ii) The conclusions hold not just for some alternatives, but for many of them.

Theorem 2 requires some additional terminology. A "continuous" probability assigns measure 0 to each point. A "singular" probability on [0, 1] concentrates on a set of Lebesgue measure 0.

Theorem 2. *Consider probabilities μ on the Borel unit interval. Consider testing the simple null hypothesis*

$N:$ $\mu = \lambda$

against the composite alternative

$A:$ μ *is continuous and singular.*

Under these circumstances, any test of size α has power at most α against some alternatives.

Proof. We identify 0 and 1, then visualize $[0, 1)$ as the additive group modulo 1 with convolution operator $*$. If μ is any probability, then $\lambda * \mu = \lambda$. Let ϕ be a randomized test function of size α. Then $\alpha \geq \int \phi \, d\lambda = \iint \phi(x + y) \, \mu(dx) \, dy$. Hence, there are y with $\alpha \geq \int \phi(x + y) \, \mu(dx) = \int \phi(x) \, \mu_y(dx)$, where μ_y is the translation of μ by y. If μ is continuous and singular, so is μ_y; but ϕ only has power α against μ_y. QED

Remarks. (i) If we restrict ϕ to be non-randomized, then $\lambda\{\phi = 0\} \geq 1 - \alpha > 0$; the trivial case $\alpha = 1$ must be handled separately. Hence, power would be 0 rather than α.

(ii) There are tests with high power against any particular alternative. Indeed, if ν is singular, it concentrates on a Borel set B with $\lambda(B) = 0$; let ϕ be the indicator function of B. This test has size 0, and power 1 at ν. The problem lies in distinguishing λ from the cloud of *all* alternatives.

A little more terminology may help. If μ and ν are two probabilities on the same σ-field, then μ is equivalent to ν if they have the same null sets. By the Radon-Nikodym theorem, this is tantamount to saying that the derivative of μ with respect to ν is positive and finite a.e. Write λ^2 for Lebesgue measure on the Borel subsets of the unit square. Let ξ_1 and ξ_2 be the coordinate functions, so that $\xi_1(x, y) = x$ and $\xi_2(x, y) = y$. More generally, we write λ^k for Lebesgue measure on the Borel subsets of $[0, 1]^k$ and ξ_i for the coordinate functions, so $\xi_i(x_1, x_2, \ldots) = x_i$.

If μ is a probability on the unit square, let ρ_μ be the correlation between ξ_1 and ξ_2, computed according to μ. This is well-defined unless μ concentrates on a horizontal or vertical line.

For the proof of Theorem 3, if f is an integrable Borel function on the unit interval, then λ-almost all $x \in (0, 1)$ are Lebesgue points, in the sense that

$$(1) \qquad \lim_{h \to 0} \frac{1}{h} \int_x^{x+h} f \, d\lambda \to f(x).$$

The result extends to k-dimensional space. See, for instance, Dunford and Schwartz (1958, p. 215).

Theorem 3. *Consider probabilities μ on the Borel unit square that are equivalent to λ^2. Consider testing the simple null hypothesis*

N: $\mu = \lambda^2$

against the composite alternative

A: $\rho_\mu \neq 0$.

Under these circumstances, a non-randomized test of size $\alpha < 1/2$ has power arbitrarily close to 0 at some alternative μ with $|\rho_\mu|$ arbitrarily close to 1; furthermore, μ can be chosen to have a continuous positive density.

Proof. Consider a non-randomized test function ϕ with size α. Let G be the subset of the unit square where $\phi = 0$. So

$$\lambda^2(G) \geq 1 - \alpha > 1/2.$$

Let G^* be the set of pairs $(1 - x, 1 - y)$ with $(x, y) \in G$. So $\lambda^2(G^*) > 1/2$ and $\lambda^2(G \cap G^*) > 0$. We can find u, v with $u \neq 1/2, v \neq 1/2$ and (u, v) a Lebesgue point of $G \cap G^*$. Thus, (u, v) is a Lebesgue point of G, and so is $(1 - u, 1 - v)$. These two points are different, and lie on a line of non-zero slope; if we put mass $1/2$ at each point, the correlation between ξ_1 and ξ_2 would be ± 1.

Now construct a continuous positive density f that puts mass nearly $1/2$ in a small neighborhood of (u, v), and mass nearly $1/2$ in a small neighborhood of $(1 - u, 1 - v)$. With respect to f, the correlation between ξ_1 and ξ_2 is essentially ± 1. Moreover, $\int_G f \, d\lambda^2$ is nearly 1, so $\int \phi f \, d\lambda^2$ is nearly 0.   QED

Remark. The correlation is used to pick out alternatives that are quite different from independence.

Let $\mathcal{D}$ be the set of densities on $[0, 1]^k$ that can be represented as a finite sum $\sum_j c_j g_j$, where $c_j \geq 0$, $\sum_j c_j = 1$, $x = (x_1, \ldots, x_k)$, and

$$(2) \qquad g_j(x) = \prod_{i=1}^{k} g_{ij}(x_i),$$

the g_{ij} being continuous densities on $[0, 1]$. Unless otherwise specified, densities are with respect to Lebesgue measure.

Lemma 1. *The uniform closure of $\mathcal{D}$ is the set of all continuous densities on $[0, 1]^k$.*

Proof. This is easily done, using k-dimensional Bernstein polynomials. See Lorentz (1986, p. 51).   QED

Lemma 2. *Let ϕ be a randomized test function on $[0, 1]^k$. Suppose $\int \phi \, d\mu \leq \alpha$ for all probabilities μ on $[0, 1]^k$ that make the coordinate functions independent with continuous positive densities. Then ess sup $\phi \leq \alpha$.*

Proof. The condition is

$$(3) \qquad \int_{[0,1]^k} \phi \times \left(\prod_{i=1}^{k} f_i \right) d\lambda^k \leq \alpha$$

for all continuous positive densities f_i on $[0, 1]$. By an easy passage to the limit, inequality (3) holds for all continuous densities; that is, $f_i \geq 0$ rather than $f_i > 0$.

A convexity argument shows that

$$(4) \qquad \int \phi \varphi \, d\lambda^k \leq \alpha$$

for any $\varphi \in \mathcal{D}$, hence for any continuous density φ on $[0, 1]^k$ by Lemma 1. A density on $[0, 1]^k$ can be approximated in L^1 by a continuous density. Thus, inequality (4) holds for all densities φ on $[0, 1]^k$. Let $B = \{\phi > \alpha\}$. Suppose $\lambda^k(B) > 0$. Let $\varphi = 1/\lambda^k(B)$ on B, and let φ vanish off B. Then $\alpha \geq \int \phi \varphi \, d\lambda^k \geq \int_B \phi \varphi \, d\lambda^k > \alpha \int_B \varphi \, d\lambda^k = \alpha$, a contradiction showing that $\lambda^k(B) = 0$.   QED

Theorem 4. *Consider probabilities μ on the Borel unit hypercube $[0, 1]^k$ that are equivalent to λ^k. Consider testing the composite null hypothesis*

> *N: the coordinate functions are independent with respect to μ, each coordinate having a continuous positive density*

against any alternative set A of μ's. Under these circumstances, any test of size α has power at most α.

Proof. This is immediate from Lemma 2.   QED

Remark. The alternative A can consist of a single density f that is positive a.e. but is otherwise badly behaved. The null hypothesis can be substantially restricted, say to polynomial densities.

Theorem 5 is couched in terms of remote alternatives, which are distant from the null hypothesis. For rigor, we would have to metrize the space of probabilities on $[0, 1]^k$. This can be done in several ways without changing the argument. Here are three possibilities. (i) Variation distance can be used; remote alternatives will be nearly singular with respect to the probability satisfying the (simple) null hypothesis. (ii) The sup norm on distribution functions is another possibility; remote alternatives will be at a distance nearly 1 from the null. (iii) Distance can also be defined so as to metrize the weak-star topology; remote alternatives will be at a substantial distance from the null, with details depending a little on the metric that is used. Recall that λ^k is Lebesgue measure on the Borel subsets of $[0, 1]^k$.

Theorem 5. *Fix α with $0 < \alpha < 1$. Fix δ with $\delta > 0$ and $\alpha + \delta < 1$. Let μ be a probability on $[0, 1]^k$, and let $\xi_1, \ldots, \xi_k$ be the coordinate functions. Consider testing the simple null hypothesis*

N: $\mu = \lambda^k$

against the composite alternative

A: *μ makes the coordinate functions independent with different distributions, each distribution having a continuous positive density on $[0, 1]$.*

There is a fixed positive integer k_0 such for any $k > k_0$, and any randomized test ϕ_k of size α, there is a remote alternative μ where power is less than $\alpha + \delta$. The alternative μ makes ξ_i independent with continuous positive density f_i, but each f_i is arbitrarily concentrated near some point c_i in $(0, 1)$. Moreover,

(i) *$c_i < 1/3$ for more than $k/4$ of the indices $i = 1, \ldots, k$, and*
(ii) *$c_i > 2/3$ for more than $k/4$ of the indices $i = 1, \ldots, k$.*

Proof. *Step 1.* Let ϕ_k be a randomized test function of size α on $[0, 1]^k$, so

$$(5) \qquad \int \phi_k \, d\lambda^k \leq \alpha.$$

Apply Markov's inequality to $1 - \phi_k$, to see that for all k,

$$(6) \qquad P(B_k) \geq \frac{\delta}{\alpha + \delta}, \text{ where } B_k = \{\phi_k < \alpha + \delta\}.$$

Step 2. Write 1_Q for the indicator function of the set Q. Let C_k be the subset of $[0, 1]^k$ where $\xi_i < 1/3$ for more than $k/4$ of the indices $i = 1, \ldots, k$. Formally,

$$(7) \qquad\qquad C_k = \left\{ \sum_{i=1}^{k} 1_{\{\xi_i < 1/3\}} > k/4 \right\}.$$

Similarly, let

$$(8) \qquad\qquad D_k = \left\{ \sum_{i=1}^{k} 1_{\{\xi_i > 2/3\}} > k/4 \right\}.$$

The ξ_i are independent and $\lambda^k\{\xi_i < 1/3\} = 1/3 > 1/4$. So $\lambda^k(C_k) \to 1$ by the law of large numbers, and likewise for D_k.

Step 3. Choose k_0 so that

$$(9) \qquad\qquad \lambda^k(C_k \cap D_k) \geq 1 - \frac{1}{2} \frac{\delta}{\alpha + \delta} \text{ for } k > k_0.$$

Then

$$(10) \qquad \lambda^k(B_k \cap C_k \cap D_k) \geq \frac{1}{2} \frac{\delta}{\alpha + \delta} \text{ for } k > k_0.$$

Step 4. There must be $c_i \in (0, 1)$ with $c = (c_1, \ldots, c_k)$ a Lebesgue point of $B_k \cap C_k \cap D_k$. For each i, we construct a continuous positive density f_i on $[0, 1]$ that is concentrated around c_i, with

$$(11) \qquad \int_{[0,1]^k} \phi_k \times \left(\prod_{i=1}^{k} f_i \right) d\lambda^k \leq \phi_k(c) + \delta \leq \alpha + \delta.$$

To get the densities, start by putting the uniform distribution on a small hypercube centered at $(c_1, \ldots, c_k)$. Then smooth each edge separately. QED

Remarks. (i) If the test function is not randomized, we get power approaching 0 at remote alternatives, by the same argument.

(ii) The alternatives are remote from the null λ^k. They are also radically different from any power measure θ^k.

(iii) What if we have a suite of diagnostics? Let ϕ be the final result: 1 if the model is rejected, 0 if accepted, and $0 < \phi < 1$ if results are ambiguous or the decision is randomized. Thus, it suffices to consider a single test function.

(iv) Lehmann and Romano (2005, section 14.6) discuss statistical literature on limits to diagnostic power, the chief result being due to Janssen (2000). Even with IID data and a simple null hypothesis, goodness-of-fit tests have little power unless the set of alternative hypotheses can be substantially restricted. The context, however, is "local alternatives" that are $\mathrm{const.}/\sqrt{k}$ away from the null; such alternatives tend to the null as the sample size k increases. There are also some relevant papers in the game-theory literature: for instance, Lehrer (2001) and Olszewski and Sandroni (2008).

(v) By way of comparison, with IID data, the Kolmogorov-Smirnov test (among others) has power approaching 1 at any fixed alternative when testing the simple null hypothesis that the common distribution is uniform on [0, 1]. In Theorem 5 above, the data are IID under the null, not the alternative, and the sample size k is fixed.

19.2. Specific models

In regression models, a key assumption is exogeneity: Variables included in the model should be independent of error terms. Theorems 3 and 4 indicate the difficulty of testing this assumption. For many purposes, errors may be required to be independent, or independent and identically distributed. Theorems 3–5 indicate the difficulties. Requiring orthogonality rather than independence will not solve the problem and may not suffice for the usual asymptotics.

Rather than pursuing such topics, we turn to the proportional-hazards model, where subjects have failure times and censoring times. These are positive random variables, and only the smaller of the two is observed. These variables are generally assumed to be independent, or conditionally independent in a suitable sense given certain information. For a review of the model, see Freedman (2008d) [Chapter 11].

As shown by Tsiatis (1975), independence of failure times and censoring times is not testable in the usual data structures. Without that assumption, marginal distributions are not identifiable. Also see Clifford (1977). Furthermore, what happens after the end of a study is plainly unknowable. This already puts severe limits on the power of diagnostics. Therefore, let us assume that all failure times are fully observable, and see what can be done in that context.

In the model, failure times τ_i are independent positive variables with absolutely continuous distribution functions F_i. The density is $f_i = F_i'$ and the hazard rate is $h_i = f_i/(1 - F_i)$. According to the model, there is a baseline hazard rate h, and $h_i = h \exp(X_i \beta)$, where β is a parameter, or parameter vector if X_i is a vector. For present purposes, the covariate X_i is allowed to depend on i but not on time, and X_i is non-stochastic; we require $0 < h_i < \infty$ a.e. More general forms of the model relax these restrictions.

Theorems 1 and 2 show that we cannot tell whether failure times have densities and hazard rates—unless we restrict the class of alternatives, or impose additional assumptions, qualifications that will not be repeated. Theorems 3 and 4 show that we cannot determine whether failure times are independent or dependent.

Let us therefore assume that failure times have continuous positive densities. Even so, Theorem 5 shows that we cannot test the proportional-hazards assumption; this takes some mathematical effort to verify. Let us begin with the null hypothesis that the baseline hazard rate is identically 1, corresponding to a standard exponential failure time, and all the covariates are identically 0.

We can reduce to the IID uniform case covered by the theorem. All it takes is a change of variables: Replace the failure time τ_i by $\exp(-\tau_i)$. The conclusion is that any test of size α will have power barely above α against certain remote alternatives; the latter make the τ_i independent with continuous positive densities on $(0, \infty)$, but highly concentrated.

More general null hypotheses follow the same pattern. Suppose the covariates X_i are linearly independent p-vectors. The parameter vector β is a p-vector too. The null hypothesis specifies that the baseline hazard rate is 1, but allows the ith subject to have the hazard rate $\exp(X_i\beta)$, where β is free to vary—although it must be constant across i's.

We can again replace τ_i by $\exp(-\tau_i)$. Theorem 5 will give independent τ_i with highly concentrated densities at which power is low. Over 1/4 of these densities will be concentrated at values larger than $\log 3$ and will be quite different from any exponential density.

Remarks. (i) According to the theorem, for each large k, there are remote alternatives that are nearly indistinguishable from the null.

(ii) Replacing the baseline hazard rate by an unknown h does not change the position.

(iii) Usual tests of the proportional-hazards model involve adding another covariate, or stratifying on that covariate; the implied alternatives are tamer than the ones constructed here.

(iv) Altman and de Stavola (1994) discuss some of the practical problems in testing proportional-hazards models, and note that power is generally limited—even with conventional alternatives.

Theorem 6. *Suppose $\tau_i : i = 1, \ldots, k$ are positive random variables, each having a distribution with a continuous positive density on $(0, \infty)$. Suppose the τ_i are independent. Under these circumstances, the τ's obey a proportional-hazards model with a pre-specified baseline hazard rate h_0. The covariates are non-stochastic but time-dependent.*

Proof. We construct the model as follows. Let h_i be the hazard rate of τ_i. The covariates are $X_{it} = \log h_i(t) - \log h_0(t)$ for $i = 1, 2, 3, \ldots$. Furthermore, $h_i(t) = h_0(t)\exp(X_{it}\beta)$ with $\beta = 1$.   QED

Remark. Unless we restrict the set of covariates, the proportional-hazards model includes all distributions for failure times.

It would seem that Theorem 5 can be modified to handle time-varying non-stochastic covariates. But that includes all distributions, according to Theorem 6. To resolve the air of paradox, let ϕ be a randomized test function on $[0, 1]^k$ with

$$(12) \qquad \int_{[0,1]^k} \phi \times \left(\prod_{i=1}^k f_i\right) d\lambda^k \leq \alpha$$

for all continuous positive densities f_i. By Lemma 2, ess sup $\phi \leq \alpha$. There is no contradiction, because there are no non-trivial tests.

19.3. Discussion

A Google search (performed on September 3, 2009) gave 54,000 hits on the phrase "regression diagnostics," so this is a topic of some interest. Amazon.com gave 1300 hits on the phrase, the two most relevant books being Belsley et al. (2004) and Fox (1991). These texts do not reach the issues discussed here. Diagnostics for the proportional-hazards model are frequently mentioned, but standard references do not indicate the limitations on power; see, for instance, Andersen and Keiding (2006).

Models are frequently used to make causal inferences from observational data. See *Scandinavian Journal of Statistics* (2004) 31(2), for a recent survey. In brief,

> Fortunately, the days of "statistics can only tell us about association, and association is not causation" seem to be permanently over. (p. 161)

For causal inference, the crucial assumption is "invariance to intervention": Statistical relationships, including parameter values, that obtain in an observational setting will also obtain under intervention. For discussion and some historical background, see Freedman (2009).

The invariance assumption is not entirely statistical. Absent special circumstances, it does not appear that the assumption can be tested with the data that are used to fit the model. Indeed, it may be difficult to test the assumption without an experiment, either planned or natural. Such tests are beyond the scope of this chapter. They are also beyond the scope of conventional diagnostic procedures.

19.4. What about forecasting?

In principle, forecasting should be easier than making causal inferences from observational data, because forecasts are more readily calibrated against outcomes. On the other hand, the system that we are forecasting may be unstable, or we may be interested in forecasting rare events, or we may need the forecast before calibration data are available.

Conventional models seem to offer abundant ways to measure forecast uncertainty, just based on the data at hand. We can compute R^2 or the standard error of regression (700,000 hits on the latter phrase in Google); we can use cross-validation (800,000 hits) and so forth. However, as recent economic history makes clear, a major source of uncertainty in forecasts is specification error in the forecasting models. Specification error is extremely difficult to evaluate using internal evidence. That is the message of the present chapter.

Standard econometric texts, like Greene (2007) or Kennedy (2003), spend many pages discussing specification error, regression diagnostics (a.k.a. specification tests, model checking), robust estimation, and similar topics. Caution is in order. Unless the relevant class of specification errors can be narrowly delimited by prior theory and experience, diagnostics have limited power, and the robust procedures may be robust only against irrelevant departures from assumptions. "Robust standard errors" are particularly misleading, since these ignore bias (Freedman, 2006a [Chapter 17]).

19.5. Recommendations

Model diagnostics are seldom reported in applied papers. My recommendation, which may seem paradoxical at first, is this. Diagnostics should be reported more often, but a skeptical attitude should be adopted toward the results. Diagnostics should be reported more often because they can yield helpful information, picking up specification errors if these are sufficiently gross. Such errors might be corrected by adding explanatory variables, or modifying assumptions about disturbance terms, or changing the functional form of the equation. Furthermore, greater transparency in model development would eventually make the whole enterprise more credible. On the other hand, skepticism about diagnostics is warranted. As shown by the theorems presented here, a model can pass diagnostics with flying colors yet be ill-suited for the task at hand.

Acknowledgments

Peter Bickel, Russ Lyons, and Philip B. Stark made helpful comments.

Part IV

Shoe Leather Revisited

20

On Types of Scientific Inquiry:
The Role of Qualitative Reasoning

ABSTRACT. *One type of scientific inquiry involves the analysis of large data sets, often using statistical models and formal tests of hypotheses. Large observational studies have, for example, led to important progress in health science. However, in fields ranging from epidemiology to political science, other types of scientific inquiry are also productive. Informal reasoning, qualitative insights, and the creation of novel data sets that require deep substantive knowledge and a great expenditure of effort and shoe leather have pivotal roles. Many breakthroughs came from recognizing anomalies and capitalizing on accidents, which require immersion in the subject. Progress means refuting old ideas if they are wrong, developing new ideas that are better, and testing both. Qualitative insights can play a key role in all three tasks. Combining the qualitative and the quantitative—and a healthy dose of skepticism—may provide the most secure results.*

One type of scientific inquiry involves the analysis of large data sets, often using statistical models and formal tests of hypotheses. A moment's

Oxford Handbook of Political Methodology. (2008) J. M. Box-Steffensmeier, H. E. Brady, and D. Collier, eds. Oxford University Press, pp. 300–18.

thought, however, shows that there must be other types of scientific inquiry. For instance, something has to be done to answer questions like the
following. How should a study be designed? What sorts of data should be
collected? What kind of a model is needed? Which hypotheses should be
formulated in terms of the model and then tested against the data?

The answers to these questions frequently turn on observations, qualitative or quantitative, that give crucial insights into the causal processes
of interest. Such observations generate a line of scientific inquiry, or
markedly shift the direction of the inquiry by overturning prior hypotheses, or provide striking evidence to confirm hypotheses. They may well
stand on their own rather than being subsumed under the systematic data
collection and modeling activities mentioned above.

Such observations have come to be called "Causal Process Observations" (CPO's). These are contrasted with the "Data Set Observations"
(DSO's) that are grist for statistical modeling (Brady and Collier 2004).
My object in this essay is to illustrate the role played by CPO's, and qualitative reasoning more generally, in a series of well-known episodes drawn
from the history of medicine.

Why is the history of medicine relevant to us today? For one thing,
medical researchers frequently confront observational data that present
familiar challenges to causal inference. For another, distance lends perspective, allowing gains and losses to be more sharply delineated. The
examples show that an impressive degree of rigor can be obtained by
combining qualitative reasoning, quantitative analysis, and experiments
when those are feasible. The examples also show that great work can be
done by spotting anomalies and trying to understand them.

20.1 Jenner and vaccination

The setting is the English countryside in the 1790's. Cowpox, as
will be clear from the name, is a disease of cows. The symptoms include
sores on the cows' teats. Those who milked the cows often also became
infected, with sores on their hands; by the standards of the time, the illness
was rarely serious. In contrast, smallpox was one of the great killers of
the eighteenth century.

In 1796, Edward Jenner took some matter from a cowpox sore on
the hand of dairymaid Sarah Nelmes, and inserted it into the arm of an
eight-year-old boy, "by means of two superficial incisions, barely pene-
trating the cutis, each about half an inch long." The boy was "perceptibly
indisposed" on the ninth day, but recovered the following day. Six weeks
later, Jenner inoculated him with matter taken from a smallpox pustule,
"but no disease followed" (Jenner 1798, Case XVII).

Jenner published twenty-three case studies to demonstrate the safety and efficacy of "vaccination," as his procedure came to be called: *vacca* is the Latin term for cow, and *vaccinia* is another term for cowpox. Despite initial opposition, vaccination became standard practice within a few years, and Jenner achieved international fame. By 1978, smallpox had been eradicated.

What led Jenner to try his experiment? The eighteenth-century view of disease was quite different from ours. The great Scottish doctor of the time, William Cullen, taught that most diseases were "caused by external influences—climate, foodstuffs, effluvia, humidity, and so on—and ... the same external factors could cause different diseases in different individuals, depending on the state of the nervous system" (Porter 1997, p. 262).

Despite such misconceptions, it was known that smallpox could somehow be communicated from one person to another; moreover a person who contracted smallpox and survived was generally immune to the disease from that point on. As a preventive measure, patients could be deliberately infected (through scratches on the skin) with minute quantities of material taken from smallpox pustules, the idea being to induce a mild case of the disease that would confer immunity later.

This procedure was called "inoculation" or "variolation." It was not free of risk: Serious disease was sometimes caused in the patient and in people who came into contact with the patient (smallpox is highly contagious). On the other hand, failure to inoculate could easily lead to death from smallpox.

By the early part of the eighteenth century, variolation had reached England. Jenner was a country doctor who performed variolations. He paid attention to two crucial facts—although these facts were not explicable in terms of the medical knowledge of his time. (i) People who had the cowpox never seemed to contract smallpox afterwards, whether they had been inoculated or not. (ii) Some of his patients who had been ill with cowpox in the past still wanted to be inoculated: Such patients reacted very little to inoculation—

> What renders the Cow-pox virus so extremely singular, is, that the person who has been thus affected is for ever after secure from the infection of the Small Pox; neither exposure to the variolous effluvia, nor the insertion of the matter into the skin, producing this distemper. (Jenner 1798, p. 6)

These two facts led him to a hypothesis: Cowpox created immunity against smallpox. That is the hypothesis he tested, observationally and experimentally, as described above. In our terminology, Jenner vaccinated a boy

(Case XVII) who showed no response to subsequent inoculation. Immunity to smallpox had been induced by the vaccination.

By "virus," Jenner probably meant "contagious matter," that being a standard usage in his time. Viruses in the modern sense were not to be discovered for another century. By a curious twist, smallpox and cowpox are viral diseases in our sense, too.

20.2 Semmelweis and puerperal fever

The time is 1844 and the place is Vienna. The discovery of microbes as the cause of infectious disease would not be made for some decades. Ignaz Semmelweis was an obstetrician in the First Division of the Lying-in Hospital, where medical students were trained. (Midwives were trained in the Second Division.) Pregnant women were admitted to one division or the other, according to the day of the week that they come to the hospital, in strict alternation. Mortality from "puerperal fever" was much higher in the First Division (Semmelweis 1941 [1861], p. 356).

Eventually, Semmelweis discovered the cause. The medical students were doing autopsies, and then examining the "puerperae" (women who were giving birth, or who had just given birth). "Cadaveric particles" were thus transferred to the women, entering the bloodstream and causing infection. In 1847, Semmelweis instituted the practice of disinfection, and mortality plummeted (Semmelweis 1941 [1861], pp. 393–94).

But how did Semmelweis make his discovery? To begin with, he had to reject conventional explanations, including "epidemic influences," which meant something different then:

> Epidemic influences ... are to be understood [as] certain hitherto inexplicable, atmospheric, cosmic, telluric changes, which sometimes disseminate themselves over whole countrysides, and produce childbed fever in individuals predisposed thereto by the puerperal state. ["Telluric" means earthly.] Now, if the atmospheric-cosmic-telluric conditions of the City of Vienna are so disposed that they cause puerperal fever in individuals susceptible thereto as puerperae, how does it happen that these atmospheric-cosmic-telluric conditions over such a long period of years have carried off individuals disposed thereto as puerperae in the First Clinic, while they have so strikingly spared others also in Vienna, even in the same building in the Second Division and similarly vulnerable as puerperae? (Semmelweis 1941 [1861], p. 357)

The reasoning was qualitative; and similar qualitative arguments disposed of other theories—diet, ventilation, use of hospital linens, and so forth.

Now he had to discover the real cause. In 1847, his revered colleague Professor Kolletschka was accidentally cut with a knife used in a medico-legal autopsy. Kolletschka became ill, with symptoms remarkably similar to puerperal fever; then he died. Again, qualitative analysis was crucial. Close attention to symptoms and their progression was used to identify Kolletschka's illness with puerperal fever (Semmelweis 1941 [1861], p. 391). Tracing of causal processes came into play as well:

> Day and night this picture of Kolletschka's disease pursued me ... I was obliged to acknowledge the identity of the disease, from which Kolletschka died, with that disease of which I saw so many puerperae die.... I must acknowledge, if Kolletschka's disease and the disease from which I saw so many puerperae die, are identical, then in the puerperae it must be produced by the self-same engendering cause, which produced it in Kolletschka. In Kolletschka, the specific agent was cadaveric particles, which were introduced into his vascular system [the bloodstream]. I must ask myself the question: Did the cadaveric particles make their way into the vascular systems of the individuals, whom I had seen die of an identical disease? This question I answer in the affirmative. (Semmelweis 1941 [1861], pp. 391–92)

The source of the infectious agent also could have been a wound in a living person (Semmelweis 1941 [1861], p. 396). Once the cause was discovered, the remedy was not far away: Eliminate the infectious particles from the hands that will examine the puerperae. Washing with soap and water was insufficient, but disinfection with chlorine compounds was sufficient (Semmelweis 1941 [1861], pp. 392–96).

Few of his contemporaries accepted Semmelweis' work, due in part to his troubled and disputatious personality, although his picture of the disease was essentially correct. Puerperal fever is a generalized infection, typically caused by bacteria in the group *Streptococcus pyogenes*. These bacteria enter the bloodstream through wounds suffered during childbirth (for instance, at the site where the placenta was attached). Puerperal fever can be—and today it generally is—avoided by proper hygiene.

20.3 Snow and cholera

John Snow was a physician in Victorian London. In 1854, he demonstrated that cholera was an infectious disease, which could be prevented by cleaning up the water supply. The demonstration took advantage of a natural experiment. A large area of London was served by two water companies. The Southwark and Vauxhall company distributed contaminated water, and households served by it had a death rate "between eight and

nine times as great as in the houses supplied by the Lambeth company,"
which supplied relatively pure water (Snow 1965 [1855], p. 86, data in
table IX).

What led Snow to design the study and undertake the arduous task
of data collection? To begin with, he had to reject the explanations of
cholera epidemics that were conventional in his time. The predominant
theory attributed cholera to "miasmas," that is, noxious odors—especially
odors generated by decaying organic material. Snow makes qualitative
arguments against such explanations:

> [Cholera] travels along the great tracks of human intercourse,
> never going faster than people travel, and generally much more
> slowly. In extending to a fresh island or continent, it always
> appears first at a sea-port. It never attacks the crews of ships
> going from a country free from cholera, to one where the disease
> is prevailing, till they have entered a port, or had intercourse
> with the shore. Its exact progress from town to town cannot
> always be traced; but it has never appeared except where there
> has been ample opportunity for it to be conveyed by human
> intercourse. (Snow 1965 [1855], p. 2)

These phenomena are easily understood if cholera is an infectious disease,
but hard to explain on the miasma theory. Similarly,

> The first case of decided Asiatic cholera in London, in the au-
> tumn of 1848, was that of a seaman named John Harnold, who
> had newly arrived by the *Elbe* steamer from Hamburgh, where
> the disease was prevailing . . . . Now the next case of cholera, in
> London, occurred in the very room in which the above patient
> died. (Snow 1965 [1855], p. 3)

The first case was infected in Hamburg; the second case was infected by
contact with dejecta from the first case, on the bedding or other furnishings
in that fatal room. The miasma theory, on the other hand, does not provide
good explanations.

Careful observation of the disease led to the conclusion "that cholera
invariably commences with the affection of the alimentary canal" (Snow
1965 [1855], p. 10). A living organism enters the body, as a contaminant
of water or food, multiplies in the body, and creates the symptoms of the
disease. Many copies of the organism are expelled from the body with
the dejecta, contaminate water or food, then infect other victims. The task
is now to prove this hypothesis.

According to Sir Benjamin Ward Richardson, who wrote the introduction to Snow's book, the decisive proof came during the Broad Street
epidemic of 1854:

> [Snow] had fixed his attention on the Broad Street pump as the
> source and centre of the calamity. He advised the removal of the
> pump-handle as the grand prescription. The vestry [in charge of
> the pump] was incredulous, but had the good sense to carry out
> the advice. The pump-handle was removed and the plague was
> stayed. (Snow 1965 [1855], p. xxxvi)

The pump-handle as the decisive test is a wonderful fable, which has beguiled many a commentator.

What are the facts? Contamination at the pump did cause the epidemic, Snow recommended closing the pump, his advice was followed, and the epidemic stopped. However, the epidemic was stopping anyway. Closing the pump had no discernible effect: The episode proves little. Snow explains this with great clarity (Snow 1965 [1855], pp. 40–55, see especially table I on p. 49 [Chapter 3, Table 3.1, p. 51] and the conclusory paragraph on pp. 51–52). Richardson's account is therefore a classic instance of post hoc, ergo propter hoc.

The reality is more interesting than the fable. Snow was intimately familiar with the Broad Street area, because of his medical practice. He says,

> As soon as I became acquainted with the situation and extent of
> this irruption of cholera, I suspected some contamination of the
> water of the much-frequented street-pump in Broad Street ...
> but on examining the water, on the evening of 3rd September, I
> found so little impurity in it of an organic nature, that I hesitated
> to come to a conclusion. (Snow 1965 [1855], pp. 38–39)

Snow had access to the death certificates at the General Register Office and drew up a list of the cholera fatalities registered shortly before his inspection of the pump. He then made a house-to-house canvass (the death certificate shows the address of the deceased) and discovered that the cases clustered around the pump, confirming his suspicion. Later, he made a more complete tally of cholera deaths in the area. His "spot map" displays the locations of cholera fatalities during the epidemic, and the clustering is apparent from the map (Snow 1965 [1855], pp. 44–45; Cholera Inquiry Committee 1855, pp. 106–09).

However, there were a number of exceptions that had to be explained. For example, there was a brewery near the pump; none of the workers contracted the disease: Why not? First, the workers drank beer; second, if water was desired, there was a pump on the premises (Snow 1965 [1855], p. 42). For another example, a lady in Hampstead contracted cholera. Why? As it turned out, she liked the taste of the water from the

Broad Street pump, and had it brought to her house (Snow 1965 [1855], p. 44). Snow gives many other such examples.

Snow's work on the Broad Street epidemic illustrates the power of case studies. His refutation of the usual explanations for cholera, and the development of his own explanation, are other indicators of the power of qualitative reasoning. The analysis of his natural experiment, referred to above, shows the power of simple quantitative methods and good research design. This was the great quantitative test of his theory that cholera was a waterborne infectious disease.

In designing the quantitative study, however, Snow made some key qualitative steps: (i) seeing that conventional theories were wrong, (ii) formulating the water hypothesis, and (iii) noticing that in 1852, the Lambeth company moved its intake pipe to obtain relatively pure water, while Southwark and Vauxhall continued to draw heavily contaminated water. It took real insight to see—a priori rather than a posteriori—that this difference between the companies allowed the crucial study to be done.

Snow's ideas gained some circulation, especially in England. However, widespread acceptance was achieved only when Robert Koch isolated the causal agent (*Vibrio cholerae*, a comma-shaped bacillus) during the Indian epidemic of 1883. Even then, there were dissenters, with catastrophic results in the Hamburg epidemic of 1892: see Evans (1987).

Inspired by Koch and Louis Pasteur, there was a great burst of activity in microbiology during the 1870's and 1880's. The idea that microscopic lifeforms could arise by spontaneous generation was cast aside, and the germ theory of disease was given solid experimental proof. Besides the cholera vibrio, the bacteria responsible for anthrax (*Bacillus anthracis*) and tuberculosis (*Mycobacterium tuberculosis*) were isolated, and a vaccine was developed against rabies (a viral disease). However, as we shall see in a moment, these triumphs made it harder to solve the riddle of beriberi. Beriberi is a deficiency disease, but the prestige of the new microbiology made investigators suspicious of any explanation that did not involve microorganisms.

20.4 Eijkman and beriberi

Beriberi was endemic in Asia, from about 1750 until 1930 or so. Today, the cause is known. People need minute amounts (about one part per million in the diet) of a vitamin called "thiamin." Many Asians eat a diet based on rice, and white rice is preferred to brown.

Thiamin in rice is concentrated in the bran—the skin that gives rice its color. White rice is obtained by polishing away the skin, and with it most of the thiamin; what is left is further degraded by cooking. The diet is then deficient in thiamin, unless supplemented by other foods rich in that substance. Beriberi is the sequel.

In 1888, knowledge about vitamins and deficiency diseases lay decades in the future. That year, Christiaan Eijkman—after studying microbiology with Koch in Berlin—was appointed director of the Dutch Laboratory for Bacteriology and Pathology in the colony of Java, near the city now called Jakarta. His research plan was to show that beriberi was an infectious disease, with Koch's methods for the proof.

Eijkman tried to infect rabbits and then monkeys with blood drawn from beriberi patients. This was unsuccessful. He then turned to chickens. He tried to infect some of the birds, leaving others as controls. After a time, many of his chickens came down with polyneuritis, which he judged to be very similar to beriberi in humans. ("Polyneuritis" means inflammation of multiple nerves.)

However, the treated chickens and the controls were equally affected. Perhaps the infection spread from the treated chickens to the controls? To minimize cross infection, he housed the treated chickens and the controls separately. That had no effect. Perhaps his whole establishment had become infected? To eliminate this possibility, he started work on another, remote experimental station—at which point the chickens began recovering from the disease.

> [Eijkman] wrote "something struck us that had escaped our attention so far." The chickens had been fed a different diet during the five months in which the disease had been developing. In that period (July through November 1889), the man in charge of the chickens had persuaded the cook at the military hospital, without Eijkman being aware of it, to provide him with leftover cooked [white] rice from the previous day, for feeding to the birds. A new cook, who started duty on 21 November, had refused to continue the practice. Thirty years later, Eijkman was to say that "[the new cook] had seen no reason to give military rice to civilian hens." (Carpenter 2000, p. 38)

In short, the chickens became ill when fed cooked, polished rice; they recovered when fed uncooked, unpolished rice. This was an accidental experiment, arranged by the cooks. One of Eijkman's great insights was paying attention to the results, because the cooks' experiment eventually changed the understanding of beriberi.

Eijkman's colleague, Adolphe Vorderman, undertook an observational study of prisons to confirm the relevance to humans. Where prisoners were fed polished rice, beriberi was common; with a diet of unpolished rice, beriberi was uncommon. Beriberi is a deficiency disease, not an infectious disease. The evidence may seem compelling, but that is because we know the answer. At the time, the picture was far from clear. Eijkman himself thought that white rice was poisonous, the bran containing the antidote. Later, he was to reverse himself: Beriberi is an infectious disease, although a poor diet makes people (and chickens) more vulnerable to infection.

In 1896, Gerrit Grijns took over Eijkman's laboratory (Eijkman suffered from malaria and had to return to Holland). Among other contributions, after a long series of careful experiments, Grijns concluded that beriberi was a deficiency disease, the missing element in the diet being concentrated in rice bran—and in other foods like mung beans.

In 1901, Grijn's colleague Hulshoff Pol ran a controlled experiment at a mental hospital, showing that mung beans prevented or cured beriberi. In three pavilions out of twelve, the patients were fed mung beans; in three pavilions, other green vegetables. In three pavilions, there was intensive disinfection, and three pavilions were used as controls. The incidence of beriberi was dramatically lower in the pavilions with mung beans.

Still, medical opinion remained divided. Some public health professionals accepted the deficiency hypothesis. Others continued to favor the germ theory, and still others thought the cause was an inanimate poison. It took another ten years or so to reach consensus that beriberi was a deficiency disease, which could be prevented by eating unpolished rice or enriching the diet in other ways. From a public health perspective, the problem of beriberi might be solved, but the research effort turned to extracting the critical active ingredient in rice bran—no mean challenge, since there is about one teaspoon of thiamin in a ton of bran.

Around 1912, Casimir Funk coined the term "vitamines," later contracted to vitamins, as shorthand for "vital amines." The claim that he succeeded in purifying thiamin may be questionable. But he did guess that beriberi and pellagra were deficiency diseases, which could be prevented by supplying trace amounts of organic nutrients.

By 1926, B. C. P. Jansen and W. F. Donath had succeeded in extracting thiamin (vitamin B_1) in pure crystal form. Ten years later, Robert R. Williams and his associates managed to synthesize the compound in the laboratory. In the 1930's there were still beriberi cases in the East—and these could be cured by injecting a few milligrams of the new vitamin B_1.

20.5 Goldberger and pellagra

> Pellagra was first observed in Europe in the eighteenth cen-
> tury by a Spanish physician, Gaspar Casal, who found that it
> was an important cause of ill-health, disability, and premature
> death among the very poor inhabitants of the Asturias. In the
> ensuing years, numerous ... authors described the same con-
> dition in northern Italian peasants, particularly those from the
> plain of Lombardy. By the beginning of the nineteenth cen-
> tury, pellagra had spread across Europe, like a belt, causing
> the progressive physical and mental deterioration of thousands
> of people in southwestern France, in Austria, in Rumania, and
> in the domains of the Turkish Empire. Outside Europe, pella-
> gra was recognized in Egypt and South Africa, and by the first
> decade of the twentieth century it was rampant in the United
> States, especially in the south . . . . (Roe 1973, p. 1)

Pellagra seemed to hit some villages much harder than others. Even
within affected villages, many households were spared, but some had pellagra cases year after year. Sanitary conditions in diseased households
were primitive: Flies were everywhere. One blood-sucking fly (*Simulium*
species) had the same geographical range as pellagra, at least in Europe; and the fly was most active in the spring, just when most pellagra cases developed. Many epidemiologists concluded the disease was
infectious, and—like malaria or yellow fever—was transmitted from one
person to another by insects.

Joseph Goldberger was an epidemiologist working for the U.S. Public
Health Service. In 1914, he was assigned to work on pellagra. Despite the
climate of opinion described above, he designed a series of observational
studies and experiments showing that pellagra was caused by a bad diet
and is not infectious. The disease could be prevented or cured by foods
rich in what Goldberger called the P-P (pellagra-preventive) factor.

By 1926, he and his associates had tentatively identified the P-P factor as part of the vitamin B complex. By 1937, C. A. Elvehjem and his
associates had identified the P-P factor as niacin, also called vitamin B_3
(this compound had been discovered by C. Huber around 1870, but its
significance had not been recognized). Since 1940, most of the flour sold
in the United States has been enriched with niacin, among other vitamins.

Niacin occurs naturally in meat, milk, eggs, some vegetables, and
certain grains. Corn, however, contains relatively little niacin. In the pellagra areas, the poor ate corn—and not much else. Some villages and
some households were poorer than others and had even more restricted

diets. That is why they were harder hit by the disease. The flies were a marker of poverty, not a cause of pellagra.

What prompted Goldberger to think that pellagra was a deficiency disease rather than an infectious disease? In hospitals and asylums, the inmates frequently developed pellagra, the attendants almost never—which is unlikely if the disease is infectious because the inmates could infect the attendants. This observation, although far from definitive, set Goldberger on the path to discovering the cause of pellagra and methods for prevention or cure. The qualitative thinking precedes the quantitative investigation. Pellagra is virtually unknown in the developed world today, although it remains prevalent in some particularly poor countries.

20.6 McKay and fluoridation

> Dental caries is an infectious, communicable, multifactorial disease in which bacteria dissolve the enamel surface of a tooth . . . . Soon after establishing his dental practice in Colorado Springs, Colorado, in 1901, Dr. Frederick S. McKay noted an unusual permanent stain or 'mottled enamel' (termed 'Colorado brown stain' by area residents) on the teeth of many of his patients. After years of personal field investigations, McKay concluded that an agent in the public water supply probably was responsible for mottled enamel. McKay also observed that teeth affected by this condition seemed less susceptible to dental caries. (Centers for Disease Control and Prevention, 1999, p. 933; internal citations omitted)

Mottling was caused by something in the drinking water: That was the main hypothesis at the time (McKay and Black 1916, p. 635). McKay and Black found that mottled teeth were endemic to specific areas. Mottling affected people born in the area, not people who moved to the area after their teeth had been formed. If mottling was prevalent in one area but not in a nearby area, the two areas had different water supplies. These observations supported the water hypothesis, the idea being that the causal agent affects the teeth as they are developing in the body.

McKay and Black (1916) could not identify the causal agent in the water, but explained that their chemical analyses (p. 904)—

> were made according to the standard quantitative form. There are present, however, in waters certain other elements of rarer varieties that exist only in traces, the determination of which requires much elaborate technique and spectroscopic and polariscopic tests, which are beyond the capacities of ordinary chemical laboratories.

As a consequence of mottling, two towns (Oakley in 1925 and Bauxite in 1928) changed the source of their water supply. After the change, newborn children in those towns developed normal teeth. This is, at least in retrospect, striking confirmation of the water hypothesis (McClure 1970, chapters 2–3).

Bauxite was a company town (Aluminum Company of America). H. V. Churchill, an ALCOA chemist, discovered in 1931 that fluorides were naturally present in the original source—a deep well—at relatively high concentrations: He had a spectrograph at the company laboratory. McKay and Churchill also found high levels of fluorides in the water at several other towns where mottling was endemic, which suggested that fluorides might cause mottling and prevent tooth decay.

H. Trendley Dean, along with others in the U.S. Public Health Service, collected more systematic data on fluorides in the water, mottling, and tooth decay. The data confirmed the associations noted by McKay and Churchill. Moreover, the data indicated that, at lower doses, fluorides in the water could prevent decay without mottling the teeth. (Mottling was unsightly and carried risks of its own.) Starting in 1945, community experiments strengthened these conclusions about the role of fluorides, although some controversy remained. Fluoridation of drinking water followed within a few years, and tooth decay in childhood declined precipitously.

20.7 Fleming and penicillin

Alexander Fleming was working at St. Mary's Hospital in London, under the direction of Sir Almroth Wright, studying the life cycle of staphylococcus (bacteria that grow in clusters, looking under the microscope like clusters of grapes). Fleming had a number of culture plates on which he was growing staphylococcus colonies. He left the plates in a corner of his office for some weeks while he was on holiday. When he returned, one of the plates had been contaminated by mold. So far, this is unremarkable. He noticed, however, "that around a large colony of a contaminating mould the staphylococcus colonies became transparent and were obviously undergoing lysis" (Fleming 1929, p. 226).

Bacteria "lyse" when their cell walls collapse. What caused the lysis? Rather than discarding the plate—the normal thing to do—Fleming thought that the lysis was worth investigating. He did so by growing the mold in broth, watching its behavior, and trying filtered broth on various kinds of bacteria. The mold, a species of *Penicillium*, generated a substance that "to avoid the repetition of the rather cumbersome phrase 'mould broth filtrate' [will be named] 'penicillin'" (Fleming 1929,

p. 227). It was the penicillin that caused the bacteria to lyse. Fleming showed that penicillin destroyed—or at least inhibited the growth of— many kinds of bacteria besides staphylococcus.

Penicillin's therapeutic potential went unrealized until Howard Florey and his associates at Oxford took up the research in 1938 and found processes for purification and larger-scale production. Due to the exigencies of World War II, much of the work was done in the U.S., where a strain of *Penicillium* that gave high yields was found on a moldy cantaloupe at a market in Peoria. (Industrial-scale development was being done at a nearby Department of Agriculture laboratory under the direction of Kenneth Raper, and people were encouraged to bring in moldy fruit for analysis.)

Penicillin was widely used to treat battlefield injuries, largely pre- venting gangrene, for example. Along with the sulfa drugs (prontosil was discovered by Gerhard Domagk in 1932) and streptomycin (discovered by Selman Waksman in 1944), penicillin was among the first of the modern antibiotics.

20.8 Gregg and German measles

Norman Gregg was a pediatric ophthalmologist in Australia. In 1941, he noticed in his practice an unusually large number of infants with cataracts and heart defects. ("Cataracts" make the lens of the eye opaque.) On investigation, he found that many of his colleagues were also treating such cases. The similarity of the cases, and their widespread geographic distribution, led him to guess that the cause must have been exposure to some infectious agent early in the mother's pregnancy, rather than genetics—which was the conventional explanation at the time for birth defects. But what was the infectious agent? This is how Gregg explained his thought process:

> The question arose whether [the cause] could have been some
> disease or infection occurring in the mother during pregnancy
> which had then interfered with the developing cells of the lens.
> By a calculation from the date of the birth of the baby, it was
> estimated that the early period of pregnancy corresponded with
> the period of maximum intensity of the very widespread and se-
> vere epidemic in 1940 of the so-called German measles. (Gregg
> 1941, p. 430)

Detailed epidemiological research showed that exposure of the mother to German measles in the first or second month of pregnancy markedly increases the risk of birth defects in the baby. The association is generally viewed as causal. Today, there is a vaccine that prevents German measles, and cataracts at birth are exceedingly rare.

20.9 Herbst and DES

Herbst and Scully described seven cases of adenocarcinoma of the vagina in adolescent girls. This is an unusual kind of cancer, especially in adolescence. What was the cause? The mother of one patient suggested diethylstilbestrol (DES), an artificial hormone often prescribed in those days to prevent miscarriage. Arthur Herbst and his associates were intrigued, but skeptical. They did a case control study and established a highly significant association, confirmed by a number of other studies, and now accepted as causal.

Two key insights precede any statistical analysis: (i) this is a cluster of cancers worth investigating; and (ii) the cause might have been exposure of the mother during pregnancy—not the daughter after birth—to some toxic substance. A priori, neither point could have been obvious.

20.10 Conclusions

In the health sciences, there have been enormous gains since the time of Jenner, many of which are due to statistics. Snow's analysis of his natural experiment shows the power of quantitative methods and good research design. Semmelweis' argument depends on statistics; so too with Goldberger, Dean, Gregg, and Herbst et al. On the other hand, as the examples demonstrate, substantial progress also derives from informal reasoning and qualitative insights. Recognizing anomalies is important; so is the ability to capitalize on accidents. Progress depends on refuting conventional ideas if they are wrong, developing new ideas that are better, and testing the new ideas as well as the old ones. The examples show that qualitative methods can play a key role in all three tasks.

In Fleming's laboratory, chance circumstances generated an anomalous observation. Fleming resolved the anomaly and discovered penicillin. Semmelweis used qualitative reasoning to reject older theories about the cause of puerperal fever, to develop a new theory from observations on a tragic accident, and to design an intervention that would prevent the disease. The other examples lead to similar conclusions.

What are the lessons for methodologists in the twenty-first century? Causal inference from observational data presents many difficulties, especially when underlying mechanisms are poorly understood. There is a natural desire to substitute intellectual capital for labor, and an equally natural preference for system and rigor over methods that seem more haphazard. These are possible explanations for the current popularity of statistical models.

Indeed, far-reaching claims have been made for the superiority of a quantitative template that depends on modeling—by those who manage to ignore the far-reaching assumptions behind the models. However, the assumptions often turn out to be unsupported by the data (Duncan 1984; Berk 2004; Brady and Collier 2004; Freedman 2009). If so, the rigor of advanced quantitative methods is a matter of appearance rather than substance.

The historical examples therefore have another important lesson to teach us. Scientific inquiry is a long and tortuous process, with many false starts and blind alleys. Combining qualitative insights and quantitative analysis—and a healthy dose of skepticism—may provide the most secure results.

20.11 Further reading

Brady, Collier, and Seawright (2004) compare qualitative and quantitative methods for causal inference in the social sciences. As they point out (pp. 9–10),

> it is difficult to make causal inferences from observational data, especially when research focuses on complex political processes. Behind the apparent precision of quantitative findings lie many potential problems concerning equivalence of cases, conceptualization and measurement, assumptions about the data, and choices about model specification . . . .

These authors recommend using a diverse mix of qualitative and quantitative techniques in order to exploit the available information; no particular set of tools is universally best. Causal process observations (including anomalies and results of accidental experiments, even experiments with $N = 1$) can be extremely helpful, as they were in the epidemiological examples discussed here.

The role of anomalies in political science is also discussed by Rogowski (2004). He suggests that scholars in that field may be excessively concerned with hypothesis testing based on statistical models. Scholars may underestimate the degree to which the discovery of anomalies can overturn prior hypotheses and open new avenues of investigation. Anomalies that matter have been discovered in case studies—even when the cases have been selected in ways that do considerable violence to large-N canons for case selection. He also suggests that failure to search for anomalies can lead to a kind of sterility in research programs.

Scientific progress often begins with inspired guesswork. On the other hand, if guesses cannot be verified, progress may be illusory. For example, by analogy with cholera, Snow (1965 [1855], pp. 125–33) theorized that plague, yellow fever, dysentery, typhoid fever, and malaria (which he calls "ague" or "intermittent fever") were infectious waterborne diseases. His supporting arguments were thin. As it turns out, these diseases are infectious; however, only dysentery and typhoid fever are waterborne.

Proof for dysentery and typhoid fever, and disproof for the other diseases, was not to come in Snow's lifetime. Although William Budd (1873) made a strong case on typhoid fever, reputable authors of the late nineteenth century still denied that such diseases were infectious (Bristowe and Hutchinson 1876, pp. 211, 629; Bristowe et al. 1879, pp. 102–03). In the following decades, evidence from epidemiology and microbiology settled the issue.

Plague is mainly spread by flea bites, although transmission by coughing is also possible in cases of pharyngitis or pneumonia. The causal agent is the bacterium *Yersinia pestis*. Yellow fever and malaria are spread by mosquitoes. Yellow fever is caused by a virus, while malaria is caused by several species of *Plasmodium* (one-celled organisms with nuclei and extravagantly complicated life cycles, spent partly in humans and partly in mosquitoes). The medieval Black Death is usually identified with modern plague, but this is still contested by some scholars (Nutton 2008).

Buck et al. (1989) reprints many of the classic papers in epidemiology; some classic errors are included too. Porter (1997) is a standard reference on the history of medicine. Jenner's papers are reprinted in Eliot (1910). Bazin (2000) discusses the history of smallpox, Jenner's work, and later developments, including the eradication of smallpox; the last recorded cases were in 1977–78. There is a wealth of additional information on the disease and its history in Fenner et al. (1988).

Inoculation was recorded in England by 1721 (Bazin 2000, p. 13; Fenner et al. 1988, pp. 214–16). However, the practice was described in the journals some years before that (Timonius and Woodward 1714). It was a common opinion in Jenner's time that cowpox created immunity to smallpox (Jenner 1801; Baron 1838, p. 122). Over the period 1798–1978, techniques for producing and administering the vaccine were elaborated. As life spans became longer, it became clear that—contrary to Jenner's teachings—the efficacy of vaccination gradually wore off. Revaccination was introduced. By 1939, the virus in the vaccines was a little different from naturally occurring cowpox virus. The virus in the vaccines is called "vaccinia" (Bazin 2000, chapter 11; Fenner et al. 1988, chapters 6–7, especially p. 278).

Bulloch (1938) reviews the history of bacteriology. Bacteria were observed by Robert Hooke and Antonie van Leeuwenhoek before 1700. Otto Friderich Müller in Denmark developed a workable classification before 1800, improved about fifty years later by Ferdinand Cohn in Germany.

Some of Koch's work on anthrax was anticipated by Pierre François Rayer and Casimir-Joseph Davaine in France. Likewise, Pasteur's experiments disproving spontaneous generation built on previous work by others, including Lazzaro Spallanzani; contemporaneous research by John Tyndall should also be mentioned.

Freedman (2009, pp. 6–9) reports on Snow and cholera [see also Chapter 3, pp. 48–53]. For detailed information on Snow's work, see Vinten-Johansen et al. (2003). Evans (1987) gives a historical analysis of the cholera years in Europe. Koch's discovery of the vibrio was anticipated by Filippo Pacini in 1854, but the implications of Pacini's work were not recognized by his contemporaries.

Henry Whitehead was a clergyman in the Soho area. He did not believe that the Broad Street pump—famous for the purity of its water—was responsible for the epidemic. He saw a gap in Snow's argument: The fatalities cluster around the pump, but what about the population in general?

Whitehead made his own house-to-house canvass to determine attack rates among those who drank water from the pump and those who did not. Then he drew up a 2 × 2 table to summarize the results. The data convinced him that Snow was correct (Cholera Inquiry Committee 1855, pp. 121–33). Snow made this kind of analysis only for his natural experiment.

William Farr, statistical superintendent of the General Register Office, was a leading medical statistician in Victorian England and a "sanitarian," committed to eliminating air pollution and its sources. He claimed that the force of mortality from cholera in an area was inversely related to its elevation. More specifically, if y is the death rate from cholera in an area and x is its elevation, Farr proposed the equation

$$y = \frac{a}{b+x}.$$

The constants a and b were estimated from the data. For 1848–49, the fit was excellent.

Farr held the relationship to be causal, explained by atmospheric changes, including attenuation of noxious exhalations from the Thames, changes in vegetation, and changes in the soil. After the London epidemic of 1866, however, he came to accept substantial parts of Snow's theory—without abandoning his own views about miasmas and elevation (Humphreys 1885, pp. 341–84; Eyler 1979, pp. 114–22; Vinten-Johansen et al. 2003, p. 394).

For better or worse, Farr's belief in mathematical symbolism had considerable influence on the development of research methods in medicine and social science. Furthermore, the tension between the pursuit of social reform and the pursuit of truth, so evident in the work of the sanitarians, is still with us.

There are two informative web sites on Snow, Whitehead, and other major figures of the era.

http://www.ph.ucla.edu/epi/snow.html
http://johnsnow.matrix.msu.edu/

Loudon (2000) is highly recommended on puerperal fever; but also see Nuland (1979) for a more sympathetic account of Semmelweis' life. Hare (1970, chapter 7) discusses efforts to control puerperal fever in a London maternity hospital in the 1930's. The strain of *Staphylococcus pyogenes* causing the disease turned out to be a common inhabitant of the human nose and throat (Loudon 2000, pp. 201–04).

A definitive source on beriberi, only paraphrased here, is Carpenter (2000). He gives a vivid picture of a major scientific advance, including discussion of work done before Eijkman arrived in Java.

The discussion of pellagra is based on Freedman, Pisani, and Purves (2007, pp. 15–16). Goldberger's papers are collected in Terris (1964). Goldberger (1914) explains the reasoning that led him to the deficiency-disease hypothesis; Goldberger et al. (1926) identifies the P-P factor as part of the vitamin B complex. Carpenter (1981) reprints papers by many pellagra researchers, with invaluable commentary. He explains why in Mexico a corn-based diet does not lead to pellagra, discusses the role of tryptophan (an amino acid that can be converted to niacin in the body), and points out the gaps in our knowledge of the disease and the reasons for its disappearance.

The primary papers on fluoridation are McKay and Black (1916), Churchill (1931), and Dean (1938). There is a considerable secondary literature; see, for instance, McClure (1970) and Centers for Disease Control and Prevention (1999). McKay (1928) is often cited, but seems mainly about another topic: whether enamel in teeth is living tissue.

An excellent source on Fleming is Hare (1970), with Goldsmith (1946) adding useful background. Today, "penicillin" refers to the active ingredient in Fleming's mold broth filtrate. What is the cell-killing mechanism? In brief, cell walls of most bacteria include a scaffolding constructed from sugars and amino acids. Components of the scaffolding have to be manufactured and assembled when the cells are dividing to form daughter cells. In many species of bacteria, penicillin interferes with the assembly process, eventually causing the cell wall to collapse (Walsh 2003).

Some species of bacteria manufacture an enzyme ("penicillinase") that disables penicillin—before the penicillin can disable the cell. There are other bacterial defense systems too, which explain the limits to the efficacy of penicillin. Penicillin inhibits cell wall synthesis by a process that is reasonably well understood, but how does inhibition cause lysis? That is still something of a mystery, although much has been learned (Walsh 2003, p. 41; Bayles 2000; Giesbrecht et al. 1998).

Penicillin only causes lysis when bacteria are dividing. For this reason among others, a rather unusual combination of circumstances was needed to produce the effect that Fleming noticed on his Petri dish (Hare 1970, chapter 3). Was Fleming merely lucky? Pasteur's epigram is worth remembering: "Dans les champs de l'observation, le hasard ne favorise que les esprits préparés."

Almroth Wright, Fleming's mentor, was one of the founders of modern immunology (Dunnill 2001). Among other accomplishments, he developed a vaccine that prevented typhoid fever. Wright was a close friend of George Bernard Shaw's and was the basis for one of the characters in *The Doctor's Dilemma.*

Material on Gregg may be hard to find, but see Gregg (1941), Lancaster (1996), and Webster (1998). Gregg (1944) discusses infant deafness following maternal rubella.

On DES, the basic papers are Herbst and Scully (1970) and Herbst et al. (1971), with a useful summary by Colton and Greenberg (1982). Also see Freedman, Pisani, and Purves (2007, pp. 9–10). DES was an unnecessary tragedy. Doctors who prescribed DES were paying attention to observational studies that showed a positive effect in preventing miscarriage. However, clinical trials showed there was no such effect. DES was banned in 1971 for use in pregnant women.

Acknowledgments

David Collier, Thad Dunning, Paul Humphreys, Erich Lehmann, and Janet Macher made many helpful comments.

References and Further Reading

Aalen, O. O. (1978). Nonparametric inference for a family of counting processes. *Annals of Statistics* 6: 701–26.

Abbott, A. (1997). Of time and space: The contemporary relevance of the Chicago school. *Social Forces* 75: 1149–82.

Abbott, A. (1998). The causal devolution. *Sociological Methods and Research* 27: 148–81.

Achen, C. H. and Shively, W. P. (1995). *Cross-Level Inference*. Chicago, IL: University of Chicago Press.

Alderman, M. H., Madhavan, S., Cohen, H. et al. (1995). Low urinary sodium is associated with greater risk of myocardial infarction among treated hypertensive men. *Hypertension* 25: 1144–52.

Alderman, M. H., Madhavan, S., Ooi, W. L. et al. (1991). Association of the renin-sodium profile with the risk of myocardial infarction in patients with hypertension. *New England Journal of Medicine* 324: 1098–1104.

Altman, D. G. and de Stavola, B. L. (1994). Practical problems in fitting a proportional hazards model to data with updated measurements of the covariates. *Statistics in Medicine* 13: 301–41.

Altman, D. G., Schulz, K. F., Moher, D. et al. (2001). The revised CONSORT statement for reporting randomized trials: Explanation and elaboration. *Annals of Internal Medicine* 134: 663–94.

Amemiya, T. (1981). Qualitative response models: A survey. *Journal of Economic Literature* 19: 1483–1536.

Amemiya, T. (1985). *Advanced Econometrics*. Cambridge, MA: Harvard University Press.

American Medical Association (1987). Radioepidemiological Tables, Council on Scientific Affairs. *Journal of the American Medical Association* 257: 806–09.

Andersen, P. K. (1991). Survival analysis 1982–1991: The second decade of the proportional hazards regression model. *Statistics in Medicine* 10: 1931–41.

Andersen, P. K., Borgan, Ø., Gill, R. D., and Keiding, N. (1996). *Statistical Models Based on Counting Processes*. Corr. 4th printing. New York: Springer-Verlag.

Andersen, P. K. and Keiding, N., eds. (2006). *Survival and Event History Analysis*. Chichester, U.K.: Wiley.

Anderson, M., Ensher, J. R., Matthews, M. R., Wieman, C. E., and Cornell, E. A. (1995). Observation of Bose-Einstein condensation in a dilute atomic vapor. *Science* 269: 198–201.

Anderson, M. and Fienberg, S. E. (1999). *Who Counts? The Politics of Census-Taking in Contemporary America*. New York: Russell Sage Foundation.

Angrist, J. D. (2001). Estimation of limited dependent variable models with binary endogenous regressors: Simple strategies for empirical practice. *Journal of Business and Economic Statistics* 19: 2–16.

Angrist, J. D., Imbens, G. W., and Rubin, D. B. (1996). Identification of causal effects using instrumental variables. *Journal of the American Statistical Association* 91: 444–72.

Angrist, J. D. and Krueger, A. B. (2001). Instrumental variables and the search for identification: From supply and demand to natural experiments. *Journal of Economic Perspectives* 15: 69–85.

Appel, L. J., Moore, T. J., Obarzanek, E. et al. (1997). A clinical trial of the effects of dietary patterns on blood pressure. DASH Collaborative Research Group. *New England Journal of Medicine* 336: 1117–24.

Arceneaux, K., Gerber, A. S., and Green, D. P. (2006). Comparing experimental and matching methods using a large-scale voter mobilization experiment. *Political Analysis* 14: 37–62.

Archer, J. (2000). Sex differences in aggression between heterosexual partners: A meta-analytic review. *Psychological Bulletin* 126: 651–80.

Aris, E. M. D., Hagenaars, J. A. P., Croon, M., and Vermunt, J. K. (2000). The use of randomization for logit and logistic models. In J. Blasius, J. Hox, E. de Leuw, and P. Smidt, eds. *Proceedings of the Fifth International Conference on Social Science Methodology*. Cologne: TT Publications.

Arminger, G. and Bohrnstedt, G. W. (1987). Making it count even more: A review and critique of Stanley Lieberson's *Making It Count: The Improvement of Social Theory and Research*. In C. Clogg, ed. *Sociological Methodology 1987*. Washington, DC: American Sociological Association, pp. 198–201.

Bahry, D. and Silver, B. D. (1987). Intimidation and the symbolic uses of terror in the USSR. *American Political Science Review* 81: 1065–98.

Bailar, J. C. (1997). The promise and problems of meta-analysis. *New England Journal of Medicine* 337: 559–61.

Bailar, J. C. (1999). Passive smoking, coronary heart disease, and meta-analysis. *New England Journal of Medicine* 340: 958–59.

Bang, H. and Robins, J. M. (2005). Doubly robust estimation in missing data and causal inference models. *Biometrics* 61: 962–72.

Baron, J. (1838). *The Life of Edward Jenner*. vol. I. London: Henry Colburn.

Barrett-Connor, E. (1991). Postmenopausal estrogen and prevention bias. *Annals of Internal Medicine* 115: 455–56.

Bayes, T. (1764). An essay towards solving a problem in the doctrine of chances. *Philosophical Transactions of the Royal Society of London* 53: 370–418.

Bayles, K. W. (2000). The bactericidal action of penicillin: New clues to an unsolved mystery. *Trends in Microbiology* 8: 274–78.

Bazin, H. (2000). *The Eradication of Smallpox*. London: Academic Press.

Beck, N., Katz, J. N., Alvarez, R. M., Garrett, G., and Lange, P. (1993). Government partisanship, labor organization, and macroeconomic performance. *American Political Science Review* 87: 945–48.

Belsley, D. A., Kuh, E., and Welsch, R. E. (2004). *Regression Diagnostics: Identifying Influential Data and Sources of Collinearity*. New York: Wiley.

Berger, J. (1985). *Statistical Decision Theory and Bayesian Analysis*. 2nd edn. New York: Springer-Verlag.

Berger, J. and Wolpert, R. (1988). *The Likelihood Principle*. 2nd edn. Hayward, CA: Institute of Mathematical Statistics.

Berk, R. A. (1988). Causal inference for statistical data. In N. J. Smelser, ed. *Handbook of Sociology*. Beverly Hills: Sage Publications, pp. 155–72.

Berk, R. A. (1991). Toward a methodology for mere mortals. In P. V. Marsden, ed. *Sociological Methodology 1991*. Washington, DC: The American Sociological Association, pp. 315–24.

Berk, R. A. (2004). *Regression Analysis: A Constructive Critique*. Thousand Oaks, CA: Sage Publications.

Berk, R. A. and Campbell, A. (1993). Preliminary data on race and crack charging practices in Los Angeles. *Federal Sentencing Reporter* 6: 36–38.

Berk, R. A. and Freedman, D. A. (2003). Statistical assumptions as empirical commitments. In T. G. Blomberg and S. Cohen, eds. *Law, Punishment, and Social Control: Essays in Honor of Sheldon Messinger*. 2nd edn. New York: Aldine de Gruyter, pp. 235–54.

Berk, R. A. and Freedman, D. A. (2008). On weighting regressions by propensity scores. *Evaluation Review* 32: 392–409.

Berkson, J. (1944). Application of the logistic function to bio-assay. *Journal of the American Statistical Association* 39: 357–65.

Bernoulli, D. (1760). Essai d'une nouvelle analyse de la mortalité causée par la petite variole, et des avantages de l'inoculation pour la prévenir. *Mémoires de Mathématique et de Physique de l'Académie Royale des Sciences*, Paris, pp. 1–45. Reprinted in *Histoire de l'Académie Royale des Sciences* (1766).

Bhattacharya, J., Goldman, D., and McCaffrey, D. (2006). Estimating probit models with self-selected treatments. *Statistics in Medicine* 25: 389–413.

Bickel, P. J. and Doksum, K. A. (1977). *Mathematical Statistics: Basic Ideas and Selected Topics*. San Francisco, CA: Holden-Day.

Black, B. and Lilienfeld, D. E. (1984). Epidemiologic proof in toxic tort litigation. *Fordham Law Review* 52: 732–85.

Blau, P. M. and Duncan, O. D. (1967). *The American Occupational Structure*. New York: Wiley.

Bluthenthal, R. N., Ridgeway, G., Schell, T. et al. (2006). Examination of the association between Syringe Exchange Program (SEP) dispensation policy and SEP client-level syringe coverage among injection drug users. *Addiction* 102: 638–46.

Brady, H. E. and Collier, D., eds. (2004). *Rethinking Social Inquiry: Diverse Tools, Shared Standards*. Lanham, MD: Rowman & Littlefield.

Brady, H. E., Collier, D., and Seawright, J. (2004). Refocusing the discussion of methodology. In Brady and Collier (2004), pp. 3–20.

Brant, R. (1996). Digesting logistic regression results. *The American Statistician* 50: 117–19.

Briggs, D. C. (2004). Causal inference and the Heckman model. *Journal of Educational and Behavioral Statistics* 29: 397–420.

Bristowe, J. S. and Hutchinson, J. S. (1876). *A Treatise on the Theory and Practice of Medicine*. Philadelphia, PA: Henry C. Lea.

Bristowe, J. S., Wardell, J. R., Begbie, J. W. et al. (1879). *Diseases of the Intestines and Peritoneum*. New York: William Wood and Company.

Brown, L. D., Eaton, M. L., Freedman, D. A. et al. (1999). Statistical controversies in Census 2000. *Jurimetrics* 39: 347–75.

Buck, C., Llopis, A., Nájera, E., and Terris, M., eds. (1989). *The Challenge of Epidemiology: Issues and Selected Readings*. Geneva: World Health Organization, Scientific Publication No. 505.

Budd, W. (1873). *Typhoid Fever: Its Nature, Mode of Spreading, and Prevention*. London: Longmans, Green, and Co. Reprinted in 1977 by Ayer Publishing, Manchester, NH (http://www.deltaomega.org/typhoid.pdf).

Bulloch, W. (1938). *The History of Bacteriology*. Oxford: Oxford University Press.

Bushway, S., Johnson, B. D., and Slocum, L. A. (2007). Is the magic still there? The use of the Heckman two-step correction for selection bias in criminology. *Journal of Quantitative Criminology* 23: 151–78.

Carmelli, D. and Page, W. F. (1996). 24-year mortality in smoking-discordant World War II U.S. male veteran twins. *International Journal of Epidemiology* 25: 554–59.

Carpenter, K. J. (1981). *Pellagra*. Stroudsberg, PA: Hutchinson Ross.

Carpenter, K. J. (2000). *Beriberi, White Rice, and Vitamin B*. Berkeley, CA: University of California Press.

Cartwright, N. (1989). *Nature's Capacities and Their Measurement*. Oxford: Clarendon Press.

Casella, G. and Berger, R. L. (1990). *Statistical Inference*. Pacific Grove, CA: Wadsworth & Brooks/Cole.

Centers for Disease Control and Prevention (1999). Fluoridation of drinking water to prevent dental caries. *Morbidity and Mortality Weekly Report*, October 22, Vol. 48, No. 41, 933–40. U.S. Department of Health and Human Services.

Chattopadhyay, R. and Duflo, E. (2004). Women as policy makers: Evidence from a randomized policy experiment in India. *Econometrica* 72: 1409–43.

Cho, W. K. Tam (1998). Iff the assumption fits . . . : A comment on the King ecological inference solution. *Political Analysis* 7: 143–63.

Chobanian, A. V. and Hill, M. (2000). National Heart, Lung, and Blood Institute Workshop on Sodium and Blood Pressure: A critical review of current scientific evidence. *Hypertension* 35: 858–63. Quotes are from the online unabridged version (http://www.nhlbi.nih.gov/health/prof/heart/hbp/salt_sum.htm).

Cholera Inquiry Committee (1855). *Report on the Cholera Outbreak in the Parish of St. James, Westminster during the Autumn of 1854*. London: Churchill.

362 REFERENCES AND FURTHER READING

REFERENCES AND FURTHER READING

Chrystal, G. (1889). *Algebra: An Elementary Text Book for the Higher Classes of Secondary Schools and for Colleges*. Part II. Edinburgh: Adam and Charles Black.

Churchill, H. V. (1931). Occurrence of fluorides in some waters of the United States. *Journal of Industrial and Engineering Chemistry* 23: 996–98.

Citro, C. F., Cork, D. L., and Norwood, J. L., eds. (2004). *The 2000 Census: Counting Under Adversity*. Washington, DC: National Academy Press.

Clifford, P. (1977). Nonidentifiability in stochastic models of illness and death. *Proceedings of the National Academy of Sciences USA* 74: 1338–40.

Clogg, C. C. and Haritou, A. (1997). The regression method of causal inference and a dilemma confronting this method. In V. McKim and S. Turner, eds. *Causality in Crisis?* Notre Dame, IN: University of Notre Dame Press, pp. 83–112.

Cochran, W. G. (1957). Analysis of covariance: Its nature and uses. *Biometrics* 13: 261–81.

Cohen, J. (1988). *Statistical Power Analysis for the Behavioral Sciences*. 2nd edn. Hillsdale, NJ: Lawrence Erlbaum.

Cohen, M. L., White, A. A., and Rust, K. F., eds. (1999). *Measuring a Changing Nation: Modern Methods for the 2000 Census*. Washington, DC: National Academy Press.

Colton, T. and Greenberg, E. R. (1982). Epidemiologic evidence for adverse effects of DES exposure during pregnancy. *The American Statistician* 36: 268–72.

Cook, N. R., Cutler, J. A., Obarzanek, E. et al. (2007). Long term effects of dietary sodium reduction on cardiovascular disease outcomes: Observational followup of the trials of hypertension prevention. *British Medical Journal* 334: 885–92.

Cook, T. D. and Campbell, D. T. (1979). *Quasi-Experimentation: Design & Analysis Issues for Field Settings*. Boston, MA: Houghton Mifflin.

Copas, J. B. and Li, H. G. (1997). Inference for non-random samples. *Journal of the Royal Statistical Society*, Series B, 59: 55–77.

Cork, D. L., Cohen, M. L., and King, B. F., eds. (2004). *Reengineering the 2010 Census: Risks and Challenges*. Washington, DC: National Academy Press.

Cornfield, J., Haenszel, W., Hammond, E. C., Lilienfeld, A. M., Shimkin, M. B., and Wynder, E. L. (1959). Smoking and lung cancer: Recent evidence and a discussion of some questions. *Journal of the National Cancer Institute* 22: 173–203.

Cournot, A. A. (1843). *Exposition de la Théorie des Chances et des Probabilités*. Paris: Hachette. Reprinted in B. Bru, ed. (1984). *Œuvres Complètes de Cournot*. Paris: J. Vrin, vol. 1.

Cox, D. R. (1956). A note on weighted randomization. *Annals of Mathematical Statistics* 27: 1144–51.

Cox, D. R. (1972). Regression models and lifetables. *Journal of the Royal Statistical Society*, Series B, 34: 187–220 (with discussion).

Crump, R. K., Hotz, V. J., Imbens, G. W., and Mitnik, O. A. (2007). Dealing with overlap in estimation of average treatment effects. *Biometrika* 96: 187–99.

Cutler, J. A., Follmann, D., and Allender, P. S. (1997). Randomized trials of sodium reduction: An overview. *American Journal of Clinical Nutrition* 65, Supplement: S643–51.

Dabrowska, D. and Speed, T. P. (1990). On the application of probability theory to agricultural experiments. Essay on principles. (English translation of Neyman 1923.) *Statistical Science* 5: 463–80 (with discussion).

Darga, K. (2000). *Fixing the Census Until It Breaks*. Lansing, MI: Michigan Information Center.

Dawid, A. P. (2000). Causal inference without counterfactuals. *Journal of the American Statistical Association* 95: 407–48.

Dean, H. T. (1938). Endemic fluorosis and its relation to dental caries. *Public Health Reports* 53: 1443–52.

de Finetti, B. (1959). *La Probabilità, la Statistica, nei Rapporti con l'Induzione, Secondo Diversi Punti di Vista*. Rome: Centro Internazionale Matematica Estivo Cremonese. English translation in de Finetti (1972).

de Finetti, B. (1972). *Probability, Induction, and Statistics*. New York: Wiley.

de Moivre, A. (1697). A method of raising an infinite multinomial to any given power, or extracting any given root of the same. *Philosophical Transactions of the Royal Society of London* 19, no. 230: 619–25.

Diaconis, P. and Freedman, D. A. (1980a). De Finetti's generalizations of exchangeability. In R. C. Jeffrey, ed. *Studies in Inductive Logic and Probability*. Berkeley, CA: University of California Press, vol. 2, pp. 233–50.

Diaconis, P. and Freedman, D. A. (1980b). Finite exchangeable sequences. *Annals of Probability* 8: 745–64.

Diaconis, P. and Freedman, D. A. (1981). Partial exchangeability and sufficiency. In *Proceedings of the Indian Statistical Institute Golden Jubilee International Conference on Statistics: Applications and New Directions*. Calcutta: Indian Statistical Institute, pp. 205–36.

Diaconis, P. and Freedman, D. A. (1986). On the consistency of Bayes' estimates. *Annals of Statistics* 14: 1–87 (with discussion).

Diaconis, P. and Freedman, D. A. (1988). Conditional limit theorems for exponential families and finite versions of de Finetti's theorem. *Journal of Theoretical Probability* 1: 381–410.

Diaconis, P. and Freedman, D. A. (1990). Cauchy's equation and de Finetti's theorem. *Scandinavian Journal of Statistics* 17: 235–50.

Dietz, K. and Heesterbeek, J. A. P. (2002). Daniel Bernoulli's epidemiological model revisited. *Mathematical Biosciences* 180: 1–21.

Doss, H. and Sethuraman, J. (1989). The price of bias reduction when there is no unbiased estimate. *Annals of Statistics* 17: 440–42.

Duch, R. M. and Palmer, H. D. (2004). It's not whether you win or lose, but how you play the game. *American Political Science Review* 98: 437–52.

Ducharme, G. R. and Lepage, Y. (1986). Testing collapsibility in contingency tables. *Journal of the Royal Statistical Society*, Series B, 48: 197–205.

Duncan, O. D. (1984). *Notes on Social Measurement*. New York: Russell Sage.

Dunford, N. and Schwartz, J. T. (1958). *Linear Operators: Part I, General Theory*. New York: Wiley.

Dunnill, M. S. (2001). *The Plato of Praed Street: The Life and Times of Almroth Wright*. London: Royal Society of Medicine Press.

Dunning, T. and Freedman, D. A. (2007). Modeling selection effects. In S. Turner and W. Outhwaite, eds. *The Handbook of Social Science Methodology*. London: Sage Publications, pp. 225–31.

Dyer, A. R., Elliott, P., Marmot, M. et al. (1996). Commentary: Strength and importance of the relation of dietary salt to blood pressure. *British Medical Journal* 312: 1663–65.

Eaton, M. L. and Freedman, D. A. (2004). Dutch book against some 'objective' priors. *Bernoulli* 10: 861–72.

Eaton, M. L. and Sudderth, W. D. (1999). Consistency and strong inconsistency of group-invariant predictive inferences. *Bernoulli* 5: 833–54.

Ebrahim, S. and Davey-Smith, G. (1998). Lowering blood pressure: A systematic review of sustained effects of non-pharmacological interventions. *Journal of Public Health Medicine* 20: 441–48.

Efron, B. (1986). Why isn't everyone a Bayesian? *The American Statistician* 40: 1–11 (with discussion).

Ehrenberg, A. S. C. and Bound, J. A. (1993). Predictability and prediction. *Journal of the Royal Statistical Society*, Series A, 156, Part 2: 167–206 (with discussion).

Eliot, C. W., ed. (1910). *Scientific Papers: Physiology, Medicine, Surgery, Geology*. vol. 38 in *The Harvard Classics*. New York: P. F. Collier & Son; originally published in 1897.

Elliott, P., Stamler, J., Nichols, R. et al. (1996). Intersalt revisited: Further analyses of 24 hour sodium excretion and blood pressure within and across populations. *British Medical Journal* 312: 1249–53.

Ellsworth, W., Matthews, M., Nadeau, R. et al. (1998). A physically-based earthquake recurrence model for estimation of long-term earthquake probabilities. In *Proceedings of the Second Joint Meeting of the UJNR Panel on Earthquake Research*, pp. 135–49.

Engle, R. F., Hendry, D. F., and Richard, J. F. (1983). Exogeneity. *Econometrica* 51: 277–304.

Erikson, R. S., McIver, J. P., and Wright, Jr., G. C. (1987). State political culture and public opinion. *American Political Science Review* 81: 797–813.

Evans, R. J. (1987). *Death in Hamburg: Society and Politics in the Cholera Years*. Oxford: Oxford University Press.

Evans, S. N. and Stark, P. B. (2002). Inverse problems as statistics. *Inverse Problems* 18: R1–43.

Evans, W. N. and Schwab, R. M. (1995). Finishing high school and starting college: Do Catholic schools make a difference? *Quarterly Journal of Economics* 110: 941–74.

Eyler, J. M. (1979). *Victorian Social Medicine: The Ideas and Methods of William Farr*. Baltimore, MD: Johns Hopkins University Press.

Fahrmeir, L. and Kaufmann, H. (1985). Consistency and asymptotic normality of the maximum likelihood estimator in generalized linear models. *The Annals of Statistics* 13: 342–68.

Fearon, J. D. and Laitin, D. D. (2008). Integrating qualitative and quantitative methods. In J. M. Box-Steffensmeier, H. E. Brady, and D. Collier, eds. *The Oxford Handbook of Political Methodology*. Oxford: Oxford University Press, pp. 756–76.

Feller, W. (1968). *An Introduction to Probability Theory and Its Applications, Vol. I*, 3rd edn. New York: Wiley.

Fenner, F., Henderson, D. A., Arita, I., Jezek, Z., and Ladnyi, I. D. (1988). *Smallpox and its Eradication*. Geneva: World Health Organization (http://whqlibdoc.who.int/smallpox/9241561106.pdf).

Ferguson, T. (1967). *Mathematical Statistics: A Decision Theoretic Approach*. New York: Academic Press.

Finlay, B. B., Heffron, F., and Falkow, S. (1989). Epithelial cell surfaces induce *Salmonella* proteins required for bacterial adherence and invasion. *Science* 243: 940–43.

Fisher, F. M. (1980). Multiple regression in legal proceedings. *Columbia Law Review* 80: 702–36.

Fisher, R. A. (1958). *Statistical Models for Research Workers*. 13th edn. Edinburgh: Oliver and Boyd.

Fleming, A. (1929). On the antibacterial action of cultures of a penicillium, with special reference to their use in the isolation of *B. influenzae*. *British Journal of Experimental Pathology* 10: 226–36.

Fleming, T. R. and Harrington, D. P. (2005). *Counting Processes and Survival Analysis*. 2nd rev. edn. New York: John Wiley & Sons.

Fox, J. (1991). *Regression Diagnostics: An Introduction*. Thousand Oaks, CA: Sage Publications.

Fraker, T. and Maynard, R. (1987). The adequacy of comparison group designs for evaluations of employment-related programs. *Journal of Human Resources* 22: 194–227.

Francesconi, M. and Nicoletti, C. (2006). Intergenerational mobility and sample selection in short panels. *Journal of Applied Econometrics* 21: 1265–93.

Freedman, D. A. (1971). *Markov Chains*. San Francisco, CA: Holden-Day. Reprinted in 1983 by Springer-Verlag, New York.

Freedman, D. A. (1983). A note on screening regression equations. *The American Statistician* 37: 152–55.

Freedman, D. A. (1985). Statistics and the scientific method. In W. M. Mason and S. E. Fienberg, eds. *Cohort Analysis in Social Research: Beyond the Identification Problem*. New York: Springer-Verlag, pp. 343–90 (with discussion).

Freedman, D. A. (1987). As others see us: A case study in path analysis. *Journal of Educational Statistics* 12: 101–28 (with discussion).

Freedman, D. A. (1991). Statistical models and shoe leather. In P. V. Marsden, ed. *Sociological Methodology* 21: 291–313 (with discussion). Washington, DC: American Sociological Association.

Freedman, D. A. (1995). Some issues in the foundation of statistics. *Foundations of Science* 1: 19–83 (with discussion). Reprinted in B. C. van Fraassen, ed. *Some Issues in the Foundation of Statistics*. Dordrecht, The Netherlands: Kluwer, pp. 19–83.

Freedman, D. A. (1997). From association to causation via regression. In V. McKim and S. Turner, eds. *Causality in Crisis: Statistical Methods and the Search for Causal Knowledge in the Social Sciences*. Notre Dame, IN: University of Notre Dame Press, pp. 113–61.

Freedman, D. A. (1999). From association to causation: Some remarks on the history of statistics. *Statistical Science* 14: 243–58. Reprinted in *Journal de la Société Française de Statistique* (1999) 140: 5–32; and in J. Panaretos, ed. (2003) *Stochastic Musings: Perspectives from the Pioneers of the Late 20th Century*. Mahwah, NJ: Lawrence Erlbaum Associates, pp. 45–71.

Freedman, D. A. (2003). Structural equation models: A critical review. Technical Report No. 651, Department of Statistics, University of California, Berkeley.

Freedman, D. A. (2004). On specifying graphical models for causation and the identification problem. *Evaluation Review* 26: 267–93. Reprinted in D. W. K. Andrews and J. H. Stock, eds. (2005). *Identification and Inference for Econometric Models: Essays in Honor of Thomas Rothenberg*. Cambridge: Cambridge University Press, pp. 56–79.

Freedman, D. A. (2006a). On the so-called "Huber Sandwich Estimator" and "robust standard errors." *The American Statistician* 60: 299–302.

Freedman, D. A. (2006b). Statistical models for causation: What inferential leverage do they provide? *Evaluation Review* 30: 691–713.

Freedman, D. A. (2007). How can the score test be inconsistent? *The American Statistician* 61: 291–95.

Freedman, D. A. (2008a). On regression adjustments to experimental data. *Advances in Applied Mathematics* 40: 180–93.

Freedman, D. A. (2008b). On regression adjustments in experiments with several treatments. *Annals of Applied Statistics* 2: 176–96.

Freedman, D. A. (2008c). Randomization does not justify logistic regression. *Statistical Science* 23: 237–50.

Freedman, D. A. (2008d). Survival analysis. *The American Statistician* 62: 110–19.

Freedman, D. A. (2008e). Diagnostics cannot have much power against general alternatives. To appear in *International Journal of Forecasting* (http://www.stat.berkeley.edu/users/census/nopower.pdf).

Freedman, D. A. (2008f). Some general theory for weighted regressions. (http://www.stat.berkeley.edu/users/census/wtheory.pdf).

Freedman, D. A. (2008g). On types of scientific enquiry: The role of qualitative reasoning. In J. M. Box-Steffensmeier, H. E. Brady, and D. Collier, eds.

The Oxford Handbook of Political Methodology. Oxford: Oxford University Press, pp. 300–18.

Freedman, D. A. (2009). *Statistical Models: Theory and Practice.* Revised edn. New York: Cambridge University Press.

Freedman, D. A. and Humphreys, P. (1999). Are there algorithms that discover causal structure? *Synthese* 121: 29–54.

Freedman, D. A., Klein, S. P., Ostland, M., and Roberts, M. R. (1998). Review of *A Solution to the Ecological Inference Problem. Journal of the American Statistical Association* 93: 1518–22.

Freedman, D. A., Klein, S. P., Sacks, J., Smyth, C. A., and Everett, C. G. (1991). Ecological regression and voting rights. *Evaluation Review* 15: 659–817 (with discussion).

Freedman, D. A. and Lane, D. (1983). A nonstochastic interpretation of reported significance levels. *Journal of Business and Economic Statistics* 1: 292–98.

Freedman, D. A. and Navidi, W. (1989). Multistage models for carcinogenesis. *Environmental Health Perspectives* 81: 169–88.

Freedman, D. A., Ostland, M., Klein, S. P., and Roberts, M. R. (1999). Response to King's comments. *Journal of the American Statistical Association* 94: 352–57.

Freedman, D. A. and Petitti, D. B. (2001). Salt and blood pressure: Conventional wisdom reconsidered. *Evaluation Review* 25: 267–87.

Freedman, D. A., Petitti, D. B., and Robins, J. M. (2004). On the efficacy of screening for breast cancer. *International Journal of Epidemiology* 33: 43–73 (with discussion). Correspondence, 1404–06.

Freedman, D. A., Pisani, R., and Purves, R. A. (2007). *Statistics.* 4th edn. New York: Norton.

Freedman, D. A. and Purves, R. A. (1969). Bayes method for bookies. *Annals of Mathematical Statistics* 40: 1177–86.

Freedman, D. A., Rothenberg, T., and Sutch, R. (1983). On energy policy models. *Journal of Business and Economic Statistics* 1: 24–36 (with discussion).

Freedman, D. A. and Stark, P. B. (1999). The swine flu vaccine and Guillain-Barré syndrome: A case study in relative risk and specific causation. *Evaluation Review* 23: 619–47.

Freedman, D. A. and Stark, P. B. (2001). The swine flu vaccine and Guillain-Barré syndrome: A case study in relative risk and specific causation. *Law and Contemporary Problems* 64: 619–47.

Freedman, D. A. and Stark, P. B. (2003). What is the chance of an earthquake? In F. Mulargia and R. J. Geller, eds. *Earthquake Science and Seismic Risk Reduction*. NATO Science Series IV: Earth and Environmental Sciences, vol. 32. Dordrecht, the Netherlands: Kluwer, pp. 201–16.

Freedman, D. A., Stark, P. B., and Wachter, K. W. (2001). A probability model for census adjustment. *Mathematical Population Studies* 9: 165–80.

Freedman, D. A. and Wachter, K. W. (1994). Heterogeneity and census adjustment for the intercensal base. *Statistical Science* 9: 458–537 (with discussion).

Freedman, D. A. and Wachter, K. W. (2003). On the likelihood of improving the accuracy of the census through statistical adjustment. In D. R. Goldstein, ed. *Science and Statistics: A Festscrift for Terry Speed*. IMS Monograph 40, pp. 197–230.

Freedman, D. A. and Wachter, K. W. (2007). Methods for Census 2000 and statistical adjustments. In S. Turner and W. Outhwaite, eds. *Handbook of Social Science Methodology*. Thousand Oaks, CA: Sage Publications, pp. 232–45.

Freedman, D. A., Wachter, K. W., Coster, D. C., Cutler, R. C., and Klein, S. P. (1993). Adjusting the Census of 1990: The smoothing model. *Evaluation Review* 17: 371–443.

Freedman, D. A., Wachter, K. W., Cutler, R. C., and Klein, S. P. (1994). Adjusting the Census of 1990: Loss functions. *Evaluation Review* 18: 243–80.

Fremantle, N., Calvert, M., Wood, J. et al. (2003). Composite outcomes in randomized trials: Greater precision but with greater uncertainty? *Journal of the American Medical Association* 289: 2554–59.

Frey, B. S. and Meier, S. (2004). Social comparisons and pro-social behavior: Testing "conditional cooperation" in a field experiment. *American Economic Review* 94: 1717–22.

Gail, M. H. (1986). Adjusting for covariates that have the same distribution in exposed and unexposed cohorts. In S. H. Moolgavkar and R. L. Prentice, eds. *Modern Statistical Methods in Chronic Disease Epidemiology*. New York: Wiley, pp. 3–18.

Gail, M. H. (1988). The effect of pooling across strata in perfectly balanced studies. *Biometrics* 44: 151–62.

Gani, J. (1978). Some problems of epidemic theory. *Journal of the Royal Statistical Society*, Series A, 141: 323–47 (with discussion).

Gartner, S. S. and Segura, G. M. (2000). Race, casualties, and opinion in the Vietnam war. *Journal of Politics* 62: 115–46.

Geiger, D., Verma, T., and Pearl, J. (1990). Identifying independence in Bayesian networks. *Networks* 20: 507–34.

Geller, N. L., Sorlie, P., Coady, S. et al. (2004). Limited access data sets from studies funded by the National Heart, Lung, and Blood Institute. *Clinical Trials* 1: 517–24.

Gertler, P. (2004). Do conditional cash transfers improve child health? Evidence from PROGRESA's control randomized experiment. *American Economic Review* 94: 336–41.

Gibson, J. L. (1988). Political intolerance and political repression during the McCarthy Red Scare. *American Political Science Review* 82: 511–29.

Giesbrecht, P., Kersten, T., Maidhof, H., and Wecke, J. (1998). Staphylococcal cell wall: Morphogenesis and fatal variations in the presence of penicillin. *Microbiology and Molecular Biology Reviews* 62: 1371–1414.

Gigerenzer, G. (1996). On narrow norms and vague heuristics. *Psychological Review* 103: 592–96.

Gilens, M. (2001). Political ignorance and collective policy preferences. *American Political Science Review* 95: 379–96.

Gill, R. D. and Robins, J. M. (2004). Causal inference for complex longitudinal data: The continuous case. *Annals of Statistics* 29: 1785–1811.

Glazerman, S., Levy, D. M., and Myers, D. (2003). Nonexperimental versus experimental estimates of earnings impacts. *Annals of the American Academy of Political and Social Science* 589: 63–93.

Godlee, F. (1996). The food industry fights for salt. *British Medical Journal* 312: 1239–40.

Goertz, G. (2008). Choosing cases for case studies: A qualitative logic. *Qualitative and Multi-Method Research* 6: 11–14.

Goldberger, J. (1914). The etiology of pellagra. *Public Health Reports* 29: 1683–86. Reprinted in Buck et al. (1989), pp. 99–102; and in Terris (1964), pp. 19–22.

Goldberger, J., Wheeler, G. A., Lillie, R. D., and Rogers, L. M. (1926). A further study of butter, fresh beef, and yeast as pellagra preventives, with consideration of the relation of factor P-P of pellagra (and black tongue of dogs) to vitamin B_1. *Public Health Reports* 41: 297–318. Reprinted in Terris (1964), pp. 351–70.

Goldsmith, M. (1946). *The Road to Penicillin.* London: Lindsay Drummond.

Goldthorpe, J. H. (1999). *Causation, Statistics and Sociology.* Twenty-ninth Geary Lecture, Nuffield College, Oxford. Published by the Economic and Social Research Institute, Dublin, Ireland.

Goldthorpe, J. H. (2001). Causation, statistics, and sociology. *European Sociological Review* 17: 1–20.

Good, I. J. (1967). The white shoe is a red herring. *The British Journal for the Philosophy of Science* 17: 322.

Good, I. J. (1968). The white shoe *qua* herring is pink. *The British Journal for the Philosophy of Science* 19: 156–57.

Goodman, L. (1953). Ecological regression and the behavior of individuals. *American Sociological Review* 18: 663–64.

Goodman, L. (1959). Some alternatives to ecological correlation. *American Journal of Sociology* 64: 610–25.

Gordis, L. (2008). *Epidemiology*. 4th edn. Philadelphia, PA: Saunders.

Gould, E. D., Lavy, V., and Passerman, M. D. (2004). Immigrating to opportunity: Estimating the effect of school quality using a natural experiment on Ethiopians in Israel. *Quarterly Journal of Economics* 119: 489–526.

Gozalo, P. L. and Miller, S. C. (2007). Predictors of mortality: Hospice enrollment and evaluation of its causal effect on hospitalization of dying nursing home patients. *Health Services Research* 42: 587–610.

Grace, N. D., Muench, H., and Chalmers, T. C. (1966). The present status of shunts for portal hypertension in cirrhosis. *Gastroenterology* 50: 684–91.

Graudal, N. A., Galløe, A. M., and Garred, P. (1998). Effects of sodium restriction on blood pressure, renin, aldosterone, catecholamines, cholesterols, and triglyceride: A meta-analysis. *Journal of the American Medical Association* 279: 1383–91.

Graunt, J. (1662). *Natural and Political Observations Mentioned in a Following Index, and Made upon the Bills of Mortality*. London. Printed by Tho. Roycroft, for John Martin, James Allestry, and Tho. Dicas, at the Sign of the Bell in St. Paul's Church-yard, MDCLXII. Reprinted in 2006 by Ayer Company Publishers, Manchester, NH (http://www.edstephan.org/Graunt/bills.html).

Green, M., Freedman, D. M., and Gordis, L. (2000). Reference guide on epidemiology. In *Reference Manual on Scientific Evidence*. 2nd edn., Washington, DC: Federal Judicial Center, §VII, pp. 333–400.

Greene, W. H. (2007). *Econometric Analysis*. 6th edn. Prentice Hall, NJ: Upper Saddle River.

Greenland, S., Pearl, J., and Robins, J. (1999). Causal diagrams for epidemiologic research. *Epidemiology* 10: 37–48.

Gregg, N. M. (1941). Congenital cataract following German measles in the mother. *Transactions of the Ophthalmological Society of Australia* 3: 35–46. Reprinted in Buck et al. (1989), pp. 426–34.

Gregg, N. M. (1944). Further observations on congenital defects in infants following maternal rubella. *Transactions of the Ophthalmological Society of Australia* 4: 119–31.

Grodstein, F., Stampfer, M. J., Colditz, G. A. et al. (1997). Postmenopausal hormone therapy and mortality. *New England Journal of Medicine* 336: 1769–75.

Grodstein, F., Stampfer, M. J., Manson, J. et al. (1996). Postmenopausal estrogen and progestin use and the risk of cardiovascular disease. *New England Journal of Medicine* 335: 453–61.

Grofman, B. (1991). Statistics without substance. *Evaluation Review* 15: 746–69.

Gross, S. R. and Mauro, R. (1989). *Death and Discrimination*. Boston, MA: Northeastern University Press.

Guo, G. H. and Geng, Z. (1995). Collapsibility of logistic regression coefficients. *Journal of the Royal Statistical Society*, Series B, 57: 263–67.

Hald, A. (2005). *A History of Probability and Statistics and Their Applications before 1750*. New York: Wiley.

Halley, E. (1693). An estimate of the mortality of mankind, drawn from curious tables of the births and funerals at the city of Breslaw; with an attempt to ascertain the price of annuities upon lives. *Philosophical Transactions of the Royal Society of London* 196: 596–610, 654–56.

Hanneman, R. L. (1996). Intersalt: Hypertension rise with age revisited. *British Medical Journal* 312: 1283–84.

Hare, R. (1970). *The Birth of Penicillin and the Disarming of Microbes*. London: Allen & Unwin.

Harsha, D. W., Lin, P. H., Obarzanek, E. et al. (1999). Dietary approaches to stop hypertension: A summary of study results. *Journal of the American Dietetic Association* 99, Supplement: 35–39.

Hart, H. L. A. and Honoré, A. M. (1985). *Causation in the Law*. 2nd edn. Oxford: Oxford University Press.

Hartigan, J. (1983). *Bayes Theory*. New York: Springer-Verlag.

He, J., Ogden, L. G., Vupputuri, S. et al. (1999). Dietary sodium intake and subsequent risk of cardiovascular disease in overweight adults. *Journal of the American Medical Association* 282: 2027–34.

Heckman, J. J. (1978). Dummy endogenous variables in a simultaneous equation system. *Econometrica* 46: 931–59.

Heckman, J. J. (1979). Sample selection bias as a specification error. *Econometrica* 47: 153–61.

Heckman, J. J. (2000). Causal parameters and policy analysis in economics: A twentieth century retrospective. *The Quarterly Journal of Economics* 115: 45–97.

Heckman, J. J. (2001a). Micro data, heterogeneity, and the evaluation of public policy: Nobel lecture. *Journal of Political Economy* 109: 673–748.

Heckman, J. J. (2001b). Econometrics and empirical economics. *Journal of Econometrics* 100: 3–5.

Heckman, J. J. and Hotz, V. J. (1989). Choosing among alternative nonexperimental methods for estimating the impact of social programs: The case of manpower training. *Journal of the American Statistical Association* 84: 862–80.

Hedges, L. V. and Olkin, I. (1985). *Statistical Methods for Meta-Analysis.* New York: Academic Press.

Hedström, P. and Swedberg, R., eds. (1998). *Social Mechanisms.* Cambridge: Cambridge University Press.

Heiss, G., Wallace, R., Anderson, G. L. et al. (2008). Health risks and benefits 3 years after stopping randomized treatment with estrogen and progestin. *Journal of the American Medical Association* 299: 1036–45.

Hempel, C. G. (1945). Studies in the logic of confirmation. *Mind* 54: 1–26, 97–121.

Hempel, C. G. (1967). The white shoe: No red herring. *The British Journal for the Philosophy of Science* 18: 239–40.

Hendry, D. F. (1980). Econometrics—alchemy or science? *Economica* 47: 387–406. Reprinted in D. F. Hendry (2000), *Econometrics—Alchemy or Science?* Oxford: Blackwell, chapter 1. Page cites are to the journal article.

Henschke, C. I., Yankelevitz, D. F., Libby, D. M. et al. (2006). The International Early Lung Cancer Action Program Investigators. Survival of patients with Stage I lung cancer detected on CT screening. *New England Journal of Medicine* 355: 1763–71.

Herbst, A. L. and Scully, R. E. (1970). Adenocarcinoma of the vagina in adolescence: A report of 7 cases including 6 clear cell carcinomas. *Cancer* 25: 745–57.

Herbst, A. L., Ulfelder, H., and Poskanzer, D. C. (1971). Adenocarcinoma of the vagina: Association of maternal stilbestrol therapy with tumor appearance in young women. *New England Journal of Medicine* 284: 878–81. Reprinted in Buck et al. (1989), pp. 446–50.

Hernán, M. A., Brumback, B., and Robins, J. M. (2001). Marginal structural models to estimate the joint causal effects of nonrandomized treatments. *Journal of the American Statistical Association* 96: 440–48.

Hill, A. B. (1961). *Principles of Medical Statistics.* 7th edn. London: The Lancet.

Hirano, K. and Imbens, G. W. (2001). Estimation of causal effects using propensity score weighting: An application to data on right heart catheterization. *Health Services and Outcomes Research Methodology* 2: 259–78.

Hodges, J. L., Jr. and Lehmann, E. (1964). *Basic Concepts of Probability and Statistics*. San Francisco, CA: Holden-Day.

Hoeffding, H. (1963). Probability inequalities for sums of bounded random variables. *Journal of the American Statistical Association* 58: 13–30.

Hoeffding, W. (1951). A combinatorial central limit theorem. *Annals of Mathematical Statistics* 22: 558–66.

Höglund, T. (1978). Sampling from a finite population: A remainder term estimate. *Scandinavian Journal of Statistics* 5: 69–71.

Holland, P. W. (1986). Statistics and causal inference. *Journal of the American Statistical Association* 8: 945–70 (with discussion).

Holland, P. W. (1988). Causal inference, path analysis, and recursive structural equation models. In C. Clogg, ed. *Sociological Methodology 1988*. Washington, DC: American Sociological Association, pp. 449–93.

Howard-Jones, N. (1975). *The Scientific Background of the International Sanitary Conferences 1851–1938*. Geneva: World Health Organization.

Hu, W.-Y. (2003). Marriage and economic incentives: Evidence from a welfare experiment. *Journal of Human Resources* 38: 942–63.

Huber, P. J. (1967). The behavior of maximum likelihood estimates under nonstandard conditions. *Proceedings of the Fifth Berkeley Symposium on Mathematical Statistics and Probability*, vol. I, pp. 221–33.

Humphreys, N. A., ed. (1885). *Vital Statistics: A Memorial Volume of Selections from the Reports and Writings of William Farr*. London: Edward Stanford.

Humphreys, P. (1997). A critical appraisal of causal discovery algorithms. In V. McKim and S. Turner, eds. *Causality in Crisis: Statistical Methods and the Search for Causal Knowledge in the Social Sciences*. Notre Dame, IN: University of Notre Dame Press, pp. 249–63.

Humphreys, P. and Freedman, D. A. (1996). The grand leap. *British Journal for the Philosophy of Science* 47: 113–23.

Hurwitz, E. S., Schonberger, L. B., Nelson, D. B., and Holman, R. C. (1981). Guillain-Barré syndrome and the 1978–1979 influenza vaccine. *New England Journal of Medicine* 304: 1557–61.

International Agency for Research on Cancer (1986). *Tobacco Smoking*. Monographs on the Evaluation of the Carcinogenic Risk of Chemicals to Humans, Vol. 38. IARC, Lyon, France.

Intersalt Cooperative Research Group (1986). Intersalt study. An international co-operative study on the relation of blood pressure to electrolyte excretion in populations. Design and methods. *Journal of Hypertension* 4: 781–87.

Intersalt Cooperative Research Group (1988). Intersalt: An international study of electrolyte excretion and blood pressure. Results for 24 hour urinary sodium and potassium excretion. *British Medical Journal* 297: 319–28.

Iyengar, S. I. and Greenhouse, J. B. (1988). Selection models and the file drawer problem. *Statistical Science* 3: 109–17.

Jacobs, D. and Carmichael, J. T. (2002). The political sociology of the death penalty. *American Sociological Review* 67: 109–31.

Janssen, A. (2000). Global power functions of goodness-of-fit tests. *Annals of Statistics* 28: 239–53.

Jeffrey, R. C. (1983). *The Logic of Decision.* 2nd edn. Chicago, IL: University of Chicago Press.

Jenner, E. (1798). *An Inquiry into the Causes and Effects of the Variolae Vaccinae, a Disease Discovered in Some of the Western Counties of England, Particularly Gloucestershire, and Known by the Name of the Cow Pox.* London: printed for the author by Sampson Low. Reprinted in Eliot (1910), pp. 151–80.

Jenner, E. (1801). *The Origin of the Vaccine Inoculation.* London: D. N. Shury. Reprinted in Fenner et al. (1988), pp. 258–61.

Jewell, N. P. (2003). *Statistics for Epidemiology.* Boca Raton, FL: Chapman & Hall/CRC.

Johnson, M. T., Drew, C. E., and Miletich, D. P. (1998). *Use of Expert Testimony, Specialized Decision Makers, and Case-Management Innovations in the National Vaccine Injury Compensation Program.* Washington, DC: Federal Judicial Center.

Johnston, J. (1984). *Econometric Methods.* New York: McGraw-Hill.

Kahneman, D., Slovic, P., and Tversky, A., eds. (1982). *Judgment Under Uncertainty: Heuristics and Biases.* Cambridge: Cambridge University Press.

Kahneman, D. and Tversky, A. (1974). Judgment under uncertainty: Heuristics and biases. *Science* 185: 1124–31.

Kahneman, D. and Tversky, A. (1996). On the reality of cognitive illusions. *Psychological Review* 103: 582–91.

Kahneman, D. and Tversky, A., eds. (2000). *Choices, Values, and Frames.* Cambridge: Cambridge University Press.

Kalbfleisch, J. D. and Prentice, R. L. (1973). Marginal likelihoods based on Cox's regression and life model. *Biometrika* 60: 267–78.

Kalbfleisch, J. D. and Prentice, R. L. (2002). *The Statistical Analysis of Failure Time Data*. 2nd edn. New York: Wiley.

Kanarek, M. S., Conforti, P. M., Jackson, L. A., Cooper, R. C., and Murchio, J. C. (1980). Asbestos in drinking water and cancer incidence in the San Francisco Bay Area. *American Journal of Epidemiology* 112: 54–72.

Kang, J. D. Y. and Schafer, J. L. (2007). Demystifying double robustness: A comparison of alternative strategies for estimating a population mean from incomplete data. *Statistical Science* 22: 523–39.

Kaplan, E. L. and Meier, P. (1958). Nonparametric estimation from incomplete observations. *Journal of the American Statistical Association* 53: 457–81.

Kaplan, J. E., Katona, P., Hurwitz, E. S., and Schonberger, L. B. (1982). Guillain-Barré syndrome in the United States, 1979–1980 and 1980–1981: Lack of an association with influenza vaccination. *Journal of the American Medical Association* 248: 698–700.

Kaprio, J. and Koskenvuo, M. (1989). Twins, smoking and mortality: A 12-year prospective study of smoking-discordant twin pairs. *Social Science and Medicine* 29: 1083–89.

Kaye, D. H. and Freedman, D. A. (2000). Reference guide on statistics. In *Reference Manual on Scientific Evidence*. 2nd edn. Washington, DC: Federal Judicial Center, pp. 83–178.

Kempthorne, O. (1952). *The Design and Analysis of Experiments*. New York: Wiley.

Kennedy, P. (2003). *A Guide to Econometrics*. 5th edn. Cambridge, MA: MIT Press.

Keynes, J. M. (1939). Professor Tinbergen's method. *The Economic Journal* 49: 558–70.

Keynes, J. M. (1940). Comment on Tinbergen's response. *The Economic Journal* 50: 154–56.

Kiiveri, H. and Speed, T. (1982). Structural analysis of multivariate data: A review. In S. Leinhardt, ed. *Sociological Methodology 1982*. San Francisco, CA: Jossey Bass, pp. 209–89.

King, G. (1997). *A Solution to the Ecological Inference Problem*. Princeton, NJ: Princeton University Press.

King, G. (1999). A reply to Freedman et al. *Journal of the American Statistical Association* 94: 352–55.

Kirk, D. (1996). Demographic transition theory. *Population Studies* 50: 361–87.

Klein, S. P. and Freedman, D. A. (1993). Ecological regression in voting rights cases. *Chance* 6: 38–43.

Klein, S. P., Sacks, J., and Freedman, D. A. (1991). Ecological regression *versus* the secret ballot. *Jurimetrics* 31: 393–413.

Koch, C. G. and Gillings, D. B. (2005). Inference, design-based vs. model-based. In S. Kotz, C. B. Read, N. Balakrishnan, and B. Vidakovic, eds. *Encyclopedia of Statistical Sciences*. 2nd edn. Hoboken, NJ: Wiley.

Koenker, R. (2005). Maximum likelihood asymptotics under nonstandard conditions: A heuristic introduction to sandwiches (http://www.econ.uiuc.edu/∼roger/courses/476/lectures/L10.pdf).

Kolata, G. (1999). *Flu*. New York: Farrar, Straus & Giroux.

Kolmogorov, A. N. (1956 [1933]). *Foundations of the Theory of Probability*. 2nd edn. New York: Chelsea. Originally published as *Grundbegriffe der Wahrscheinlichkeitstheorie. Ergebnisse Mathematische* 2 no. 3.

Kreps, D. (1988). *Notes on the Theory of Choice*. Boulder, CO: Westview Press.

Kruskal, W. (1988). Miracles and statistics, the casual assumption of independence. *Journal of the American Statistical Association* 83: 929–40.

Kumanyika, S. K. and Cutler, J. A. (1997). Dietary sodium reduction: Is there cause for concern? *Journal of the American College of Nutrition* 16: 192–203.

LaLonde, R. J. (1986). Evaluating the econometric evaluations of training programs with experimental data. *American Economic Review* 76: 604–20.

Lancaster, P. A. L. (1996). Gregg, Sir Norman McAlister (1892–1966), Ophthalmologist. In J. Ritchie, ed. *Australian Dictionary of Biography*. Melbourne: Melbourne University Press, vol. 14, pp. 325–27 (http://www.adb.online.anu.edu.au/biogs/A140370b.htm).

Lane, P. W. and Nelder, J. A. (1982). Analysis of covariance and standardization as instances of prediction. *Biometrics* 38: 613–21.

Langmuir, A. D. (1979). Guillain-Barré syndrome: The swine influenza virus vaccine incident in the United States of America, 1976–77: Preliminary communication. *Journal of the Royal Society of Medicine* 72: 660–69.

Langmuir, A. D., Bregman, D. J., Kurland, L. T., Nathanson, N., and Victor, M. (1984). An epidemiologic and clinical evaluation of Guillain-Barré syndrome reported in association with the administration of swine influenza vaccines. *American Journal of Epidemiology* 119: 841–79.

Laplace, P. S. (1774). Mémoire sur la probabilité des causes par les événements. *Mémoires de Mathématique et de Physique Présentés à l'Académie Royale des Sciences, par Divers Savants, et Lus dans ses Assemblées* 6.

Reprinted in Laplace's *Œuvres Complètes* 8: 27–65. English translation by S. Stigler (1986). *Statistical Science* 1: 359–78.

Lassen, D. D. (2005). The effect of information on voter turnout: Evidence from a natural experiment. *American Journal of Political Science* 49: 103–18.

Lauritzen, S. L. (1996). *Graphical Models*. Oxford: Oxford University Press.

Lauritzen, S. L. (2001). Causal inference in graphical models. In O. E. Barndorff-Nielsen, D. R. Cox, and C. Klüppelberg, eds. *Complex Stochastic Systems*. Boca Raton, FL: Chapman & Hall/CRC, pp. 63–108.

Law, M. (1996). Commentary: Evidence on salt is consistent. *British Medical Journal* 312: 1284–85.

Lawless, J. F. (2003). *Statistical Models and Methods for Lifetime Data*. 2nd edn. New York: Wiley.

Leamer, E. (1978). *Specification Searches*. New York: Wiley.

Le Cam, L. M. (1977). A note on metastatistics or "An essay toward stating a problem in the doctrine of chances." *Synthese* 36: 133–60.

Le Cam, L. M. (1986). *Asymptotic Methods in Statistical Decision Theory*. New York: Springer-Verlag.

Le Cam, L. M. and Yang, G. L. (1990). *Asymptotics in Statistics: Some Basic Concepts*. New York: Springer-Verlag.

Lee, E. T. and Wang, J. W. (2003). *Statistical Methods for Survival Data Analysis*. 3rd edn. New York: Wiley.

Lee, L. F. (1981). Simultaneous equation models with discrete and censored dependent variables. In C. Manski and D. McFadden, eds. *Structural Analysis of Discrete Data with Economic Applications*. Cambridge, MA: MIT Press, pp. 346–64.

Legendre, A. M. (1805). *Nouvelles Méthodes pour la Détermination des Orbites des Comètes*. Paris: Courcier. Reprinted in 1959 by Dover, New York.

Lehmann, E. L. (1986). *Testing Statistical Hypotheses*. 2nd edn. New York: Wiley.

Lehmann, E. L. (1998). *Elements of Large-Sample Theory*. New York: Springer-Verlag.

Lehmann, E. L. and Casella, G. (2003). *Theory of Point Estimation*. 2nd edn. New York: Springer-Verlag.

Lehmann, E. and Romano, J. (2005). *Testing Statistical Hypotheses*. 3rd edn. New York: Springer-Verlag.

Lehrer, E. (2001). Any inspection rule is manipulable. *Econometrica* 69: 1333–47.

Leslie, S. and Theibaud, P. (2007). Using propensity scores to adjust for treatment selection bias. *SAS Global Forum 2007: Statistics and Data Analysis*, paper 184-2007.

Lichtman, A. (1991). Passing the test. *Evaluation Review* 15: 770–99.

Lieberson, S. (1985). *Making It Count: The Improvement of Social Theory and Research*. Berkeley, CA: University of California Press.

Lieberson, S. (1988). Asking too much, expecting too little. *Sociological Perspectives* 31: 379–97.

Lieberson, S. and Lynn, F. B. (2002). Barking up the wrong branch: Alternatives to the current model of sociological science. *Annual Review of Sociology* 28: 1–19.

Lieberson, S. and Waters, M. (1988). *From Many Strands: Ethnic and Racial Groups in Contemporary America*. New York: Russell Sage Foundation.

Lim, W. (1999). Estimating impacts on binary outcomes under random assignment. Technical report, MDRC, New York.

Lipsey, M. W. (1992). Juvenile delinquency treatment: A meta-analysis inquiry into the variability of effects. In T. C. Cook, D. S. Cooper, H. Hartmann et al., eds. *Meta-Analysis for Explanation*. New York: Russell Sage, pp. 83–127.

Lipsey, M. W. (1997). What can you build with thousands of bricks? Musings on the cumulation of knowledge in program evaluation. *New Directions for Evaluation* 76: 7–24.

Lipsey, M. W. and Wilson, D. (2001). *Practical Meta-Analysis*. Newbury Park, CA: Sage Publications.

Littlewood, J. (1953). *A Mathematician's Miscellany*. London: Methuen & Co. Ltd.

Liu, T. C. (1960). Underidentification, structural estimation, and forecasting. *Econometrica* 28: 855–65.

Lombard, H. L. and Doering, C. R. (1928). Cancer studies in Massachusetts, 2. Habits, characteristics and environment of individuals with and without lung cancer. *New England Journal of Medicine* 198: 481–87.

Lorentz, G. G. (1986). *Bernstein Polynomials*. 2nd edn. New York: Chelsea.

Loudon, I. (2000). *The Tragedy of Childbed Fever*. Oxford: Oxford University Press.

Louis, P. (1986 [1835]). *Researches on the Effects of Bloodletting in Some Inflammatory Diseases, and the Influence of Emetics and Vesication in Pneumonitis*. Translated and reprinted by Classics of Medicine Library, Birmingham, AL.

Lucas, R. E., Jr. (1976). Econometric policy evaluation: A critique. In K. Brunner and A. Meltzer, eds. *The Phillips Curve and Labor Markets*. The Carnegie-Rochester Conferences on Public Policy, supplementary series to the *Journal of Monetary Economics*. Amsterdam: North-Holland, vol. 1, pp. 19–64 (with discussion).

Lunceford, J. K. and Davidian, M. (2004). Stratification and weighting via the propensity score in estimation of causal treatment effects: A comparative study. *Statistics in Medicine* 23: 2937–60.

MacGregor, G. A. and Sever, P. S. (1996). Salt—overwhelming evidence but still no action: Can a consensus be reached with the food industry? *British Medical Journal* 312: 1287–89.

MacKenzie, D. L. (1991). The parole performance of offenders released from shock incarceration (boot camp prisons): A survival time analysis. *Journal of Quantitative Criminology* 7: 213–36.

Mahoney, J. and Goertz, G. (2004). The possibility principle: Choosing negative cases in comparative research. *The American Political Science Review* 98: 653–69.

Mahoney, J. and Rueschemeyer, D. (2003). *Comparative Historical Analysis in the Social Sciences*. Cambridge: Cambridge University Press.

Manski, C. F. (1995). *Identification Problems in the Social Sciences*. Cambridge, MA: Harvard University Press.

Marini, M. M. and Singer, B. (1988). Causality in the social sciences. In C. Clogg, ed. *Sociological Methodology 1988*. Washington, DC: American Sociological Association, pp. 347–409.

Marks, J. S. and Halpin, T. J. (1980). Guillain-Barré syndrome in recipients of A/New Jersey influenza vaccine. *Journal of the American Medical Association* 243: 2490–94.

Massey, D. S. (1981). Dimensions of the new immigration to the United States and the prospects for assimilation. *Annual Review of Sociology* 7: 57–85.

Massey, D. S. and Denton, N. A. (1985). Spatial assimilation as a socioeconomic outcome. *American Sociological Review* 50: 94–105.

McCarron, D. A. and Reusser, M. E. (1999). Finding consensus in the dietary calcium-blood pressure debate. *Journal of the American College of Nutrition* 18, Supplement: S398–405.

McClure, F. J. (1970). *Water Fluoridation*. Bethesda, MD: National Institute of Dental Research.

McCue, K. F. (1998). Deconstructing King: Statistical problems. In *A Solution to the Ecological Inference Problem*. Technical report, California Institute of Technology, Pasadena, CA.

McKay, F. S. (1928). Relation of mottled enamel to caries. *Journal of the American Dental Association* 15: 1429–37.

McKay, F. S. and Black, G. V. (1916). An investigation of mottled teeth: An endemic developmental imperfection of the enamel of the teeth, heretofore unknown in the literature of dentistry. *Dental Cosmos* 58: 477–84, 627–44, 781–92, 894–904.

McNiel, D. E. and Binder, R. L. (2007). Effectiveness of mental health court in reducing recidivism and violence. *American Journal of Psychiatry* 164: 1395–1403.

Meehl, P. E. (1954). *Clinical Versus Statistical Prediction: A Theoretical Analysis and a Review of the Evidence.* Minneapolis, MN: University of Minnesota Press.

Meehl, P. E. (1978). Theoretical risks and tabular asterisks: Sir Karl, Sir Ronald, and the slow progress of soft psychology. *Journal of Consulting and Clinical Psychology* 46: 806–34.

Meehl, P. E. and Waller, N. G. (2002). The path analysis controversy: A new statistical approach to strong appraisal of verisimilitude. *Psychological Methods* 7: 283–337 (with discussion).

Middleton, J. (2007). Even for randomized experiments, logistic regression is not generally consistent. Technical report, Political Science Department, Yale University, New Haven, CT.

Midgley, J. P., Matthew, A. G., Greenwood, C. M., and Logan, A. G. (1996). Effect of reduced dietary sodium on blood pressure. *Journal of the American Medical Association* 275: 1590–97.

Miller, D. P., Neuberg, D., De Vivo, I. et al. (2003). Smoking and the risk of lung cancer: Susceptibility with GSTP1 polymorphisms. *Epidemiology* 14: 545–51.

Miller, J. F., Mekalanos, J. J., and Falkow, S. (1989). Coordinate regulation and sensory transduction in the control of bacterial virulence. *Science* 243: 916–22.

Miller, R. G., Jr. (1998). *Survival Analysis.* New York: Wiley.

Mills, J. P. (1926). Table of the ratio: Area to boundary ordinate, for any portion of the normal curve. *Biometrika* 18: 395–400.

Moore, T. J., Vollmer, W. M., Appel, L. J. et al. (1999). Effect of dietary patterns on ambulatory blood pressure: Results from the Dietary Approaches

382   REFERENCES AND FURTHER READING

to Stop Hypertension (DASH) Trial. DASH Collaborative Research Group. *Hypertension* 34: 472–77.

Müller, F. H. (1939). Tabakmissbrauch und Lungcarcinom (Tobacco abuse and lung cancer). *Zeitschrift für Krebsforschung* 49: 57–84.

Muthen, B. (1979). A structural probit model with latent variables. *Journal of the American Statistical Association* 74: 807–11.

Nagin, D. S. and Paternoster, R. (1993). Enduring individual differences and rational choice theories of crime. *Law and Society Review* 27: 467–96.

Nakachi, K., Ima, K., Hayashi, S.-I., and Kawajiri, K. (1993). Polymorphisms of the CYP1A1 and glutathione S-transferase genes associated with susceptibility to lung cancer in relation to cigarette dose in a Japanese population. *Cancer Research* 53: 2994–99.

Nathanson, N. and Alexander, E. R. (1996). Infectious disease epidemiology, *American Journal of Epidemiology* 144: S34, S37.

National Research Council (1997). *Possible Health Effects of Exposure to Residential Electric and Magnetic Fields*. Washington, DC: National Academy of Science.

Netto, E. (1927). *Lehrbuch der Combinatorik*. Leipzig: B. G. Teubner.

Neustadt, R. E. and Fineberg, H. V. (1981). *The Epidemic That Never Was: Policy-Making and the Swine Flu Affair*. New York: Random House.

Neyman, J. (1923). Sur les applications de la théorie des probabilités aux experiences agricoles: Essai des principes. *Roczniki Nauk Rolniczych* 10: 1–51, in Polish. English translation by D. M. Dabrowska and T. P. Speed (1990). *Statistical Science* 5: 465–80.

Neyman, J., Kolodziejczyk, S., and Iwaszkiewicz, K. (1935). Statistical problems in agricultural experimentation. *Journal of the Royal Statistical Society* 2, Supplement: 107–54.

Ní Bhrolcháin, M. (2001). Divorce effects and causality in the social sciences. *European Sociological Review* 17: 33–57.

Nicod, J. (1930). *Foundations of Geometry and Induction*. in French. English translation by P. P. Wiener. New York: Harcourt Brace.

Nuland, S. (1979). The enigma of Semmelweis—An interpretation. *Journal of the History of Medicine and Allied Sciences* 34: 255–72.

Nutton, V., ed. (2008). *Pestilential Complexities*. London: Wellcome Trust.

Oakes, M. (1990). *Statistical Inference*. Chestnut Hill, MA: Epidemiology Resources.

Olszewski, W. and Sandroni, A. (2008). Manipulability of future-independent tests. *Econometrica* 76: 1437–66.

Ono, H. (2007). Careers in foreign-owned firms in Japan. *American Sociological Review* 72: 267–90.

Pargament, K. I., Koenig, H. G., Tarakeshwar, N., and Hahn, J. (2001). Religious struggle as a predictor of mortality among medically ill patients. *Archives of Internal Medicine* 161: 1881–85.

Pasteur, L. (1878). *La Théorie des Germes et ses Applications à la Médecine et à la Chirurgie,* lecture faite à l'Académie de Médecine le 30 avril 1878, par M. Pasteur en son nom et au nom de MM. Joubert et Chamberland. Paris: G. Masson.

Pate, A. M. and Hamilton, E. E. (1992). Formal and informal deterrents to domestic violence: The Dade County spouse assault experiment. *American Sociological Review* 57: 691–97.

Patz, E. F., Jr., Goodman, P. C., and Bepler, G. (2000). Screening for lung cancer. *New England Journal of Medicine* 343: 1627–33.

Pearl, J. (1988). *Probabilistic Reasoning in Intelligent Systems.* San Mateo, CA: Morgan Kaufmann Publishers.

Pearl, J. (1995). Causal diagrams for empirical research. *Biometrika* 82: 669–710 (with discussion).

Pearl, J. (2000). *Causality: Models, Reasoning, and Inference.* Cambridge: Cambridge University Press.

Peikes, D. N., Moreno, L., and Orzol, S. M. (2008). Propensity score matching: A note of caution for evaluators of social programs. *The American Statistician* 62: 222–31.

Petitti, D. B. (1994). Coronary heart disease and estrogen replacement therapy: Can compliance bias explain the results of observational studies? *Annals of Epidemiology* 4: 115–18.

Petitti, D. B. (1996). Review of "Reference Guide on Epidemiology," *Jurimetrics* 36:159–68.

Petitti, D. B. (1998). Hormone replacement therapy and heart disease prevention: Experimentation trumps observation. *Journal of the American Medical Association* 280: 650–52.

Petitti, D. B. (1999). *Meta-Analysis, Decision Analysis, and Cost-Effectiveness Analysis.* 2nd edn. New York: Oxford University Press.

Petitti, D. B. (2002). Hormone replacement therapy for prevention. *Journal of the American Medical Association* 288: 99–101.

Petitti, D. B. and Chen, W. (2008). Statistical adjustment for a measure of healthy lifestyle doesn't yield the truth about hormone therapy. In D. Nolan and T. Speed, eds. *Probability and Statistics: Essays in Honor of*

David A. Freedman. Beachwood, OH: Institute of Mathematical Statistics, pp. 142–52.

Petitti, D. B. and Freedman, D. A. (2005). Invited commentary: How far can epidemiologists get with statistical adjustment? *American Journal of Epidemiology* 162: 415–18.

Phillips, S. and Grattet, R. (2000). Judicial rhetoric, meaning-making, and the institutionalization of hate crime law. *Law and Society Review* 34: 567–606.

Port, S., Demer, L., Jennrich, R., Walter, D., and Garfinkel, A. (2000). Systolic blood pressure and mortality. *Lancet* 355: 175–80.

Porter, R. (1997). *The Greatest Benefit to Mankind*. New York: Norton.

Prakasa Rao, B. L. S. (1987). *Asymptotic Theory of Statistical Inference*. New York: Wiley.

Pratt, J. W. (1981). Concavity of the log likelihood. *Journal of the American Statistical Association* 76: 103–06.

Pratt, J. W. and Schlaifer, R. (1984). On the nature and discovery of structure. *Journal of the American Statistical Association* 79: 9–33 (with discussion).

Pratt, J. W. and Schlaifer, R. (1988). On the interpretation and observation of laws. *Journal of Econometrics* 39: 23–52.

Prewitt, K. (2000). Accuracy and coverage evaluation: Statement on the feasibility of using statistical methods to improve the accuracy of Census 2000. *Federal Register* 65: 38, 373–98.

Psaty, B. M., Weiss, N. S., Furberg, C. D. et al. (1999). Surrogate end points, health outcomes, and the drug-approval process for the treatment of risk factors for cardiovascular disease. *Journal of the American Medical Association* 282: 786–90.

Ramsey, F. P. (1926). in R. B. Braithwaite (1931). *The Foundations of Mathematics and other Logical Essays*. London: Routledge and Kegan Paul.

Rao, C. R. (1973). *Linear Statistical Inference and its Applications*. 2nd edn. New York: Wiley.

Redfern, P. (2004). An alternative view of the 2001 census and future census taking. *Journal of the Royal Statistical Society*, Series A, 167: 209–48 (with discussion).

Reif, F. (1965). *Fundamentals of Statistical and Thermal Physics*. New York: McGraw-Hill.

Resnick, L. M. (1999). The role of dietary calcium in hypertension: A hierarchical overview. *American Journal of Hypertension* 12: 99–112.

Retailliau, H. F., Curtis, A. C., Storr, G., Caesar, et al. (1980). Illness after influenza vaccination reported through a nationwide surveillance system, 1976–1977. *American Journal of Epidemiology* 111: 270–78.

Ridgeway, G., McCaffrey, D., and Morral, A. (2006). Toolkit for weighting and analysis of nonequivalent groups: A tutorial for the TWANG package. RAND Corporation, Santa Monica, CA.

Rindfuss, R. R., Bumpass, L., and St. John, C. (1980). Education and fertility: Implications for the roles women occupy. *American Sociological Review* 45: 431–47.

Rivers, D. and Vuong, Q. H. (1988). Limited information estimators and exogeneity tests for simultaneous probit models. *Journal of Econometrics* 39: 347–66.

Robins, J. M. (1986). A new approach to causal inference in mortality studies with a sustained exposure period— application to control of the healthy worker survivor effect. *Mathematical Modelling* 7: 1393–1512.

Robins, J. M. (1987a). A graphical approach to the identification and estimation of causal parameters in mortality studies with sustained exposure periods. *Journal of Chronic Diseases* 40, Supplement 2: 139S–61.

Robins, J. M. (1987b). Addendum to "A new approach to causal inference in mortality studies with a sustained exposure period—application to control of the healthy worker survivor effect." *Computers and Mathematics with Applications* 14: 923–45.

Robins, J. M. (1995). Discussion. *Biometrika* 82: 695–98.

Robins, J. M. (1999). Association, causation, and marginal structural models. *Synthese* 121: 151–79.

Robins, J. M. and Greenland, S. (1989). The probability of causation under a stochastic model for individual risk. *Biometrics* 45: 1125–38.

Robins, J. M. and Rotnitzky, A. (1992). Recovery of information and adjustment for dependent censoring using surrogate markers. In N. Jewell, K. Dietz, and V. Farewell, eds. *AIDS Epidemiology—Methodological Issues*. Boston, MA: Birkhäuser, pp. 297–331.

Robins, J. M. and Rotnitzky, A. (1995). Semiparametric efficiency in multivariate regression models with missing data. *Journal of the American Statistical Association* 90: 122–29.

Robins, J. M., Rotnitzky, A., and Zhao, L. P. (1994). Estimation of regression coefficients when some regressors are not always observed. *Journal of the American Statistical Association* 89: 846–66.

Robins, J. M., Sued, M., Lei-Gomez, Q., and Rotnitzky, A. (2007). Performance of double-robust estimators when "inverse probability" weights are highly variable. *Statistical Science* 22: 544–59.

Robinson, L. D. and Jewell, N. P. (1991). Some surprising results about covariate adjustment in logistic regression models. *International Statistical Review* 58: 227–40.

Robinson, W. S. (1950). Ecological correlations and the behavior of individuals. *American Sociological Review* 15: 351–57.

Roe, D. A. (1973). *A Plague of Corn*. Ithaca, NY: Cornell University Press.

Rogowski, R. (2004). How inference in the social (but not the physical) sciences neglects theoretical anomaly. In Brady and Collier (2004), pp. 75–82.

Rosenbaum, P. R. (2002). Covariance adjustment in randomized experiments and observational studies. *Statistical Science* 17: 286–327 (with discussion).

Rosenberg, C. E. (1962). *The Cholera Years*. Chicago, IL: University of Chicago Press.

Rosenblum, M. and van der Laan, M. J. (2009). Using regression models to analyze randomized trials: Asymptotically valid hypothesis tests despite incorrectly specified models. *Biometrics* 65: 937–45 (http://www.bepress.com/ucbbiostat/paper219).

Rosenthal, R. (1979). The "file drawer" and tolerance for null results. *Psychological Bulletin* 86: 638–41.

Rossouw, J. E., Anderson, G. L., Prentice, R. L. et al. (2002). Risks and benefits of estrogen plus progestin in healthy postmenopausal women: Principal results from the Women's Health Initiative randomized controlled trial. *Journal of the American Medical Association* 288: 321–33.

Rotnitzky, A., Robins, J. M., and Scharfstein, D. O. (1998). Semiparametric regression for repeated outcomes with nonignorable nonresponse. *Journal of the American Statistical Association* 93: 1321–39.

Rubin, D. (1974). Estimating causal effects of treatments in randomized and nonrandomized studies. *Journal of Educational Psychology* 66: 688–701.

Rudin, W. (1976). *Principles of Mathematical Analysis*. 3rd. edn. New York: McGraw-Hill.

Sacks, F. M., Svetkey, L. P., Vollmer, W. M. et al. (2001). Effects on blood pressure of reduced dietary sodium and the dietary approaches to stop hypertension (DASH) diet. *New England Journal of Medicine* 344: 3–10.

Sampson, R. J., Laub, J. H., and Wimer, C. (2006). Does marriage reduce crime? A counterfactual approach to within-individual causal effects. *Criminology* 44: 465–508.

Savage, L. J. (1972 [1954]). *The Foundations of Statistics*. 2nd rev. edn. New York: Dover Publications.

Scharfstein, D. O., Rotnitzky, A., and Robins, J. M. (1999). Adjusting for nonignorable drop-out using semiparametric nonresponse models. *Journal of the American Statistical Association* 94: 1096–1146.

Scheffé, H. (1956). Alternative models for the analysis of variance. *Annals of Mathematical Statistics* 27: 251–71.

Schonberger, L. B., Bregman, D. J., Sullivan-Bolyai, J. Z. Keenlyside, R. A., et al. (1979). Guillain-Barré syndrome following vaccination in the National Influenza Immunization Program, United States, 1976–1977. *American Journal of Epidemiology* 110: 105–23.

Schonlau, M. (2006). Charging decisions in death-eligible federal cases (1995–2005): Arbitrariness, capriciousness, and regional variation. In S. P. Klein, R. A. Berk, and L. J. Hickman, eds. *Race and the Decision to Seek the Death Penalty in Federal Cases*. Technical report #TR-389-NIJ, RAND Corporation, Santa Monica, CA, pp. 95–124.

Semmelweis, I. (1981 [1861]). *The Etiology, Concept, and Prophylaxis of Childbed Fever*. English translation by F. P. Murphy. Birmingham, AL: *The Classics of Medicine Library*, pp. 350–773. Originally published as *Die Aetiologie, der Begriff und die Prophylaxis des Kindbettfiebers*. Pest, Wien und Leipzig, C. A. Hartleben's Verlags-Expedition.

Sen, A. K. (2002). *Rationality and Freedom*. Cambridge, MA: Harvard University Press.

Shadish, W. R., Cook, T. D., and Campbell, D. T. (2002). *Experimental and Quasi-Experimental Designs for Generalized Causal Inference*. Boston: Houghton Mifflin.

Shaffer, J. P. (1995). Multiple hypothesis testing. *Annual Review of Psychology* 46: 561–84.

Shapiro, S. (1994). Meta-analysis, shmeta-analysis. *American Journal of Epidemiology* 140: 771–91 (with discussion).

Shapiro, S., Venet, W., Strax, P., and Venet L. (1988). *Periodic Screening for Breast Cancer: The Health Insurance Plan Project and its Sequelae, 1963–1986*. Baltimore, MD: Johns Hopkins University Press.

Sherman, L. W., Gottfredson, D., MacKenzie, D. et al. (1997). *Preventing Crime: What Works, What Doesn't, What's Promising?* Washington, DC: U.S. Department of Justice.

Shields, P. G., Caporaso, N. E., Falk, K. T., Sugimura, H. et al. (1993). Lung cancer, race, and a CYP1A1 genetic polymorphism. *Cancer Epidemiology, Biomarkers and Prevention* 2: 481–85.

Silverstein, A. M. (1981). *Pure Politics and Impure Science: The Swine Flu Affair*. Baltimore, MD: Hopkins University Press.

Simon, H. (1957). *Models of Man*. New York: Wiley.

Sims, C. A. (1980). Macroeconomics and reality. *Econometrica* 48: 1–47.

Singer, B. and Marini, M. M. (1987). Advancing social research: An essay based on Stanley Lieberson's *Making It Count: The Improvement of Social Theory and Research*. In C. Clogg, ed. *Sociological Methodology 1987*. Washington, DC: American Sociological Association, pp. 373–91.

Skerry, P. (1995). *Mexican Americans: The Ambivalent Minority*. Cambridge, MA: Harvard University Press.

Skerry, P. (2000). *Counting on the Census? Race, Group Identity, and the Evasion of Politics*. Washington, DC: Brookings Institution Press.

Smith, G. D. and Phillips, A. N. (1996). Inflation in epidemiology: "The proof and measurement between two things" revisited. *British Medical Journal* 312: 1659–63.

Smith, W. C., Crombie, I. K., Tavendale, R. T. et al. (1988). Urinary electrolyte excretion, alcohol consumption, and blood pressure in the Scottish heart health study. *British Medical Journal* 297: 329–30.

Snow, J. (1965 [1855]). *On the Mode of Communication of Cholera*. 2nd edn. London: Churchill. Reprinted as part of *Snow on Cholera* in 1965 by Hafner, New York. Page cites are to the 1965 edition.

Sobel, M. E. (1998). Causal inference in statistical models of the process of socioeconomic achievement—A case study. *Sociological Methods and Research* 27: 318–48.

Sobel, M. E. (2000). Causal inference in the social sciences. *Journal of the American Statistical Association* 95: 647–51.

Spirtes, P., Glymour, C., and Scheines, R. (1993). *Causation, Prediction, and Search*. New York: Springer. 2nd edn., Cambridge, MA: MIT Press (2000).

Spirtes, P., Scheines, R., Glymour, C., and Meek, C. (1993). TETRAD II. Documentation for Version 2.2. Technical report, Department of Philosophy, Carnegie Mellon University, Pittsburgh, PA.

Stamler, J. (1997). The Intersalt study: Background, methods, findings, and implications. *American Journal of Clinical Nutrition* 65, Supplement: S626–42.

Stamler, J., Elliott, P., Dyer, A. R. et al. (1996). Commentary: Sodium and blood pressure in the Intersalt study and other studies—in reply to the Salt Institute. *British Medical Journal* 312: 1285–87.

Stark, P. B. (2001). Review of *Who Counts? Journal of Economic Literature* 39: 592–95.

Stata (2005). *Stata Base Reference Manual*. Stata Statistical Software. Release 9. Vol. 1. College Station, TX: StataCorp LP.

Steiger, J. H. (2001). Driving fast in reverse. *Journal of the American Statistical Association* 96: 331–38.

Stigler, S. M. (1986). *The History of Statistics*. Cambridge, MA: Harvard University Press.

Stolzenberg, R. M. and Relles, D. A. (1990). Theory testing in a world of constrained research design. *Sociological Methods and Research* 18: 395–415.

Stone, R. (1993). The assumptions on which causal inferences rest. *Journal of the Royal Statistical Society*, Series B, 55: 455–66.

Stoto, M. A. (1998). A solution to the ecological inference problem: Reconstructing individual behavior from aggregate data. *Public Health Reports* 113: 182–83.

Svetkey, L. P., Sacks, F. M., Obarzanek, E. et al. (1999). The DASH diet, sodium intake and blood pressure trial (DASH-sodium): Rationale and design. *Journal of the American Dietetic Association* 99, Supplement: 96–104.

Swales J. (2000). Population advice on salt restriction: The social issues. *American Journal of Hypertension* 13: 2–7.

Taleb, N. T. (2007). *The Black Swan*. New York: Random House.

Tauber, S. (1963). On multinomial coefficients. *American Mathematical Monthly* 70: 1058–63.

Taubes, G. (1998). The (political) science of salt. *Science* 281: 898–907.

Taubes, G. (2000). A DASH of data in the salt debate. *Science* 288: 1319.

Temple, R. (1999). Are surrogate markers adequate to assess cardiovascular disease drugs? *Journal of the American Medical Association* 282: 790–95.

Terris, M., ed. (1964). *Goldberger on Pellagra*. Baton Rouge, LA: Louisiana State University Press.

Thiébaut, A. C. M. and Bénichou, J. (2004). Choice of time-scale in Cox's model analysis of epidemiologic cohort data: A simulation study. *Statistics in Medicine* 23: 3803–20.

Timberlake, M. and Williams, K. (1984). Dependence, political exclusion and government repression: Some cross national evidence. *American Sociological Review* 49: 141–46.

Timonius, E. and Woodward, J. (1714). An account, or history, of the procuring the small pox by incision, or inoculation; as it has for some time been practised at Constantinople. *Philosophical Transactions* 29: 72–82.

Tinbergen, J. (1940). Reply to Keynes. *The Economic Journal* 50: 141–54.

Tita, G. and Ridgeway, G. (2007). The impact of gang formation on local pattern of crime. *Journal of Research on Crime and Delinquency* 44: 208–37.

Tong, Y. L. (1980). *Probability Inequalities in Multivariate Distributions.* New York: Academic Press.

Tropfke, J. (1903). *Geschichte der Elementar-Mathematik in Systematischer Darstellung.* Leipzig: Verlag von Veit & Comp.

Truett, J., Cornfield, J., and Kannel, W. (1967). A multivariate analysis of the risk of coronary heart disease in Framingham. *Journal of Chronic Diseases* 20: 511–24.

Tsiatis, A. (1975). A nonidentifiability aspect of the problem of competing risks. *Proceedings of the National Academy of Sciences, USA* 72: 20–22.

U.S. Census Bureau (2001a). *Report of the Executive Steering Committee for Accuracy and Coverage Evaluation Policy.* With supporting documentation, Reports B1–24. Washington, DC (http://www.census.gov/dmd/www/EscapRep.html).

U.S. Census Bureau (2001b). *Report of the Executive Steering Committee for Accuracy and Coverage Evaluation Policy on Adjustment for Non-Redistricting Uses.* With supporting documentation, Reports 1–24. Washington, DC (http://www.census.gov/dmd/www/EscapRep2.html).

U.S. Census Bureau (2003). *Technical Assessment of A.C.E. Revision II.* Washington, DC (http://www.census.gov/dmd/www/ace2.html).

U.S. Department of Commerce (1991). Office of the Secretary. *Decision on Whether or Not a Statistical Adjustment of the 1990 Decennial Census of Population Should Be Made for Coverage Deficiencies Resulting in an Overcount or Undercount of the Population, Explanation.* Three volumes, Washington, DC Reprinted in part in *Federal Register* 56: 33, 582–642 (July 22).

U.S. Department of Health and Human Services (1990). *The Health Benefits of Smoking Cessation: A Report of the Surgeon General.* Washington, DC

U.S. Geological Survey (1999). Working group on California earthquake probabilities. Earthquake probabilities in the San Francisco Bay Region: 2000–2030—A summary of findings. Technical Report Open-File Report 99-517, USGS, Menlo Park, CA.

U.S. Preventive Services Task Force (1996). *Guide to Clinical Preventive Services.* 2nd edn. Baltimore, MD: Williams & Wilkins.

U.S. Public Health Service (1964). *Smoking and Health. Report of the Advisory Committee to the Surgeon General.* Washington, DC: U.S. Government Printing Office.

van der Vaart, A. (1998). *Asymptotic Statistics.* Cambridge: Cambridge University Press.

van de Ven, W. P. M. M. and van Praag, B. M. S. (1981). The demand for deductibles in private health insurance: A probit model with sample selection. *Journal of Econometrics* 17: 229–52.

Verhulst, P. F. (1845). Recherches mathématiques sur la loi d'accroissement de la population. *Nouveaux Mémoires de l'Académie Royale des Sciences et Belles-Lettres de Bruxelles* 18: 1–38.

Verma, T. and Pearl, J. (1990). Causal networks: Semantics and expressiveness. In R. Shachter, T. S. Levitt, and L. N. Kanal, eds. *Uncertainty in AI 4.* Elsevier Science Publishers, pp. 69–76.

Victora, C. G., Habicht, J. P., and Bryce, J. (2004). Evidence-based public health: Moving beyond randomized trials. *American Journal of Public Health* 94: 400–405.

Vinten-Johansen, P., Brody, H., Paneth, N., and Rachman, S. (2003). *Cholera, Chloroform, and the Science of Medicine.* New York: Oxford University Press.

von Mises, R. (1964). *Mathematical Theory of Probability and Statistics.* H. Geiringer, ed. New York: Academic Press.

von Neumann, J. and Morgenstern, O. (1944). *Theory of Games and Economic Behavior.* Princeton, NJ: Princeton University Press.

Wachter, K. W. and Freedman, D. A. (2000). The fifth cell. *Evaluation Review* 24: 191–211.

Wainer, H. (1989). Eelworms, bullet holes, and Geraldine Ferraro: Some problems with statistical adjustment and some solutions. *Journal of Educational Statistics* 14: 121–40 (with discussion). Reprinted in J. Shaffer, ed. (1992). *The Role of Models in Nonexperimental Social Science.* Washington, DC: AERA/ASA, pp. 129–207.

Wald, A. (1940). The fitting of straight lines if both variables are subject to error. *The Annals of Mathematical Statistics* 11: 284–300.

Wald, A. and Wolfowitz, J. (1950). Bayes solutions of sequential decision problems. *The Annals of Mathematical Statistics* 21: 82–99.

Walsh, C. (2003). *Antibiotics: Actions, Origins, Resistance.* Washington, DC: ASM Press.

Webster, W. S. (1998). Teratogen update: Congenital rubella. *Teratology* 58: 13–23.

Weisberg, S. (1985). *Applied Linear Regression*. New York: Wiley.

Welch, H. G., Woloshin, S., Schwartz, L. M. et al. (2007). Overstating the evidence for lung cancer screening: The International Early Lung Cancer Action Program (I-ELCAP) study. *Archives of Internal Medicine* 167: 2289–95.

White, H. (1980). A heteroskedasticity-consistent covariance matrix estimator and a direct test for heteroskedasticity. *Econometrica* 48: 817–38.

White, H. (1994). *Estimation, Inference, and Specification Analysis*. Cambridge: Cambridge University Press.

White, M. D. (2000). Assessing the impact of administrative policy on the use of deadly force by on- and off-duty police. *Evaluation Review* 24: 295–318.

Wilde, E. T. and Hollister, R. (2007). How close is close enough? Evaluating propensity score matching using data from a class size reduction experiment. *Journal of Policy Analysis and Management* 26: 455–77.

Winship, C. and Mare, R. D. (1992). Models for sample selection bias. *Annual Review of Sociology* 18: 327–50.

Woodward, J. (1997). Causal models, probabilities, and invariance. In V. McKim and S. Turner, eds. *Causality in Crisis?* Notre Dame, IN: University of Notre Dame Press, pp. 265–315.

Woodward, J. (1999). Causal interpretation in systems of equations. *Synthese* 121: 199–247.

Wright, P. G. (1928). *The Tariff on Animal and Vegetable Oils*. New York: MacMillan.

Wright, S. (1921). Correlation and causation. *Journal of Agricultural Research* 20: 557–85.

Yee, T. W. (2007). The VGAM Package. (http://www.stat.auckland.ac.nz/~yee/VGAM)

Ylvisaker, D. (2001). Review of *Who Counts? Journal of the American Statistical Association* 96: 340–41.

Yule, G. U. (1899). An investigation into the causes of changes in pauperism in England, chiefly during the last two intercensal decades. *Journal of the Royal Statistical Society* 62: 249–95.

Yule, G. U. (1925). The growth of population and the factors which control it. *Journal of the Royal Statistical Society* 88: 1–62 (with discussion).

Zaslavsky, A. M. (1993). Combining census, dual system, and evaluation study data to estimate population shares. *Journal of the American Statistical Association* 88: 1092–1105.

Index

Accuracy and Coverage Evaluation
    Survey. *see* census
additive error. *see* error term
adjustment, 13, 53, 55, 65–66, 69–76, 78–
    82, 141, 179, 204–06, 209, 260
    census. *see* census
    regression, 137, 195
alternative hypothesis, 229, 323–24, 330
assumption, modeling, xi, xiv–xvi, 10,
    12–16, 20, 24–39, 41–43, 46, 52, 56–62,
    69–70, 72, 77, 80, 82, 84–85, 87–93,
    95–99, 101–03, 110–13, 122, 136, 139,
    141, 144, 147, 152, 159, 162, 169–70,
    172–73, 175–77, 185, 187–89, 191,
    196–98, 203–05, 207–8, 217, 219, 221–
    22, 224, 230, 235, 237, 242, 245–50,
    252–53, 256–68, 271–76, 284–85, 290–
    92, 296, 299–306, 316, 320, 323, 330–
    33, 352
    behavioral, 24, 29, 84, 92, 96, 99, 256,
        301, 315
    causal Markov condition, 245–46,
        251–53
    constancy, 84, 90–91, 93, 95–96, 103,
        196–98, 257–58
    diagnostic test, 83–84, 87, 89, 95, 98,
        100, 323–34
    exogeneity, 46, 59, 221–22, 258, 261,
        263–66, 271, 285, 291–92, 306, 330
    faithfulness, 245–47, 351–53
    homogeneity, 70–71, 77, 80

independence, 26, 29–31, 34, 36, 46,
    69, 84, 92, 95–96, 170, 172–73, 175,
    191, 221, 224, 257–58, 260–61,
    263–64, 266, 271, 292, 299, 304,
    306, 330
independent and identically distributed
    (IID), 12–13, 15, 32–34, 52, 59, 92,
    101, 112, 204, 224, 257, 263, 265–
    66, 268, 285, 291–92, 303, 306
invariance, 188, 255–56, 258–59,
    261–62, 264–67, 271, 274, 277–78,
    332–33
proportional-hazard, 169, 176, 180,
    187–88, 323, 331
random-sampling, 24–29, 36–41, 43
stationarity, 169, 172–73, 176
asymptotic bias. *see* bias
asymptotic standard error. *see* standard
    error
asymptotic variance. *see* variance

Bayes, Thomas, 4–5
    procedure, 9–10, 19
    rule, 124, 126–27, 129
Bayesian. *see also* subjectivist, 4, 7–9, 15,
    103, 112, 117, 126–28
beriberi. *see also* Eijkman, 344–46, 355
Bernoulli, Daniel, 170, 173, 190, 234
bias, xiv, 28–29, 31–32, 36, 40–42, 58–
    59, 69, 71–73, 80, 84–85, 88, 90, 92,
    100, 132, 143–44, 152–53, 161, 173,

Printed in the United States
By Bookmasters